# Advances in Industrial Control

Springer
*Berlin*
*Heidelberg*
*New York*
*Barcelona*
*Budapest*
*Hong Kong*
*London*
*Milan*
*Paris*
*Tokyo*

*Other titles published in this Series:*

*Parallel Processing for Jet Engine Control*
Haydn A. Thompson

*Iterative Learning Control for Deterministic Systems*
Kevin L. Moore

*Parallel Processing in Digital Control*
D. Fabian, Garcia Nocetti and Peter J Fleming

*Intelligent Seam Tracking for Robotic Welding*
Nitin Nayak and Asok Ray

*Identification of Multivariable Industrial Process for Simulation, Diagnosis and Control*
Yucai Zhu and Ton Backx

*Nonlinear Process Control: Applications of Generic Model Control*
Edited by Peter L. Lee

*Microcomputer-Based Adaptive Control Applied to Thyristor-Driven D-C Motors*
Ulrich Keuchel and Richard M. Stephan

*Expert Aided Control System Design*
Colin Tebbutt

*Modeling and Advanced Control for Process Industries, Applications to Paper Making Processes*
Ming Rao, Qijun Xia and Yiquan Ying

*Robust Multivariable Flight Control*
Richard J. Adams, James M. Buffington, Andrew G. Sparks and Siva S. Banda

*Modelling and Simulation of Power Generation Plants*
A.W. Ordys, A.W. Pike, M.A. Johnson, R.M. Katebi and M.J. Grimble

*Model Predictive Control in the Process Industry*
E.F. Camacho and C. Bordons

*$H_\infty$ Aerospace Control Design: A VSTOL Flight Application*
R.A. Hyde

*Neuro-Control and its Applications*
Sigeru Omatu, Marzuki Khalid and Rubiyah Yusof

ANDERSONIAN LIBRARY
WITHDRAWN
FROM
LIBRARY
STOCK
UNIVERSITY OF STRATHCLYDE

Kenneth Hunt, George Irwin and Kevin Warwick (Eds.)

# Neural Network Engineering in Dynamic Control Systems

With 122 Figures

Springer

Kenneth J. Hunt
Daimler-Benz AG
Alt Moabit 96 a, D-10559 Berlin, Germany

George R. Irwin
Department of Electrical and Electronic Engineering
Queen's University of Belfast, Belfast, BT9 5AH, UK

Kevin Warwick
Department of Cybernetics
School of Engineering and Information Sciences
University of Reading
P.O. Box 225, Whiteknights, Reading RG6 2AY, UK

*Series Editors*

Michael J. Grimble, Professor of Industrial Systems and Director
Michael A. Johnson, Reader in Control Systems and Deputy Director

Industrial Control Centre, Department of Electronic and Electrical Engineering, Graham Hills Building, 60 George Street, Glasgow G1 1QE, UK

ISBN 3-540-19973-X Springer-Verlag Berlin Heidelberg New York

British Library Cataloguing in Publication Data
Neural Network Engineering in Dynamic Control Systems. - (Advances in Industrial Control Series)
I. Hunt, K. J. II. Series
006.3
ISBN 3-540-19960-8

Library of Congress Cataloging-in-Publication Data
Neural network engineering in dynamic control systems / Kenneth Hunt, George Irwin, and Kevin Warwick, eds.
p. cm. -- (Advances in industrial control)
Includes bibliographical references and index.
ISBN 3-540-19973-X (alk. paper)
1. Neural networks (Computer science) 2. Automatic control.
I. Hunt, K. J. (Kenneth J.), 1963- . II. Irwin, George (George R.), 1950- .
III. Warwick, K. IV. Series.
QA76.87.A386 1995 95-30315
629.8'9--dc20 CIP


Printed in Great Britain



Typesetting: Camera ready by authors
Printed and bound at the Athenæum Press Ltd., Gateshead, Tyne and Wear
69/3830-543210 Printed on acid-free paper

# Series Editors' Foreword

The series *Advances in Industrial Control* aims to report and encourage technology transfer in control engineering. The rapid development of control technology impacts all areas of the control discipline. New theory, new controllers, actuators, sensors, new industrial processes, computer methods, new applications, new philosophies,...., new challenges. Much of this development work resides in industrial reports, feasibility study papers and the reports of advanced collaborative projects. The series offers an opportunity for researchers to present an extended exposition of such new work in all aspects of industrial control for wider and rapid dissemination.

Within the control community there has been much discussion of and interest in the new Emerging Technologies and Methods. Neural networks along with Fuzzy Logic and Expert Systems is an emerging methodology which has the potential to contribute to the development of intelligent control technologies.

This volume of some thirteen chapters edited by Kenneth Hunt, George Irwin and Kevin Warwick makes a useful contribution to the literature of neural network methods and applications. The chapters are arranged systematically progressing from theoretical foundations, through the training aspects of neural nets and concluding with four chapters of applications. The applications include problems as diverse as oven temperature control, and energy/load forecasting routines. We hope this interesting but balanced mix of material appeals to a wide range of readers from the theoretician to the industrial applications engineer.

M.J. Grimble and M.A. Johnson
Industrial Control Centre
Glasgow, Scotland, U.K.

# Series Editors' Preface

# Preface

Neural networks have been seen for some years now as providing considerable promise for application in nonlinear control and systems problems. This promise stems from the theoretical ability of networks of various types to approximate arbitrarily well continuous nonlinear mappings.

This book is based upon a very successful colloquium of the same title held in May 1994 at the Daimler-Benz Systems Engineering Research Centre in Berlin. The aim in this book is to evaluate the state-of-the-art in this very popular field from the engineering perspective. The book covers both theoretical and applied aspects. A major goal of the book is to examine ways of improving the engineering involved in neural network modelling and control, so that the theoretical power of learning systems can be harnessed for practical applications. This includes questions such as: which network architecture for which application? Can constructive learning algorithms capture the underlying dynamics while avoiding overfitting? How can we introduce a priori knowledge or models into neural networks? Can experiment design and active learning be used to automatically create 'optimal' training sets? How can we validate a neural network model?

The structuring of the book into four parts reflects these aims. In part 1 (chapters 1–3) the theoretical foundations of neural networks with specific focus on control systems is examined. Part 2 is the engineering core of the book and consists of chapters 4–6. Here, a variety of very recent constructive learning methods are presented. The main motivation for these techniques is to move towards a systematic engineering design procedure for neural modelling and control systems. Further technical aspects of the learning process are considered in part 3 (chapters 7–9).

In line with the goal of better engineering methods, the book places emphasis on real industrial applications of the technology; Part 4, consisting of chapters 10–13, describes in detail a number of concrete implementations.

## Contents Overview

Part 1: Theoretical Foundations

1. **Neural Approximation: A Control Perspective**
   *R. Żbikowski and A. Dzieliński*
   This chapter discusses the theoretical foundations of the modelling of non-linear control systems with neural networks. The major approaches based on approximation and interpolation theories are presented. These are compared within a unified framework and the relevance for neural control is stressed.

2. **Dynamic Systems in Neural Networks**
*K. Warwick, C. Kambhampati, P.C. Parks and J. Mason*
This contribution considers progress in the understanding of the dynamics of weight adjustment in neural networks when they are used as function approximators. This is seen as a vital step towards the development of a more complete stability analysis for neural control.
3. **Adaptive Neurocontrol of a Certain Class of MIMO Discrete-time Processes Based on Stability Theory**
*J-M. Renders, M. Saerens and H. Bersini*
A stability proof is provided for a class of non-linear multivariable processes controlled by a multilayer neural network. The result is based on a Lyapunov analysis and is local in nature. A simple weight adaptation strategy underlying the stability theorem is discussed.

Part 2: Constructive Training Methods

4. **Local Model Architectures for Nonlinear Modelling and Control**
*R. Murray-Smith and K.J. Hunt*
Local model networks are based upon the interpolation of simple, locally valid dynamic models. This contribution shows how the transparency of the local model network structure supports the integration of existing knowledge and a priori models. Extension of the model structure to local controller networks is described and applied to an automotive control example.
5. **On ASMOD – An Algorithm for Empirical Modelling using Spline Functions**
*T. Kavli and E. Weyer*
This chapter describes the theoretical foundations and principles of the ASMOD algorithm – a spline-based method for building dynamic models based on observed data. An incremental refinement procedure automatically adapts the model to the dependencies observed in the data. Case studies are included.
6. **Semi-Empirical Modeling of Non-linear Dynamic Systems through Identification of Operating Regimes and Local Models**
*T.A. Johansen and B.A. Foss*
Off-line algorithms for automatically determining the structure of local model networks are proposed and discussed. The algorithm searches for an optimal decomposition of the operating space into operating regimes, and local model structures. The transparency of the resulting model and the flexibility with respect to incorporation of prior knowledge is discussed.

Part 3: Further Issues in Network Learning

7. **On Interpolating Memories for Learning Control**
*H. Tolle, S. Gehlen and M. Schmitt*
General aspects of learning control are outlined and discussed. The presentation in this chapter is focussed on interpolating memories, and the CMAC structure in particular. Recent improvements in the approach are presented and results of applications in biotechnology

and automotive control are given. Finally, a critical assessment of the status of learning control is presented.

8. **Construction and Design of Parsimonious Neurofuzzy Systems**
   *K.M. Bossley, D.J. Mills, M. Brown and C.J. Harris*
   This chapter is concerned with the development of adaptive neural networks which can learn to perform ill-defined and complex tasks. The focus is on neurofuzzy systems where the transparent representation of fuzzy systems is fused with the adaptive capabilities of neural networks. The chapter concentrates on the question of how an appropriate structure for the rule base may be determined directly from the training data.
9. **Fast Gradient Based Off-line Training of Multilayer Perceptrons**
   *S. McLoone and G.R. Irwin*
   Fast off-line training of multilayer perceptrons using gradient-based algorithms is discussed. The inefficiencies of standard backpropagation algorithms are highlighted and this leads to a discussion of second-order optimisation techniques. Very significant speed-ups can be achieved, depending on problem size and convergence criterion. Parallel implementation of the algorithms is discussed.

Part 4: Applications

10. **Kohonen Network as a Classifier and Predictor for the Qualification of Metal-Oxide Surfaces**
    *W. Kessler and R.W. Kessler*
    The problem of predicting the future corrosion behaviour of low carbon steel is considered. The approach presented is based upon diffuse reflectance spectroscopy and evaluation of the spectra by a Kohonen self-organising map. This method is fast and reliable and can be applied on-line. Examples of corrosion prediction in car body steel are presented.
11. **Analysis and Classification of Energy Requirement Situations Using Kohonen Feature Maps within a Forecasting System**
    *S. Heine and I. Neumann*
    The Kohonen network is again applied, this time to the analysis and classification of energy requirements and load forecasting. For such problems a forecast model must be built for each application, and this can be a very demanding procedure. The efficiency of the Kohonen approach in a stepwise automation of model building is discussed by means of two case studies.
12. **A Radial Basis Function Network Model for Adaptive Control of Drying Oven Temperature**
    *O. Dubois, J-L. Nicolas and A. Billat*
    This contribution describes a neural control scheme for temperature control in a drying oven. The control strategy used is internal model control in which the plant is modelled by a radial basis function network. The process was identified both on- and off-line and experimental results of control trials with a real oven are presented.
13. **Hierarchical Competitive Net Architecture**
    *T. Long and E. Hanzevack*
    Development of hypersonic aircraft requires a high degree of system integration. Design tools are needed to provide rapid and accurate

calculations of complex fluid flow patterns. This chapter demonstrates that neural networks can be successfully applied to calculation of fluid flow distribution and heat transfer in a six leg heat exchanger panel, of the type envisioned for use in hypersonic aircraft.

## Acknowledgements

Most of the contributors to this book came together at the Berlin colloquium in May 1994. We'd like to gratefully acknowledge the Institution of Electrical Engineers and Daimler-Benz Systems Technology Berlin, whose support made this meeting possible. Special thanks are due to Cap'n Natho who managed the local arrangements in Berlin.

## Dedication

It is with deep sadness that we note the recent passing of our friend and colleague Professor Patrick C Parks. Patrick gave a presentation at the Berlin colloquium upon which this book is based, and is a co-author of chapter 2. For more than three decades Patrick Parks made fundamental contributions to the theory and stability analysis of nonlinear control systems. In recent years he contributed very actively to neural network applications in this area. As a token of our respect we dedicate this book to his memory.

*Kenneth Hunt, George Irwin, Kevin Warwick*
Berlin, Belfast, Reading: April 11, 1995

# Table of Contents

# List of Contributors

**H. Bersini**
Free University of Brussels

**A. Billat**
University of Reims Champagne Ardenne

**K.M. Bossley**
University of Southampton

**M. Brown**
University of Southampton

**O. Dubois**
University of Reims Champagne Ardenne

**A. Dzieliński**
University of Glasgow

**B.A. Foss**
Norwegian Institute of Technology, Trondheim

**S. Gehlen**
Centre for Neurocomputing GmbH, Bochum

**E. Hanzevack**
University of South Carolina

**C.J. Harris**
University of Southampton

**S. Heine**
Leipzig Polytechnic

**K.J. Hunt**
Daimler-Benz AG, Berlin

**G.R. Irwin**
Queen's University of Belfast

**T.A. Johansen**
Norwegian Institute of Technology, Trondheim

**C. Kambhampati**
University of Reading

**T. Kavli**
SINTEF, Oslo

**R.W. Kessler**
Reutlingen Polytechnic

**W. Kessler**
Reutlingen Polytechnic

**T. Long**
Neurodyne Inc., Williamsburg, VA

**J. Mason**
University of Reading

**S. McLoone**
Queen's University of Belfast

**D.J. Mills**
University of Southampton

**R. Murray-Smith**
Daimler-Benz AG, Berlin

**I. Neumann**
Best Data Engineering GmbH, Berlin

**J-L. Nicolas**
University of Reims Champagne Ardenne

**P.C. Parks**
University of Oxford

**J-M. Renders**
Free University of Brussels

**M. Saerens**
Free University of Brussels

**M. Schmitt**
Robert Bosch GmbH, Schwieberdingen

**H. Tolle**
Technical University of Darmstadt

# Neural Approximation: A Control Perspective

Rafał Żbikowski and Andrzej Dzieliński

Control Group, Department of Mechanical Engineering, James Watt Building,
Glasgow University, Glasgow G12 8QQ, Scotland, UK

**Abstract.** This chapter discusses theoretical foundations of modelling of nonlinear control systems with neural networks. Both feedforward and recurrent networks are described with emphasis on the practical implications of the mathematical results. The major approaches based on approximation and interpolation theories are presented: Stone-Weierstrass' theorem, Kolmogorov's theorem and multidimensional sampling. These are compared within a unified framework and the relevance for neural modelling of nonlinear control systems is stressed. Also, approximation of functionals with feedforward networks is briefly explained. Finally, approximation of dynamical systems with recurrent networks is described with emphasis on the concept of differential approximation.

## 1 Introduction

One of the reasons for the remarkable success of theory and practice of linear time-invariant (LTI) systems is their genericity [44]. This means that we have *universal* control design methods for the whole class of LTI plants. In other words, given *any* LTI model (say, a transfer function) the methods (e.g. pole placement) will always work. This is to say that LTI systems admit *generic models*, which allow the familiar formulation of the control design problem: "Given a transfer function $G(s)\ldots$", as $G(s)$ is all we need to know.

An extension of this highly desirable feature to nonlinear systems has so far eluded control theorists. The main reason is the complex behaviour associated with nonlinearity and its intrinsic locality. Hence the search for a universal nonlinear model, a 'nonlinear transfer function', is highly nontrivial, as is the underlying problem of classification of nonlinear systems.

An important feature of a candidate for the 'ultimate black-box' is that it be parameterised to allow finite-dimensional identification techniques to be applied. The model should itself be a tractable system from the control viewpoint, as it is only an auxiliary step—a representation of plant's dynamics—in the overall closed-loop scheme.

It is in this context that we attempt to analyse and apply neural networks for control (neurocontrol) [19], [45]. We treat them as a candidate for a generic, parametric, nonlinear model of a broad class of nonlinear plants. As explained below, neural networks have modelling capabilities to a desired accuracy. It should, however, be borne in mind that an important question is *how* neural models represent the plant's system properties. That is, how controllability, observability, stability etc. of neural models represent these properties of the real object. These are still open questions and here we do not address them, but give an account of remarkable progress in theoretical investigations on the fundamental representational capabilities of neural networks. Not only validate they the use of the models, but also give interesting and practical suggestions for further research.

The use of feedforward neural networks (FNNs) for nonlinear control is based on the input-output discrete-time description of systems

$$y(t+1) = f\Big(y(t), \ldots, y(t-n+1); u(t), \ldots, u(t-m+1)\Big), \tag{1}$$

where the minimal assumption is that $f$ is continuous. Even when $y$ and $u$ are properly bandlimited, there remains a nontrivial question what (1) actually represents with regard to the underlying differential equation. Leaving this question aside, the issue is (in the SISO case) to approximate the continuous mapping $f: \mathcal{R}^{m+n} \to \mathcal{R}$. The essence of the neural approach is to use an FNN model for this purpose. Thus, the neural modelling is a problem of Approximation Theory.

A similar existence problem arises for recurrent neural networks. Here the issue is if one can approximate a dynamic system with a recurrent network. The relevant results will be discussed in the context of differential approximation described in section 7.1.

The chapter is organised as follows. Section 2 defines the approximation problem to be considered and two main issues to be addressed. Each of the following three sections is devoted to one of the major approaches and is structured in the same way. First the mathematical meaning of an approach is explained and then its relevance and use for neural modelling is investigated. Thus section 3 describes the ideas related to the Stone-Weierstrass theorem. Section 4 analyses the approach based on Kolmogorov's solution to Hilbert's 13th problem. Section 5 gives an account of the application of multidimensional sampling. The following section (section 6) addresses the more general problem of functional (infinite-dimensional) approximation with FNNs and its use for neural modelling. The section ends with section 7 (followed by conclusions) discussing approximation with recurrent neural networks in the context of differential approximation.

## 2 Approximation with Feedforward Networks

From the mathematical viewpoint, the problem of feedforward neural modelling can be formulated as follows. Let $f: K \to \mathcal{R}^q$ be a continuous mapping, where $K$ is an uncountable compact subset of $\mathcal{R}^{pm+qn}$ (here $p$ is the number of inputs, $q$ the number of outputs of (1)). Compactness of $K$ means here that it is closed and bounded. The mapping $f$ is not given explicitly, but by a *finite* number of pairs $(U_k, Y_k) \in K \times \mathcal{R}^q$, $k = 1, \ldots, s$ (here $s$ is the number of observed input-output pairs).[1] The problem is: find an approximation of $f$ by known functions and a finite number of real parameters, such that the representation yields uniform approximation of $f$ over $K$. Thus we ask for:

[1] For example, if $t+1$ is replaced with $t$ in (1), then $U_k = [y(t-1), \ldots, y(t-n); u(t-1), \ldots, u(t-m)]^T$ is a $q \times (pm+qn)$ vector and $Y_k = y(t)$ a $q \times 1$ vector.

1. uniform approximation of $f$ on $K$;
2. interpolation of the continuum $f(K)$ from the samples $(U_k, Y_k)$.

The question arises if the problems 1 and 2 are equivalent, as uniform approximation of $f$ on $K$ gives an interpolation of the hypersurface $f(K)$ and a well chosen interpolation implies a solution of the problem 1. From the practical point of view the equivalence does not hold.

The problem 1 is, in general, an existence result independent of the form in which $f$ is given. All that is needed to establish a possibility of approximating $f$ by some known functions are qualitative properties of $f$ (e.g. continuity).

Issue 2, on the other hand, by definition deals with functions given by a finite number of argument-value pairs. Thus it is a more specific problem, intrinsically orientated towards construction.

Hence, a solution to problem 1 gives meaning to our efforts, while 2 brings us closer practical learning algorithms (see also section 5).

We now proceed to describe the main mathematical tools an their relevance for these problems.

# 3 Stone-Weierstrass Theorem

The purpose of this section is to present the Stone-Weierstrass theorem and its applicability to problem 1 of section 2.

We give an account of the relevant results not in their full generality, but in the form suitable for our discussion and notation introduced in section 2.

## 3.1 Weierstrass' Theorem

The original theorem due to Weierstrass [3] shows that an arbitrary continuous function $f: [a, b] \to \mathcal{R}$ can be uniformly approximated by a sequence of polynomials $\{p_n(x)\}$ to within a desired accuracy. Thus, given $\varepsilon > 0$, one can always find $N \in \mathcal{N}$, such that for any $n > N \quad |f(x) - p_n(x)| < \varepsilon$ uniformly on $[a, b]$. This clearly is a model for problem 1 of section 2 with a number of real parameters (coefficients of $p_{N+1}$) finite for a given $\varepsilon$.

## 3.2 Main Result

The Weierstrass theorem was analysed by M. H. Stone [40], who tried to find the general properties of approximating functions, not necessary polynomials. He has done it rather abstractly, so we first give a motivating example of his reasoning.

The reals $\mathcal{R}$ are composed of rational and irrational numbers, $\mathcal{Q}$ and $\mathcal{R} - \mathcal{Q}$, respectively. All computations involving real numbers are performed on rationals. This is because it is very convenient and any real can be approximated to a desired accuracy by a sequence in $\mathcal{Q}$. This is formalised by saying that $\mathcal{Q}$ is *dense* in $\mathcal{R}$, or, equivalently, that $\mathcal{R}$ is the *closure* of $\mathcal{Q}$. In other words, $\mathcal{R}$ is the smallest set in which all rational Cauchy sequences have limits. This means that any number which can be approximated by a sequence with terms in $\mathcal{Q}$ is a real number.

Stone considered the converse problem [7], [32]. Given the set $B$ (playing the role of $\mathcal{R}$ above) of *all* continuous functions from a compact $K$ to $\mathcal{R}$, find a *proper* subset $A \subset B$, such that $B$ is the closure of $A$ (here $A$ plays the role of $\mathcal{Q}$ above). Thus, if $\{p_n\}$ is a sequence of functions from $A$, such that $p_n \rightarrow f$, we want its limit $f$ to be in $B$. Since we are talking about *function* approximation (as opposed to *number* approximation), it is desirable to perform simple algebraic operations on $A$, e.g. forming of linear combinations (this is like arithmetic in $\mathcal{Q}$). Also, as sequences of functions may have two modes of convergence, we insist that those with terms in $A$ converge uniformly on $K$. This motivates the following definitions, valid for any functions (not necessarily continuous) and any set $K$ (not necessarily compact).

**Definition 1.** *A set $A$ of functions from $K \subset \mathcal{R}^{pm+qn}$ to $\mathcal{R}$ is called an* algebra of functions *iff $\forall f, g \in A$ and $\forall \alpha \in \mathcal{R}$*

*(i)* $f + g \in A$;
*(ii)* $fg \in A$;
*(iii)* $\alpha f \in A$. $\diamond$

**Definition 2.** *Let $B$ be the set of all functions which are limits of uniformly convergent sequences with terms in $A$, a set of functions from $K \subset \mathcal{R}^{pm+qn}$ to $\mathcal{R}$. Then $B$ is called the* uniform closure *of $A$.* $\diamond$

We need only two more conditions of nondegeneracy; see [28, page 142].

**Definition 3.** *A set $A$ of functions from $K \subset \mathcal{R}^{pm+qn}$ to $\mathcal{R}$ is said to* separate points on $K$ *iff $\forall x_1, x_2 \in K \quad x_1 \neq x_2 \Rightarrow \exists f \in A,\ f(x_1) \neq f(x_2)$.* $\diamond$

In other words, for any two distinct points of $K$ there exists a function from $A$ having distinct values at the points.

**Definition 4.** *Let $A$ be a set of functions from $K \subset \mathcal{R}^{pm+qn}$ to $\mathcal{R}$. We say that $A$* vanishes at no point of $K$ *iff $\forall x \in K\ \exists f \in A$, such that $f(x) \neq 0$.* $\diamond$

The main result is the following.

**Theorem 5 Stone-Weierstrass.** *Let $A$ be an algebra of* some *continuous functions from a compact $K \subset \mathcal{R}^{pm+qn}$ to $\mathcal{R}$, such that $A$ separates points on $K$ and vanishes at no point of $K$. Then the uniform closure $B$ of $A$ consists of* all *continuous functions from $K$ to $\mathcal{R}$.* $\diamond$

The original formulation [32], [7] is for $f: \mathcal{R}^{pm+qn} \rightarrow \mathcal{R}$ due to condition (ii) of Definition 1. But the codomain of a vector-valued function is the cartesian product of its components, so the result remains valid for $f: \mathcal{R}^{pm+qn} \rightarrow \mathcal{R}^q$.

### 3.3 Relevance for Neural Modelling

The theorem is a criterion given functions should satisfy in order to uniformly approximate arbitrary continuous functions on compacts.

This is the essence of several publications demonstrating approximation capabilities of multilayer FNNs, e.g. [18], [6] to name a few. Before discussing the details let us make precise the terminology.

**Definition 6.** *A $C^k$-sigmoid function $\sigma: \mathcal{R} \to \mathcal{R}$ is a nonconstant, bounded, and monotone increasing function of class $C^k$ (continuously differentiable up to order $k$).* ◇

Informally, the sigmoid is a (smooth) nonlinearity with saturation.

**Definition 7.** *A $C^k$-radial basis function (RBF) $g_{c,m}: \mathcal{R}^n \to \mathcal{R}$, with $c \in \mathcal{R}^n$ and $m \in \mathcal{R}_+$, is a $C^k$ function constant on spheres $\{x \in \mathcal{R}^n \mid \|x - c\|/m = r\}$, centre $c$, radius $rm$, $r \in \mathcal{R}_+$, where $\|\cdot\|$ is the Euclidean norm on $\mathcal{R}^n$.* ◇

The RBF has spherical (radial) symmetry with the centre $c$ and the positive number $m$ controlling the size of the spherical neighbourhood of $c$. It can be alternatively defined as the function $g: \mathcal{R}_+ \cup \{0\} \to \mathcal{R}$, where the argument is taken to be the radius of spheres centered at $c$.

Neural networks often have layered structures. The sigmoidal model

$$y_i = \sum_{j=1}^{N} a_{ij}\sigma(b_{ij}^T U + d_{ij}) \quad i = 1, \ldots, q, \tag{2}$$

where $a_{ij}, d_{ij} \in \mathcal{R}$, $b_{ij} \in \mathcal{R}^{pm+qn}$ and $^T$ means transposition ($U$ is the input; see footnote 1 in section 2), has the identity input layer, sigmoidal hidden layer and linear output layer. A similar structure is obtained for the RBF network

$$\begin{aligned} y_i &= \sum_{j=1}^{N} a_{ij} g_{c_{ij},m_{ij}}(U) \\ &= \sum_{j=1}^{N} a_{ij} g(\|U - c_{ij}\|/m_{ij}) \quad i = 1, \ldots, q. \end{aligned} \tag{3}$$

According to theorem 5 it suffices to show that the set of all finite linear combinations of sigmoids and RBFs, respectively, is a nonvanishing algebra separating points on a compact $K \subset \mathcal{R}^{pm+qn}$, as specified in Definitions 1, 3, 4. This is relatively straightforward, with the additional condition of convexity of $K$ for Gaussians [35]. Thus, according to the Stone-Weierstrass theorem, both sigmoids and RBFs are suitable for uniform approximation of an arbitrary continuous mapping. However, their interpolation properties (see problem 2 of section 2) are different, for RBFs are more suitable for interpolation (see section 5).

# 4 Kolmogorov's Theorem

This section describes Kolmogorov's theorem and its applicability to problem 1 of section 2. It concerns representation of continuous functions defined on an $n$-dimensional cube by sums and superpositions of continuous functions of one variable. Kolmogorov's results are presented and their applicability in neural context is addressed.

## 4.1 Hilbert's 13th Problem

The thirteenth of the famous 23 Hilbert's problems posed at the Second International Congress of Mathematicians in 1900 was the following:

*Is every analytic function of three variables a superposition of continuous functions of two variables? Is the root $x(a, b, c)$ of the equation*

$$x^7 + ax^3 + bx^2 + cx + 1 = 0$$

*a superposition of continuous functions of two variables?*

It is important to consider the class of functions of which the superposition is constituted. Although any function of three variables can be represented as a superposition of functions of two variables by a proper choice of the latter functions [1], this may not be possible within a given class of smoothness. Hilbert proved the following:

**Theorem 8.** *There is an analytic function of three variables which is not a superposition of infinitely differentiable functions of two variables.* ◇

This motivates the problem of the possibility of reduction to superpositions of continuous functions. Hilbert's hypothesis was that a reduction of this kind would not, in general, be possible. From now on, we refer to this conjecture as Hilbert's problem.

## 4.2 Kolmogorov's Solution to Hilbert's Problem

Vitushkin showed [42] that it is not possible to represent an arbitrary smooth function of three variables by a superposition of functions of two variables and the same degree of smoothness. Hilbert's problem, however, is concerned not with smooth but with continuous functions. In this domain the results of Kolmogorov have disproved Hilbert's hypothesis.

Initially, Kolmogorov demonstrated that any continuous function given on the unit ($E = [0, 1]$) $n$-dimensional cube $E^n$ has, for $n \geq 3$, the representation

$$f(x_1, x_2, \cdots, x_n) = \sum_{r=1}^{n} h^r\Big(x_n, g_1^r(x_1, \cdots, x_{n-1}), g_2^r(x_1, \cdots, x_{n-1})\Big), \quad (4)$$

where the functions $g_1^r$, $g_2^r$ of $n-1$ variables and the functions $h^r$ of 3 variables are real and continuous. Applying this representation several times it may be noticed [22] that any continuous function of $n \geq 4$ variables is a superposition of continuous functions of 3 variables.

The Hilbert problem was formulated for functions of three variables, and the above result of Kolmogorov is valid for four. This result, however, may be restated in such a way that one can represent any continuous function given on the three dimensional cube in the form

$$f(x_1, x_2, x_3) = \sum_{i=1}^{3} \sum_{j=1}^{3} h_{ij}\Big(\phi_{ij}(x_1, x_2), x_3\Big), \tag{5}$$

where $h$ and $\phi$ are real continuous functions of two variables. This disproves Hilbert's hypothesis, and shows that any continuous function of $n \geq 3$ variables can be represented as the superposition of continuous functions of two variables.

The most important and relevant of Kolmogorov's results is his representation theorem [23]:

**Theorem 9 Kolmogorov.** *Any function continuous on the n-dimensional cube $E^n$ can be represented in the form*

$$f(x_1, \cdots, x_n) = \sum_{i=1}^{2n+1} \chi_i\Big(\sum_{j=1}^{n} \phi_{ij}(x_j)\Big), \tag{6}$$

*where $\chi_i$ and $\phi_{ij}$ are real continuous functions of one variable.* ◇

This way, we may *exactly* represent every continuous function as a superposition of a *finite* number of continuous functions of one variable and of a single particular function of two variables, viz. addition. The functions $\phi_{ij}$ are standard and independent of $f(x_1, \cdots, x_n)$. Only the functions $\chi_i$ are specific for the given function $f$.

### 4.3 Relevance for Neural Modelling

Kolmogorov's result has been improved by several other authors to make it more applicable. Lorentz [29] showed that the functions $\chi_i$ may be replaced by only one function $\chi$. Sprecher [39] replaced the functions $\phi_{ij}$ by $\alpha^{ij}\phi_j$, where $\alpha^{ij}$ are constants and $\phi_j$ are monotonic increasing functions. The latter has been reformulated by Hecht-Nielsen [17] to make the use of neural networks plausible. Thus Sprecher's version of Kolmogorov's representation theorem in the form directly relevant to neural networks is as follows.

**Theorem 10.** *Any continuous function defined on the n-dimensional cube $E^n$ can be implemented* exactly *by a three-layered network having $2n+1$ units in the hidden layer with transfer functions $\alpha^{ij}\phi_j$ from the input to the hidden layer and $\chi$ from all of the hidden units to the output layer.* ◇

Girosi and Poggio [14] criticised Hecht-Nielsen's method, pointing to two main drawbacks of the approach:

1. the functions $\phi_{ij}$ are highly nonsmooth,
2. the functions $\chi_i$ depend on the specific function $f$ and are not representable in a parameterised form.

Both these difficulties have been resolved by Kůrková [25], [26] by the use of staircase-like functions of a sigmoidal type in a sigmoidal feedforward neural network. The highly nonsmooth functions are constructed as limits or sums of infinite series of smooth functions. Since in the context of neural networks we are interested only in approximations of functions, the problems reported by Girosi and Poggio can be reduced to the following question. Can Kolmogorov's theorem be modified in such a way that all of the single-variable functions are limits of sequences of smooth functions used in sigmoidal feedforward networks (i.e. sigmoidal functions)? This question has a positive answer when staircase-like functions of sigmoidal type are used in the neural network. This type of function has the property that it can approximate any continuous function on any closed interval with an arbitrary accuracy. Taking advantage of this fact we can reformulate Kolmogorov's representation theorem in the way given by Kůrková:

**Theorem 11.** *Let $n \in \mathcal{N}$ with $n \geq 2$, $\sigma: \mathcal{R} \to E$ be a sigmoidal function, $f \in C^0(E^n)$, and $\varepsilon$ be a positive real number. Then there exist $k \in \mathcal{N}$ and staircase-like functions $\chi_i$, $\phi_{ij} \in S(\sigma)$ such that for every $(x_1, \cdots, x_n) \in E^n$*

$$\left| f(x_1, \cdots, x_n) - \sum_{i=1}^{k} \chi_i \Big( \sum_{j=1}^{n} \phi_{ij}(x_j) \Big) \right| < \varepsilon, \tag{7}$$

*where $S(\sigma)$ is the set of all staircase-like functions of the form $\sum_{i=1}^{k} a_i \sigma(b_i x + c_i)$.* $\diamond$

The theorem implies that any continuous function can be approximated arbitrarily well by a four-layer sigmoidal feedforward neural network. However, comparing (7) with (6) we see that the original *exact* representation in Kolmogorov's result (Theorem 9) is replaced by an *approximate* one. It is worth noting that it has been established, not necessarily by Kolmogorov's argument, that even three layers are sufficient for approximation of general continuous functions [6], [11], [18].

In practice one has to answer the question whether an arbitrary given multivariate function $f$ can be represented by an approximate realisation of the corresponding functions $\chi_i$ of one variable. The answer [27] is that an *approximate* implementation of $\chi_i$ does not, in general, guarantee an approximate implementation of the original function $f$, i.e. $\chi_i$ must be *exactly* realised. This highlights the limitations of the applicability of Kolmogorov's theorem to neural networks for approximation of mappings. However, the efforts toward applying this theorem are important, as some useful neural networks like the sigmoidal feedforward network can be described by Kolmogorov's representation theorem.

## 5 Multidimensional Sampling

This section considers multidimensional sampling applied to problem 2 of section 2. It is argued that this approach is the most feasible, because it also solves problem 1, as seen from the results of section 3.

### 5.1 Motivation and Background

It follows from section 3 that both sigmoids and RBFs are suitable for solving problem 1 of section 2. The question arises which of these is better for tackling problem 2.

The hypersurface $f(K)$ is, in general, a continuum and so is the domain $K$. One of practicable approches is the discretisation of $K$. Thus a *countably infinite* subset (infinite sequence) $U = \{U_k\}_{k=1}^{\infty}$ of $K$ is chosen and $f$ evaluated at these points, $Y = \{Y_k\}_{k=1}^{\infty}$. The problem is to interpolate values in $K - U$ with accuracy to within a prescribed error. The one-dimensional case, $K \subset \mathcal{R}$, is the classical [36] problem of sampling of time signals. The original formulation is treated in section 5.2, and its extension to $n$ dimensions in section 5.3. It is shown that, under certain assumptions, exact interpolation (reconstruction) of $f$ is possible.

Our problem, however, is more restricted, as we are given a *finite* subset of $K$ (finite sequence), viz. $U = \{U_k\}_{k=1}^{s}$ (and the corresponding $Y = \{Y_k\}_{k=1}^{s}$). We cannot expect the perfect reconstruction of $f$ in this case, but we seek an approximate interpolation. The essence of the approach is to use the multidimensional sampling [31] and system theories [8] to obtain bounds for accuracy of the interpolation from the finite set (see [34]).

### 5.2 Sampling in One Dimension

Shannon [36] was the first to bring to the attention of wide scientific community the problem (and solution) of reconstruction of a real continuous function from its samples, or values at a countably infinite subset of $\mathcal{R}$. The following is the main result.

**Theorem 12 Sampling Theorem.** *Let $f\colon \mathcal{R} \to \mathcal{R}$ be such that both its direct, $F$, and inverse Fourier transforms are well-defined. If the spectrum $F(\omega)$ vanishes for $|\omega| > 2\pi\nu_N$, then $f$ can be exactly reconstructed from samples $\{f(t_k)\}_{k\in\mathcal{Z}}$, $t_k = k/2\nu_N$.* $\diamond$

The essential assumption is that $f$ is bandlimited, i.e. its spectrum is zero outside the interval $[-2\pi\nu_N, 2\pi\nu_N]$. For the one-dimensional case this is ensured by the use of lowpass filters. Then, with $t_k = k/2\nu_N$,

$$f(t) = \sum_{k=-\infty}^{\infty} f(t_k)\frac{\sin\pi(2\nu_N t - k)}{\pi(2\nu_N t - k)} = \sum_{k=-\infty}^{\infty} f(t_k)g(t - t_k), \tag{8}$$

where $g(t-t_k) = \sin[2\pi\nu_N(t-t_k)]/2\pi\nu_N(t-t_k)$. This is the canonical interpolation *exact* between all samples. If the sampling period is $T$, so that $t_k = kT$, then

$$F(\omega) = \frac{G(\omega)}{T} \sum_{k=-\infty}^{\infty} F\left(\omega + \frac{2\pi k}{T}\right). \tag{9}$$

Thus $F$ is periodic and the repetitive portions of the spectrum will not overlap iff

$$T \leq 1/2\nu_N, \tag{10}$$

as stipulated in Theorem 12.

The ideal spectrum of $G$ is

$$G(\omega) = \begin{cases} T, & |\omega| < 2\pi\nu_N; \\ 0, & \text{elsewhere,} \end{cases} \tag{11}$$

which corresponds to $g(t-t_k)$ in (8). Approximate interpolation occurs when the spectrum $G(\omega)$ of (11) is realised approximately, e.g. by a suitably scaled Gaussian.

This reasoning can, in principle, be carried over to $n$ dimensions [31], i.e. when $f: \mathcal{R}^n \to \mathcal{R}$ (it remains true if $f$ is defined on a compact subset $K$ of $\mathcal{R}^n$). A fact worth noting is the necessity of use of multidimensional [21] lowpass antialiasing filters to bandlimit the spectrum of $f: \mathcal{R}^n \to \mathcal{R}$.

### 5.3 Sampling in $n$ Dimensions

The main result of multidimensional sampling theory is a generalisation of the Sampling Theorem (Theorem 12) to many variables. This may be summarised as follows.

**Theorem 13 $n$-Dimensional Sampling Theorem.** *Let $f: \mathcal{R}^n \to \mathcal{R}$ be such that both its n-dimensional direct, $F$, and inverse Fourier transforms are well-defined. If the spectrum $F(\omega_1, \ldots, \omega_n)$ vanishes outside a bounded subset of n-dimensional space, then $f$ can be everywhere exactly reconstructed from its samples $f(x_{1,k_1}, \ldots, x_{n,k_n})$, taken over a lattice of points $\{k_1 v_1 + k_2 v_2 + \ldots + k_n v_n\}, k_1, k_2, \ldots, k_n = 0, \pm 1, \pm 2, \ldots$, provided that the vectors $\{v_i\} \in \mathcal{R}^n$, $i = 1, \ldots, n$, are small enough to ensure nonoverlapping of the spectrum $F(\omega_1, \ldots, \omega_n)$ with its periodic images on the lattice defined by the vectors $\{u_j\}$, $j = 1, \ldots, n$, with $v_i \cdot u_j = 2\pi\delta_{ij}$, where $\delta_{ij}$ is the Kronecker's symbol and $\cdot$ the inner product.*

To establish the above result let us consider a set of real functions in $n$ dimensions $f(x_1, x_2, \ldots, x_n)$ defined over $\mathcal{R}^n$. Moreover, let us assume that the Fourier transforms of these functions exist and are given by [37]

$$F(\omega_1, \ldots, \omega_n) = \int_{-\infty}^{\infty} \cdots \int_{-\infty}^{\infty} f(x_1, \ldots, x_n) e^{-i(\omega_1 x_1 + \ldots + \omega_n x_n)} dx_1 \ldots dx_n. \tag{12}$$

As in the one-dimensional case, we call the subset $\mathcal{S}$ of these functions "bandlimited" if the Fourier transform of every member of $\mathcal{S}$ vanishes outside a bounded subset $\Omega$ of the set over which function $F(\omega_1, \omega_2, \ldots, \omega_n)$ is defined. It is not required that $\Omega$ be symmetrical in any way or even that it be connected.

We would like to represent the function $f(x_1, x_2, \ldots, x_n)$ as a series of terms whose coefficients are the values of the function at a set of multidimensional periodic sampling points. Therefore let us define the $n$-dimensional periodic sampling basis of the whole space over which functions $f$ are defined. This basis is a set of linearly independent vectors in $\mathcal{R}^n$

$$\{v_i\} = \{v_1, v_2, \ldots, v_n\}. \tag{13}$$

Then, all sampling lattice points are all vectors which may be represented using this basis as

$$k_1 v_1 + k_2 v_2 + \ldots + k_n v_n, \tag{14}$$

where $k_1, k_2, \ldots, k_n = 0, \pm 1, \pm 2, \ldots$

Our goal is to find a suitable interpolation filter with the multidimensional impulse response $g(x_1, \ldots, x_n)$ such that we may reconstruct $f(x_1, \ldots, x_n)$ from its samples, i.e.

$$f(x_1, \ldots, x_n) = \sum_{k_1=-\infty}^{\infty} \cdots \sum_{k_n=-\infty}^{\infty} f(x_{1,k_1}, \ldots, x_{n,k_n}) g(x_1 - x_{1,k_1}, \ldots, x_n - x_{n,k_n}). \tag{15}$$

Similarly to the one-dimensional case, the $n$-dimensional samples are generated by $n$-dimensional impulse modulation, i.e. using $n$-dimensional Dirac deltas:

$$\begin{aligned} f(x_1, \ldots, x_n) = & \sum_{k_1=-\infty}^{\infty} \cdots \sum_{k_n=-\infty}^{\infty} g(x_1 - x_{1,k_1}, \ldots, x_n - x_{n,k_n}) \times \\ & \times \int_{-\infty}^{\infty} \cdots \int_{-\infty}^{\infty} f(\tau_1, \ldots, \tau_n) \delta(\tau_1 - x_{1,k_1}, \ldots, \tau_n - x_{n,k_n}) d\tau_1 \ldots d\tau_n. \end{aligned} \tag{16}$$

Changing the order of summation and integration in (16) we obtain

$$\begin{aligned} f(x_1, \ldots, x_n) = & \int_{-\infty}^{\infty} \cdots \int_{-\infty}^{\infty} f(\tau_1, \ldots, \tau_n) g(x_1 - \tau_1, \ldots, x_n - \tau_n) \times \\ & \times \sum_{k_1=-\infty}^{\infty} \cdots \sum_{k_n=-\infty}^{\infty} \delta(\tau_1 - x_{1,k_1}, \ldots, \tau_n - x_{n,k_n}) d\tau_1 \ldots d\tau_n \end{aligned} \tag{17}$$

and, after performing the Fourier series expansion of multidimensional series of Dirac deltas, we finally get

$$f(x_1, \ldots, x_n) = \frac{1}{Q} \sum_{k_1=-\infty}^{\infty} \cdots \sum_{k_n=-\infty}^{\infty} \int_{-\infty}^{\infty} \cdots \int_{-\infty}^{\infty} f(\tau_1, \ldots, \tau_n) \times$$

$$\times \exp\left(-i\left(\tau_1 \sum_{s=1}^{n} k_s u_1^s + \ldots + \tau_n \sum_{s=1}^{n} k_s u_n^s\right)\right) \times$$

$$\times g(x_1 - \tau_1, \ldots, x_n - \tau_n) d\tau_1 \ldots d\tau_n.$$

$$(18)$$

Here $Q$ is the hypervolume of the parallelepiped, whose edges are the basis vectors $\{v_i\}$. Also $u_1^s, \ldots, u_n^s$ are components of a vector $u_s$ belonging to the set $\{u_j\}$, whose members fulfil a reciprocal relation to the vectors $\{v_i\}$, i.e.

$$v_i \cdot u_j = 2\pi\delta_{ij}, \quad i = 1, \ldots, n, \quad j = 1, \ldots, n, \tag{19}$$

$\delta_{ij}$ being the Kronecker's symbol and $v_i \cdot u_j$ the inner product of the vectors $v_i$ and $u_j$. Condition (19) is a multidimensional generalisation of the Shannon condition (10). Thus, we may look at (18) as a convolution of

$$\frac{g(x_1, \ldots, x_n)}{Q} \quad \text{and} \quad \sum_{k_1=-\infty}^{\infty} \ldots \sum_{k_n=-\infty}^{\infty} f(\tau_1, \ldots, \tau_n) e^{-i\left(\tau_1 \sum_{s=1}^{n} k_s u_1^s + \ldots + \tau_n \sum_{s=1}^{n} k_s u_n^s\right)}$$

Taking the Fourier transform of both sides of (18) yields

$$F(\omega_1, \ldots, \omega_n) = \frac{G(\omega_1, \ldots, \omega_n)}{Q} \times$$

$$\times \sum_{k_1=-\infty}^{\infty} \ldots \sum_{k_n=-\infty}^{\infty} F(\omega_1 + \sum_{s=1}^{n} k_s u_1^s, \ldots, \omega_n + \sum_{s=1}^{n} k_s u_n^s),$$

$$(20)$$

where the components of the vector $u_j$, viz. $u_1^j, \ldots, u_n^j$, can be uniquely expressed by the components of the vector $v_i$, viz. $v_1^i, \ldots, v_n^i$, using (19). It can be seen from (20) that the multidimensional spectrum of sampled continuous function $f(x_1, \ldots, x_n)$ is periodic in the appropriate $n$-dimensional space. This phenomenon is a direct analogue of sampled functions spectrum periodicity encountered in one-dimensional space (see Section 5.2). In order to reconstruct exactly the continuous $n$-dimensional functions from its samples, the nonoverlapping of this spectrum portions must be ensured. This is only possible when the $n$-dimensional "bandwidth" of the continuous function is limited, i.e. the original function belongs to the set $\mathcal{S}$. In such a case we seek for a universal function $g(x_1, \ldots, x_n)$ which will be able to reconstruct any function $f(x_1, \ldots, x_n)$ belonging to $\mathcal{S}$ from its discrete samples. It follows from (20) that this may be achieved under certain assumptions. First of all, the vectors $\{u_j\}$ must be large enough to prevent the neighbouring repetitive spectra of $F(\omega_1 + \sum_{s=1}^{n} k_s u_1^s, \ldots, \omega_n + \sum_{s=1}^{n} k_s u_n^s)$ from overlapping. Secondly, the spectrum of the reconstructing filter $G(\omega_1, \ldots, \omega_n)$ has to be equal to the constant

$Q$ all over the region in which $F(\omega_1, \ldots, \omega_n)$ is nonzero and has to be equal zero where the repetitive spectra $F(\omega_1 + \sum_{s=1}^{n} k_s u_1^s, \ldots, \omega_n + \sum_{s=1}^{n} k_s u_n^s)$ are nonzero. The value of this spectrum is arbitrary in these regions where both spectrum $F(\omega_1, \ldots, \omega_n)$ and its repetitive images are zero. It is worth mentioning that unlike in the one-dimensional case, where such arbitrariness is easily avoidable and usually unnecessary (oversampling), in many dimensions it is a general feature. There exists, in general, some region in which the spectrum of the reconstruction filter is arbitrary even though the closest possible nonoverlapping packing of the repetitive spectra of the original sampled function has been chosen.

A direct corollary of the $n$-Dimensional Sampling Theorem, leading to the equation of the $n$-dimensional reconstruction filter similar to the one-dimensional case (see (8)), may be stated as follows.

**Corollary 14.** *A function of $n$ variables, $f(x_1, \ldots, x_n)$, whose $n$-dimensional Fourier transform is equal to zero for $\omega_1 > \omega_{1s}, \ldots, \omega_n > \omega_{ns}$ is uniquely reconstructable from its sampled values taken at uniformly spaced points in the $x_1, \ldots, x_n$ hyperplane if the spacings $T_1, \ldots, T_n$ satisfy the conditions $T_1 \leq \pi/\omega_{1s}, \ldots, T_n \leq \pi/\omega_{ns}$.*

The scalar spacings $T_1, \ldots, T_n$ correspond to the vectors $v_1, \ldots, v_n$ of Theorem 13, when the latter lie along the axes $x_1, \ldots, x_n$. Then the only nontrivial information contained in $v_1, \ldots, v_n$ are their lengths, i.e. the spacings $T_1, \ldots, T_n$. It is evident that the conditions imposed in Corollary 14 are a particular form of the more general conditions of the $n$-Dimensional Sampling Theorem. However, they are more practical and therefore are of particular interest. For function $f(x_1, \ldots, x_n)$ sampled according to the above corollary all the spectrum replicas are disjoint and it is possible to recover the spectrum of the continuous function using an $n$-dimensional ideal reconstruction filter. This filter can have for instance a spectrum equal to 1 in the parallelepiped with dimensions $\omega_{1s}, \ldots, \omega_{ns}$, that is a spectrum for $\omega_{1s} = \bar{\omega}_1, \ldots, \omega_{ns} = \bar{\omega}_n$, where $\bar{\omega}_1 = (2\pi)/T_1, \ldots, \bar{\omega}_n = (2\pi)/T_n$

$$G(\omega_1, \ldots, \omega_n) = \text{rect}(\frac{\omega_1}{\bar{\omega}_1}) \ldots \text{rect}(\frac{\omega_n}{\bar{\omega}_n}) = \prod_{i=1}^{n} \text{rect}(\frac{\omega_i}{\bar{\omega}_i}), \tag{21}$$

where

$$\text{rect}(x) = \begin{cases} 1, & -1 \leq x \leq 1, \\ 0, & \text{elsewhere.} \end{cases}$$

If the sampled version of the function $f$ has the form $f(k_1 T_1, \ldots, k_n T_n)$ then its discrete Fourier transform is

$$F_s(\omega_1, \ldots, \omega_n) = \frac{1}{T_1} \cdots \frac{1}{T_n} \sum_{k_1=-\infty}^{\infty} \cdots \sum_{k_n=-\infty}^{\infty} F(\omega_1 + k_1 \bar{\omega}_1, \ldots, \omega_n + k_n \bar{\omega}_n), \tag{22}$$

where $F$ is the continuous spectrum of $f$. This spectrum can be obtained from the $F_s$ by means of the following relation

$$F(\omega_1, \ldots, \omega_n) = F_s(\omega_1, \ldots, \omega_n) G(\omega_1, \ldots, \omega_n), \tag{23}$$

which in $x_1, \ldots, x_n$ domain corresponds to

$$f(x_1, \ldots, x_n) = T_1 \ldots T_n f(k_1 T_1, \ldots, k_n T_n) * g(x_1, \ldots, x_n), \tag{24}$$

where $*$ stands for convolution, and

$$\begin{aligned} g(x_1, \ldots, x_n) &= \frac{1}{(2\pi)^n} \int_{-\infty}^{\infty} \ldots \int_{-\infty}^{\infty} \text{rect}(\frac{\omega_1}{\bar{\omega}_1}) \ldots \text{rect}(\frac{\omega_n}{\bar{\omega}_n}) e^{i(\omega_1 x_1 + \ldots + \omega_n x_n)} d\omega_1 \ldots d\omega_n \\ &= \frac{1}{T_1} \cdots \frac{1}{T_n} \frac{\sin(\bar{\omega}_1 x_1)}{\bar{\omega}_1 x_1} \cdots \frac{\sin(\bar{\omega}_n x_n)}{\bar{\omega}_n x_n}. \end{aligned} \tag{25}$$

Substituting (25) to (24), we obtain the following interpolation relation for this particular $n$-dimensional case:

$$\begin{aligned} f(x_1, \ldots, x_n) = \sum_{k_1=-\infty}^{\infty} \cdots \sum_{k_n=-\infty}^{\infty} f(k_1 T_1, \ldots, k_n T_n) \times \\ \times \frac{\sin(\bar{\omega}_1 (x_1 - k_1 T_1))}{\bar{\omega}_1 (x_1 - k_1 T_1)} \cdots \frac{\sin(\bar{\omega}_n (x_n - k_n T_n))}{\bar{\omega}_n (x_n - k_n T_n)}, \end{aligned} \tag{26}$$

which is a generalisation of the familiar sinc function, see (8).

Another important corollary of the $n$-Dimensional Sampling Theorem is presented in [2] and [5] in the form of multidimensional Generalised Sampling Expansion. This approach basing on the use of $m$ parallel processing $n$-dimensional channels instead of one as in original sampling theorem enables us to reduce the sampling frequency $m$ times. This relaxes significantly constraints imposed on the $n$-dimensional sampling procedure.

However, we should bear in mind that the actual problem of nonlinear dynamic control system modelling is somewhat different from "pure" nonlinear function reconstruction from its samples. Since some of the "independent" variables in $f$ are values of $y$ and $u$ in previous instants of time, they are not spread arbitrarily over their domain. In fact, their distribution is related to the dynamics of the system under consideration and it cannot be expected that this distribution be uniform or regular in any sense. Thus we have to face the problem of nonlinear function reconstruction from irregular samples. In this problem we ask whether—and how—it is possible to reconstruct a multidimensional bandlimited function $f$ from its nonuniformly sampled values.

Unlike the uniform sampling, there is no guarantee of the uniqueness of a bandlimited function recovery from arbitrary nonuniform samples. We must ensure that the nonuniform samples locations not only satisfy the Nyquist rate on average, but also the sample locations are not the zero-crossings of a bandlimited signal of the same bandwidth. This is always fulfilled if the average sampling rate of a set of sample locations is higher than the Nyquist rate. However, it is interesting to note that—unlike in the uniform sampling theory—even if the average sampling rate is less than the Nyquist rate it is still possible to uniquely reconstruct the functions. These and other similar results found in mathematical and signal processing literature provide several uniqueness results (see [30] and references therein for detailed information on the subject). Unfortunately such results have had little consequences for applied sciences, because they were not constructive. Only recently there have emerged some results of practical solutions to the irregular sampling problem (see [15] and [9] for a detailed quantitative analysis of known approaches). These methods—although very well theoretically established—are still at a fairly early stage of their practical implementation. The questions of the relations between sampling density and the iterative method convergence to the original function $f$ are still a matter of intensive research.

### 5.4 Relevance for Neural Modelling

As we have seen in section 5.2, there are essentially two steps in the approximate interpolation of $f$ via multidimensional sampling:

1. $n$-dimensional bandlimiting of $f$;
2. design of an interpolating filter with characteristic approximating (21).

The possibility of 1 is assumed to be available either due to properties of $f$ or implicitly by design (using for example the Generalised Sampling Expansion theorem). The essence of the neural approach is step 2. It is based on the observation [34] that the Fourier transform of a Gaussian is also Gaussian and may approximate (21) with quantifiable accuracy.

Thus, after a realistic bandlimiting of $f$, it can be approximately interpolated by linear combinations of Gaussians with series of the form (26), where $g(x_1 - x_{1,k_1}, \ldots, x_n - x_{n,k_n})$ will correspond to the approximation of (21). Since the formulation of our problem (section 2) allows only a *finite* number of samples, the series has to be truncated. It is important that an upper bound for the truncation error is available [34].

These results solve problem 2 of section 2. Problem 1 was also shown (section 3) to have a positive solution for Gaussians. The potential of Gaussian feedforward neural networks thus motivates intense research on their applications to neurocontrol.

However, it should be pointed out that the method bears similarity (on the conceptual level) to the direct methods of optimisation [24]. No *a priori* knowledge about $f$ (except for its continuity and perhaps Fourier transformability) is incorporated and the problem scales exponentially with the dimension of $K$. This phenomenon, known in optimisation as the 'curse of dimensionality' [10], led to the abandonment of direct methods in favour of the ones using some *a priori* knowledge about $f$.

By assuming little about $f$ the approach gains generality, but the price is the problem of dimensionality. In most control applications some nontrivial *a priori* information is usually available (it is in any case necessary for the design of antialiasing filters). On the other hand, the computational (real-time) constraints may be severe.

# 6 Approximation of Continuous Functionals with Feedforward Networks

All approximation approaches mentioned in previous sections are concerned with continuous functions defined on a compact set in $\mathcal{R}^n$, a space of *finite* dimension. However, there arise in practice situations where *functionals* have to be computed. These are defined on some set of functions, a space of *infinite* dimension.

This is especially interesting in the context of neural modelling as the output of a dynamic system at any particular time can be viewed as a functional [4]. This is why such an approach may be recognised as an alternative formulation of Problem 1 of section 2. Also, this approach is a generalisation of results obtained elsewhere for approximation of functions by neural networks (see [6], [11], [18], [33]).

## 6.1 Approximation of Continuous Functionals

To understand the essence of the problem, we first look at the relation between functionals and functions in many variables.

Let $C[a,b]$ denote the space of all continuous real functions on $[a,b]$ with the norm

$$\|f\|_{C[a,b]} = \sup_{x\in[a,b]} |f(x)|$$

and let us consider a functional $F: C[a,b] \to \mathcal{R}$. That is, to each function $f \in C[a,b]$ a unique real number $r \in \mathcal{R}$ is assigned, or $F: f \mapsto r$. To find a related function of the sort considered in finite-dimensional analysis, we may proceed as follows. Using the points:

$$a = x_0, x_1, \ldots, x_{m-1}, x_m = b$$

we divide the interval $[a,b]$ into $m$ equal parts. Then we replace $f$ by its values in these points i.e. $f(x_0)$, $f(x_1), \ldots$ Now we may approximate the functional $F(f)$ by the function of $m-1$ variables $F(f(x_1), \ldots, f(x_{m-1}))$. This idea, first adopted by Euler (see [13]) in the framework of the calculus of variations, serves as a background of the approach presented by Chen & Chen [4] to approximation of continuous functionals.

A set $U$ in $C[a,b]$ is compact if for any sequence $f_n \in U$, there exists a function $f$ in $U$ and a subsequence $f_{n_k}$ of $f_n$ such that $\|f - f_{n_k}\|_{C[a,b]} \to 0$. Then, we call $\sigma(\cdot)$ a generalised sigmoidal function if $\sigma: \mathcal{R} \to \mathcal{R}$ and satisfies:

$$\sigma(x) \to \begin{cases} \to 1, & \text{if } x \to +\infty; \\ \to 0, & \text{if } x \to -\infty. \end{cases}$$

It may be noted that all monotone, increasing sigmoidal functions belong to this class. It is also worth mentioning that continuity of $\sigma$ is not required either in the definition or in later results.

The main result given by Chen & Chen [4] is summarised in the form of the following theorem for the space $C[a,b]$.

**Theorem 15 Chen & Chen.** *Suppose that $U$ is a compact set in $C[a,b]$, $F$ is a continuous functional defined on $U$, and $\sigma(\cdot)$ is a bounded generalised sigmoidal function. Then for any $\varepsilon > 0$, there exist $m+1$ points $a = x_0 < \ldots < x_m = b$, a positive integer $N$ and constants $c_i$, $\theta_i$, $\xi_{ij}$, $i = 1, \ldots, N$, $j = 0, 1, \ldots, m$ such that:*

$$\left| F(u) - \sum_{i=1}^{N} c_i \sigma\Big(\sum_{j=0}^{m} \xi_{ij} u(x_j) + \theta_i\Big) \right| < \varepsilon, \quad \forall u \in U. \diamond \tag{27}$$

The significance of this theorem may be summarised in the following points:

1. the theorem provides a theoretical basis for approximation of continuous functionals defined on some compact subset in an infinite dimensional space of functions, which is more general (and more difficult) then usual problem setting in the finite dimensional case;
2. the theorem, apart from showing the representation capability of single-hidden-layer neural networks, gives an explicit form of the approximant; this result is much stronger then the purely existential ones obtained from the Stone-Weierstrass theorem;
3. it can be shown by the straightforward substitution

$$F(u_x) = f(x_1, \ldots, x_n)$$

where:

$$u_x(t) = x_i + \frac{x_{i+1} - x_i}{t_{i+1} - t_i}, \quad u_x \in U$$

$$t_i < t < t_{i+1} \qquad i = 1, \ldots, n-1,$$

$$t_i = a + (i-1)\frac{b-a}{n-1},$$

$$a = t_1 < t_2 < \ldots < t_n = b$$

(here $u_x$ is a piecewise linear function taking values $x_j$ at points $t_j$), that the function approximation problem is a special case of Theorem 15 when all functions in $U$ are piecewise linear functions with $n$ nodes.

### 6.2 Relevance for Neural Modelling

A potential advantage of the presented approach is the reformulation of the problem of approximation of dynamic systems in terms of approximation of continuous functionals defined on a compact set (of input functions). The approximation of discrete-time systems by neural networks methods is due to Sandberg [33]. With the functional approximation we may use the Sandberg approach equally effectively for continuous- and discrete-time systems.

This is achieved in the following input-output setting. Let us define $X_1$ as a set of functions $u: \mathcal{R} \to \mathcal{R}^p$ and $X_2$ as a set of functions $y: \mathcal{R} \to \mathcal{R}^q$. Then a dynamic system $G$ can be viewed as a map $G: X_1 \to X_2$, such that $\forall u \in X_1$,

$Gu = y \in X_2$. On the other hand, this map may be considered to be a set of functionals from $X_1$ to $\mathcal{R}$, i.e. correspondences of $u \in X_1$ to the values of $y(t) = (Gu)(t) \in \mathcal{R}$ for a given time $t$.

For any $x \in X$ let us also define a "windowing" operator $W$ as:

$$(W_{\alpha,a}x)(t) = \begin{cases} x(t), & \text{if } t \in \Gamma_{\alpha,a}; \\ 0, & \text{if } t \notin \Gamma_{\alpha,a}, \end{cases}$$

where $\alpha \in \mathcal{R}$ and $\Gamma_{\alpha,a} = \{r \in \mathcal{R}, |r-\alpha| \leq a\}$. That is, $W_{\alpha,a}x$ is a "windowed" version of $x$ with the window centred at $\alpha$ and its width $2a$.

If $U$ is a nonempty set in $X_1$, define $U_{\alpha,a} = u|\Gamma_{\alpha,a}$, $u \in U$, where $u|\Gamma_{\alpha,a}$ is the restriction of $u$ to $\Gamma_{\alpha,a}$, that is $u|\Gamma_{\alpha,a} = W_{\alpha,a}u$.

A map $G$ from $X_1$ to $X_2$ is said to be of approximately finite memory, if for all $\varepsilon > 0$, there is an $a > 0$ such that

$$\left|(Gu)_j(t) - (GW_{\alpha,a}u)_j(t)\right| < \varepsilon, \quad j = 1, \ldots, q$$

holds for any $\alpha \in \mathcal{R}$, $u \in U$.

For each $t_1 \in \mathcal{R}$, define $T_{t_1}: X_1 \to X_1$ to be the shift operator given by $(T_{t_1}x)(t) = x(t-t_1)$ for all $t \in \mathcal{R}$. A map $G: X_1 \to X_2$ is shift invariant if $(GT_{t_1}u)(t) = (Gu)(t-t_1)$ for any pair $(t, t_1)$, $t_1 \in \mathcal{R}$, $u \in X_1$.

Moreover, if we assume that the set $U$, i.e. the domain of $G$ in which we deal with the approximation problem, satisfies the following:

1. if $u \in U$, then $u|\Gamma_{\alpha,a} \in U$ for any $\alpha \in \mathcal{R}$; $a > 0$ (this is a nontrivial assumption; for $\sin(x)$ is periodic on $\mathcal{R}$, but aperiodic on $[-a, a]$, where $a < \pi$);
2. for all $\alpha \in \mathcal{R}$, $a > 0$, $U_{\alpha,a}$ is a compact set in $C[\alpha - a, \alpha + a]$ (here we assume that $U_{\alpha,a}$ is a set of functions $f : [\alpha - a, \alpha + a] \to \mathcal{R}$) ;
3. let $(Gu)(t) = ((Gu)_1(t), \ldots, (Gu)_q(t))$, then each $(Gu)_j(t)$ is a continuous functional defined over $U_{\alpha,a}$ with the corresponding topology in $C[\alpha-a, \alpha+a]$ (see [7]),

then following theorem holds:

**Theorem 16 [4].** *If $U$ and $G$ satisfy all the assumptions 1–3 above, and $G$ is of approximately finite memory, then for any $\varepsilon > 0$, there exist $a > 0$, a positive integer $N$, a positive integer $m$, $m+1$ points in $[\alpha - a, \alpha + a]$ and constants $c_i(G, \alpha, a)$ depending on $G$, $\alpha$, $a$ only, $\theta_i$, and $\xi_{i,j,k}$, $i = 1, \ldots, N$, $j = 0, 1, \ldots, m$, $k = 1, \ldots, p$ such that:*

$$\left|(Gu)_l(t) - \sum_{i=1}^{N} c_i(G,\alpha,a)\sigma\Big(\sum_{j=0}^{m}\sum_{k=1}^{p} \xi_{i,j,k}u_k((t_j)) + \theta_i\Big)\right| < \varepsilon \tag{28}$$

*with*

$$\begin{aligned} l &= 1, 2, \ldots, q \\ u &= (u_1, \ldots, u_p) \\ u_k &= (u_k(t_0), \ldots, u_k(t_m)). \ \diamond \end{aligned}$$

This result is a more general form of dynamic system modelling (both continuous-time and discrete-time) using neural networks than the one given by Sandberg ([33]).

# 7 Approximation with Recurrent Networks

This section analyses the possibility of approximating (29) with (31) and/or related recurrent models and discuss it in the context of differential approximation as defined in section 7.1.

## 7.1 The Problem of Differential Approximation

Consider an unknown plant with an accessible state $x$ (i.e. $y = x$) given by

$$\dot{x} = f(x, u), \quad x(t_0) = x^0, \tag{29}$$

where $x \in X \subseteq \mathcal{R}^n$, $u \in U \subseteq \mathcal{R}^p$. The problem of its identification is unsolved in general [16], [43]. There are two main neural approaches [44]. One [41] is to approximate (29) with

$$\dot{\hat{x}} = \sum_{i=1}^{N} w_i g\left(\frac{\|\bar{x} - c^i\|^2}{v_i}\right), \tag{30}$$

where $g(\cdot)$ is a radial basis function, typically Gaussian, $w_i, v_i$ are scalar and $c^i$ vector parameters. Here $c^i$ and $v_i$ define the grid and $w_i$ is adaptive. Also, $\bar{x} = [\hat{x}^T \vdots u^T]^T \in \mathcal{R}^{n+p}$ with $\hat{x} \in X \subseteq \mathcal{R}^n$ being the approximation of the plant's state $x \in X \subseteq \mathcal{R}^n$. Roughly speaking, the right-hand side of (29) is expanded into a finite weighted sum (linear combination) of nonlinear functions retaining the dimension of the original state, i.e. both $x$ and $\hat{x}$ are $n$-vectors.

The other method [44] postulates a network

$$\begin{aligned} \dot{\chi} &= -\chi + \sigma(W\chi) + u, \\ \hat{x} &= C\chi, \end{aligned} \tag{31}$$

where $\chi \in H \subseteq \mathcal{R}^N, \hat{x} \in X \subseteq \mathcal{R}^n, W \in V \subseteq \mathcal{R}^{N^2}$ with $N \geq n$, so that $\hat{x}$, the approximation of $x$ from (29), is a projection of $\chi$ to $n$ dimensions. Roughly speaking, (31) fixes its right-hand side and expands its order. The form of (31) is, however, arbitrary, in the sense that it is not motivated by any mathematical reasoning. Thus, there naturally arises a question about the sense of investigating (31) and not some other dynamic, parametric model. The following definition attempts to address the problem.

**Definition 17 Differential Approximation.** *Given a nonlinear dynamic system*

$$\Sigma\colon\ \dot{x} = f(x, u),$$

*where* $x(t_0) = x^0$, $x \in X \subseteq \mathcal{R}^n$, $u \in U \subseteq \mathcal{R}^p$, *and* $f(\cdot,\cdot)$ *Lipschitz in* $X \times U$, *find a dynamic parametric system*

$$\Sigma_W\colon\ \dot{\chi} = \phi(\chi, u, W),$$

*with* $\chi(t_0) = \chi^0$, $\chi \in H \subseteq \mathcal{R}^N, N \geq n$, *parameters* $W \in V \subseteq \mathcal{R}^r$, *and* $\phi(\cdot,\cdot,\cdot)$ *Lipschitz in* $H \times U \times V$, *such that*

$$\forall\varepsilon > 0 \quad \exists r \in \mathcal{N} \quad \exists W \in V \quad \forall t \geq 0 \quad \|x - \hat{x}\| < \varepsilon,$$

*where* $\hat{x} = \Pi(\chi)$, *with* $\Pi$ *the projection from* $H$ *to* $X$. ◇

Definition 17 postulates, strictly speaking, *local* differential approximation (for a given $x^0$). The global definition would have the additional quantifier $\forall x^0 \in X$ (between $\exists W \in V$ and $\forall t \geq 0$).

The definition leaves room for manoeuvre with the choice of the norm $\|\cdot\|$, the sets $X, H, V$ and the numbers $N$ and $r$, but obviously $X \subseteq H$. Finally, the projection $\Pi(\cdot)$ may be replaced by a less trivial mapping if this would lead to more useful results.

The insistence of parameterisation is motivated by the possibility of employing algebraic tools along with the analytic ones it offers for investigation of nonlinear dynamics. Fixing the model's structure allows focussing on $W$, simplifying identification and design, as we then deal with the finite-dimensional space $V$. Apart from the existence and uniqueness questions, the issue of interest is that the parameterisation of $\Sigma_W$ leads to generic nonlinear control. This means that if a nonlinear system $\Sigma$ can be replaced by its parametric equivalent $\Sigma_W$, then given the special, yet generic, form of $\phi(\cdot,\cdot,\cdot)$ of $\Sigma_W$ the control problem should be soluble with existing techniques or a new one. An obvious suggestion is to look for a parametric version of the affine model [20].

### 7.2 Feedforward Approach

An obvious approach to approximation of (29) with (31), already mentioned in section 7.1, is to apply the results for feedforward networks (see sections 3–4) to the right-hand side of (29). However, (29) with the expanded right-hand side is not of the form (31). Hence the need for defining extra variables, so that the resulting system of differential equations has the required form.

This procedure was independently considered by Sontag [38] and Funahashi & Nakamura [12]. In the first case the recurrent model for (29) is

$$\dot{\chi} = \boldsymbol{\sigma}(A\chi + Bu),$$

where $\boldsymbol{\sigma}$ is the vector of sigmoids and the matrices $A, B$ comprise the adjustable weights. The other authors considered

$$\dot{\chi} = -\frac{1}{\tau}\chi + W\boldsymbol{\sigma}(\chi),$$

with $\tau \in \mathcal{R} - \{0\}$, $W$ an $N \times N$ matrix, which they propose as an approximation of

$$\dot{x} = f(x),$$

where $f$ is continuously differentiable.

The Funahashi-Nakamura approach can be extended for the control model (29). Since the essence of theirs and Sontag's approach is the same, we shall give a brief account of Sontag's method to clarify the issues with a minimum of technicalities.

### 7.3 Main Results

To begin with, we formally state the results [38], [12].

**Theorem 18 Sontag.** *Let a control system*

$$\dot{x} = f(x, u), \quad x(t_0) = x^0, \tag{32}$$

$$y = h(x) \tag{33}$$

*be given with $x \in \mathcal{R}^n, u \in \mathcal{R}^p, y \in \mathcal{R}^q$ with $f$ and $h$ continuously differentiable, such that for all $u: [0, T] \to \mathcal{R}^p$ the solution of (32) exists and is unique for all $t \in [0, T]$ and some compact sets $K_1 \subset \mathcal{R}^n, K_2 \subset \mathcal{R}^p$, while $x \in K_1, u \in K_2$. Then there exists a recurrent neural network of the form*

$$\dot{\chi} = \boldsymbol{\sigma}(A\chi + Bu), \tag{34}$$

$$y = C\chi, \tag{35}$$

*where $\chi \in \mathcal{R}^N$, $N \geq n$, $y \in \mathcal{R}^q$, $\boldsymbol{\sigma}$ is a vector of sigmoids and $A, B$ are matrices such that, on $K_1 \times K_2$,*

$$\forall \varepsilon > 0 \quad \forall t \in [0, T] \quad \|x(t) - M(\chi(t))\| < \varepsilon \quad \textit{and} \quad \|h(x(t)) - C\chi(t)\| < \varepsilon,$$

*where $M$ is a differentiable map.* $\diamond$

**Theorem 19 Funahashi-Nakamura.** *Let a system*

$$\dot{x} = f(x), \quad x(t_0) = x^0$$

*be given with $x \in \mathcal{R}^n$ and $f$ continuously differentiable, such that $\forall t \in [0, T]$ $x(t) \in K_1$, a compact subset of $\mathcal{R}^n$. Then there exists a recurrent neural network of the form*

$$\dot{\chi} = -\frac{1}{\tau}\chi + W\boldsymbol{\sigma}(\chi), \tag{36}$$

*where $\chi \in \mathcal{R}^N$, $N \geq n$, $\tau \in \mathcal{R} - \{0\}$, $\boldsymbol{\sigma}$ is a vector of $N$ sigmoids and $W$ is a real matrix, such that, on $K_1$,*

$$\forall \varepsilon > 0 \quad \forall t \in [0, T] \quad \|x(t) - \Pi(\chi(t))\| < \varepsilon,$$

*where $\Pi: \mathcal{R}^N \to \mathcal{R}^n$ is the projection $\Pi: (\chi_1, \ldots, \chi_n, \ldots, \chi_N) \mapsto (\chi_1, \ldots, \chi_n)$.* $\diamond$

Note that Sontag's result involves not only the control signal $u$ in (32), but also output $y$, i.e. postulates equation (33), as well. Also (34) and (36) are *different* neural models. However, the proof methodology is the same in both cases and will be briefly summarised here for Theorem 18 as the starting point for further discussion.

We can rewrite (32)–(33) as

$$\dot{x} = f(x, u), \quad x(t_0) = x^0,$$
$$\dot{y} = (\partial h(x)/\partial x)^T f(x, u)$$

and, with the new state $\bar{x} = [x^T \vdots y^T]^T \in \mathcal{R}^{n+q}$, we can write $y = \bar{C}\bar{x}$, where $\bar{C} = [0 \; I_{q\times q}]^T$. Thus, without loss of generality, we can take $h$ in (33) to be linear, and hence the form of (35) will suffice.

Using the results for feedforward networks (see e.g. section 3) we can represent $f(x, u)$ on $K_1 \times K_2$ as

$$f(x, u) \approx W\boldsymbol{\sigma}(Ax + Bu + \gamma),$$

where the matrices are $W \in \mathcal{R}^{n\times N}$, $A \in \mathcal{R}^{N\times n}$, $B \in \mathcal{R}^{N\times p}$ and the vector $\gamma \in \mathcal{R}^n$ (see (2)). Thus the approximation of (32) is given by

$$\dot{x} = W\boldsymbol{\sigma}(Ax + Bu + \gamma), \quad x \in \mathcal{R}^n \tag{37}$$

with $y = Cx$ (recall that we may assume that $h(x) \equiv Cx$). For simplicity, let $\text{rank}(W) = n$. Then by $x = W\chi$, we get

$$W\dot{\chi} = W\boldsymbol{\sigma}(AW\chi + Bu + \gamma)$$

which is equivalent to

$$\dot{\chi} = \boldsymbol{\sigma}(AW\chi + Bu + \gamma), \quad \chi \in \mathcal{R}^N \tag{38}$$

and the elimination of $\gamma$ is easy, so that the required form (34) is achieved, complemented with $y = CW\chi$, as in (35).

Note that the passage from (37) to (38) via $x = W\chi$ is equivalent to adding an extra $N - n \geq 0$ differential equations, so that the dimension of (34) exceeds that of (32).

### 7.4 Relevance for Differential Approximation

In section 7.1 we presented two alternatives for approximation with recurrent neural networks. One (involving feedforward networks) was to develop the right-hand side of (29) in a series and truncate it to, say, $N_1$ terms, but preserve the system order, $n$, in the model. The other approach, using recurrent neural networks, consisted in fixing the right-hand side of a recurrent neural model, but expanding its order, to say $N_2$. One may wonder if, for a given system (29), $N_1$ and $N_2$ are comparable in some sense, e.g. the implementation cost.

The results of Sontag and Funahashi-Nakamura described in section 7.3 expand *both* the right-hand side of (29) *and* the order of the neural model, which means building a large feedforward network with $N_1$ units and transforming it into a recurrent network of order $N_2 = N_1$. From the engineering point of view it is the costliest and least convenient solution. It also is exactly what the definition of differential approximation was set up to avoid, so the problem in Definition 17 remains open. However, as pure existence results Theorems 18 and 19 are perfectly valid.

# 8 Conclusions

Multivariate continuous function approximation and interpolation were identified as fundamental problems for neurocontrol with feedforward networks. In this framework the major approaches to the questions were explained and evaluated from the point of view of practical relevance.

The methods stemming from the Stone-Weierstrass and Kolmogorov theorems address only the general approximation problem, i.e. they establish pure existence results. In particular, they imply that the multidimensional sampling approach is feasible. However, the latter also gives, unlike the others, a constructive, quantifiable methodology for carrying out the interpolation.

An interesting new approach to dynamic systems' modelling is the one based on functional approximation. Not only does it give an existence result to the functional (infinite-dimensional) approximation, but also allows neural modelling of both continuous- and discrete-time dynamic systems in a consistent manner.

Since the existence results are now firmly established, it seems advantageous to do more in-depth research on the practical control aspects of the multidimensional sampling approach. This must primarily address the issues of dimensionality and the practicality of bandlimiting.

Approximation with recurrent neural networks builds on the results for feedforward ones and therefore fails to solve the problem of differential approximation. The construction used in the proofs of relevant theorems is not practicable from the engineering point of view. Thus the approach gives pure existence results only.

# References


1. V. I. Arnol'd. Some questions of approximation and representation of functions. In *Proceedings of the International Congress of Mathematicians*, pages 339–348, 1958. (English translation: American Mathematical Society Translations, Vol. 53).
2. J. L. Brown. Sampling reconstruction of $n$-dimensional bandlimited images after multilinear filtering. *IEEE Trans. Circuits & Systems*, CAS-36:1035–1038, 1989.
3. J. C. Burkill and H. Burkill. *A Second Course in Mathematical Analysis.* Cambridge University Press, Cambridge, England, 1970.
4. T. Chen and H. Chen. Approximation of continuous functionals by neural networks with application to dynamic systems. *IEEE Transactions on Neural Networks*, 4:910–918, 1993.

5. K. F. Cheung. A multidimensional extension of Papoulis' Generalised Sampling Expansion with the application in minimum density sampling. In Robert J. Marks II, editor, *Advanced Topics in Shannon Sampling and Interpolation Theory.* Springer Verlag, New York, 1993.
6. G. Cybenko. Approximation by superposition of a sigmoidal function. *Mathematics of Control, Signals, and Systems*, 2:303–314, 1989.
7. J. Dugundji. *Topology.* Allyn and Bacon, Boston, 1966.
8. A. Dzieliński. *Optimal Filtering and Control of Two-dimensional Linear Discrete-Time Systems.* Ph.D. Thesis, Faculty of Electrical Engineering, Warsaw University of Technology, Warsaw, Poland, February 1992.
9. H. G. Feichtinger and K. Gröchenig. Theory and practice of irregular sampling. In J. Benedetto and M. Frazier, editors, *Wavelets: Mathematics and Applications.* CRC Press, 1993.
10. R. Fletcher. *Practical Methods of Optimization. Second Edition.* Wiley, Chichester, 1987.
11. K. Funahashi. On the approximate realization of continuous mappings by neural networks. *Neural Networks*, 2:183–192, 1989.
12. K. Funahashi and Y. Nakamura. Approximation of dynamical systems by continuous time recurrent neural networks. *Neural Networks*, 6:801–806, 1993.
13. I. M. Gelfand and S. V. Fomin. *Calculus of Variations.* Prentice-Hall, Englewood Cliffs, N.J., 1963.
14. F. Girosi and T. Poggio. Representation properties of networks: Kolmogorov's theorem is irrelevant. *Neural Computation*, 1:465–469, 1989.
15. K. Gröchenig. Reconstruction algorithms in irregular sampling. *Mathematics of Computations*, 59:181–194, 1992.
16. R. Haber and H. Unbehauen. Structure identification of nonlinear dynamic systems—a survey on input/output approaches. *Automatica*, 26:651–677, 1990.
17. R. Hecht-Nielsen. Kolmogorov's mapping neural network existence theorem. In *Proceedings of the International Joint Conference on Neural Networks*, volume 3, pages 11–14, New York, 1987. IEEE Press.
18. K. Hornik, M. Stinchcombe, and H. White. Multilayer feedforward networks are universal approximators. *Neural Networks*, 2:359–366, 1989.
19. K. J. Hunt, D. Sbarbaro, R. Żbikowski, and P. J. Gawthrop. Neural networks for control systems: A survey. *Automatica*, 28(6):1083–1112, November 1992.
20. A. Isidori. *Nonlinear Control Systems: An Introduction.* Springer-Verlag, New York, Second edition, 1989.
21. T. Kaczorek. *Two-Dimensional Linear Systems.* Springer-Verlag, Berlin, 1985.
22. A. N. Kolmogorov. On the representation of continuous functions of several variables by superpositions of continuous functions of a smaller number of variables. *Dokl. Akad. Nauk SSSR*, 108:179–182, 1956. (in Russian).
23. A. N. Kolmogorov. On the representation of continuous functions of many variables by superposition of continuous functions of one variable and addition. *Dokl. Akad. Nauk SSSR*, 114:953–956, 1957. (English translation: American Mathematical Society Translations, Vol. 28).
24. J. S. Kowalik and M. R. Osborne. *Methods for Unconstrained Optimization Problems.* Elsevier, New York, 1968.
25. V. Kůrková. Kolmogorov's theorem is relevant. *Neural Computation*, 3:617–622, 1991.
26. V. Kůrková. Kolmogorov's theorem and multilayer neural networks. *Neural Networks*, 5:501–506, 1992.

27. J. N. Lin and R. Unbehauen. On the realization of Kolmogorov's network. *Neural Computation*, 5:18–20, 1993.
28. S. Lipschutz. *General Topology.* McGraw-Hill, New York, 1965.
29. G. G. Lorentz. *Approximation of Functions.* Holt, Reinhart and Winston, New York, 1966.
30. F. Marvasti. Nonuniform sampling. In Robert J. Marks II, editor, *Advanced Topics in Shannon Sampling and Interpolation Theory.* Springer Verlag, New York, 1993.
31. D. P. Petersen and D. Middleton. Sampling and reconstruction of wave-number-limited functions in $n$-dimensional euclidean spaces. *Information and Control*, 5:279–323, 1962.
32. W. Rudin. *Principles of Mathematical Analysis, Third Edition.* McGraw-Hill, Auckland, 1976.
33. I. W. Sandberg. Approximation theorems for discrete-time systems. *IEEE Transactions on Circuits and Systems*, 38:564–566, 1991.
34. R. M. Sanner and J.-J. E. Slotine. Gaussian networks for direct adaptive control. *IEEE Transactions on Neural Networks*, 3:837–863, 1992.
35. D. Sbarbaro. *Connectionist Feedforward Networks for Control of Nonlinear Systems.* Ph.D. Thesis, Department of Mechanical Engineering, Glasgow University, Glasgow, Scotland, October 1992.
36. C. E. Shannon. Communication in the presence of noise. *Proceedings of the IRE*, 37:10–21, 1949.
37. I. N. Sneddon. *Fourier Transforms.* McGraw-Hill, New York, 1951.
38. E. Sontag. Neural nets as systems models and controllers. In *Proc. Seventh Yale Workshop on Adaptive and Learning Systems*, pages 73–79. Yale University, 1992.
39. D. A. Sprecher. On the structure of continuous functions of several variables. *Transactions of the American Mathematical Society*, 115:340–355, 1965.
40. M. H. Stone. The generalized Weierstrass approximation theorem. *Mathematics Magazine*, 21:167–184, 237–254, 1948.
41. E. Tzirkel-Hancock. *Stable Control of Nonlinear Systems Using Neural Networks.* Ph.D. thesis, Trinity College, Cambridge University, Cambridge, England, August 1992.
42. A. G. Vitushkin. On Hilbert's thirteenth problem. *Dokl. Akad. Nauk SSSR*, 95:701–704, 1954. (in Russian).
43. R. Żbikowski. The problem of generic nonlinear control. In *Proc. IEEE/SMC International Conference on Systems, Man and Cybernetics, Le Touquet, France*, volume 4, pages 74–79, 1993.
44. R. Żbikowski. *Recurrent Neural Networks: Some Control Problems.* Ph.D. Thesis, Department of Mechanical Engineering, Glasgow University, Glasgow, Scotland, May 1994.
45. R. Żbikowski, K. J. Hunt, A. Dzieliński, R. Murray-Smith, and P. J. Gawthrop. A review of advances in neural adaptive control systems. Technical Report of the ESPRIT NACT Project TP-1, Glasgow University and Daimler-Benz Research, 1994.

# Dynamic Systems in Neural Networks

Kevin Warwick[1], Chandrasekhar Kambhampati[1], Patrick Parks[2], Julian Mason[1]

[1] Department of Cybernetics, University of Reading, Reading RG6 2AY, UK.
[2] Mathematical Institute, University of Oxford, Oxford OX1 3LB, UK

## 1 Introduction

Many schemes for the employment of neural networks in control systems have been proposed [9] and some practical applications have also been made [2]. It is possible to apply a neural network to just about every conceivable control problem, however in many cases, although of interest, the network might not be the best or even a good solution, due to its relatively complex nonlinear operation. A neural network is in essence a nonlinear mapping device and in this respect, at the present time, most of the reported work describing the use of neural networks in a control environment is concerned solely with the problem of process modelling or system identification.

As a stand-alone, the system identification exercise produces an open-loop mode of operation for a neural network in that the network outputs are aimed at modelling the identified plant's outputs. The network output themselves can then be used for such as prediction of plant output values, fault analysis or inferencing. In the system identification exercise an error measurement is readily available in the form of the simple difference between the plant and neural network outputs. Further, because the network is operated in open-loop mode, network stability, in all senses, is not a consideration.

For closed-loop adaptive control systems a very different picture exists in that an outstanding problem that requires a solution is how to guarantee their stability when such systems contain one or more neural networks, the networks being used not only to identify the plant but also to provide an adaptive controller. Some progress has been made in understanding the dynamics of weight adjustments in neural networks used as function approximators, especially where "local" approximating functions are employed as in the "CMAC" devices of J.S. Albus [1, 18, 3]. Progress in this area is reported here and is seen as a vital contribution to the development of a more complete stability analysis. Some simple models suggest how such a comprehensive theory may develop, where restrictions are imposed in the dynamics of an otherwise "eventually stable" closed-loop.

One of the most popular forms of neural network at the present time is the Radial Basis Function network, which has similar weight training properties to CMAC. This particular type of network is discussed here with particular reference being made to the problem of basis function centre selection and its effect on the stability properties of a closed-loop adaptive control system.

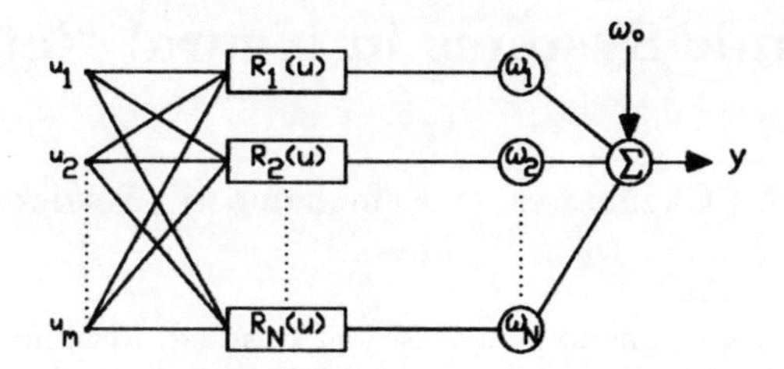


**Fig. 1.** A Radial Basis Function Network

## 2 Radial Basis function Networks for Identification and Control

Of the many types of neural networks possible, the Multilayer Perceptron (MLP), despite its problems due to slow weight convergence through nonlinear learning, has perhaps received most attention. More recently, radial basis function networks have been gaining popularity as an alternative to multilayer perceptrons. These networks have been shown to have the ability to model any arbitrary nonlinear function [7]. However radial basis functions require a large number of basis functions to achieve the required approximation properties. This is similar to the problem of estimating the number of layers and nodes in the multilayer perceptron.

With only one hidden layer, as shown in Figure 1, the radial basis function output is merely a linear combination of the hidden layer signals, such that, given an input vector, $u$ and a set of network weights $w_i (i = 1, 2, ..., N)$, the network output is found from:

$$f(u) = a_0 + \sum_{i=1}^{N} a_i \phi_i(||u - r_i||) \tag{1}$$

where $N$ is the number of radial basis functions (neurons), $a_0$ is a bias term, $\mathbf{r}$ is the set of centres and $\phi(\cdot)$ is a function of the distance between the current input point, $u$, and the centre $r_i$. The distance metric, or norm, usually employed is the Euclidean Distance. It is usual, but not necessary, for the function $\phi(\cdot)$ to be a common function type throughout the $N$ neurons, typical examples being:

(i) $\phi(t) = \exp(-t^2/2\sigma_i^2)$ Gaussian
(ii) $\phi(t) = t^2 \log(t)$ Thin Plate Spline
(iii) $\phi(t) = (t^2 + c^2)^{\frac{1}{2}}$ Multiquadratic
(iv) $\phi(t) = (t^2 + c^2)^{-\frac{1}{2}}$ Inverse Multiquadratic

in which $\sigma$ and $c$ are simply scaling or "width" parameters.

Of the selections possible, and it should be stressed that many other options exist [4], the Gaussian function is the most intuitive, providing a peak for each function which coincides with the input vector being equal to the basis function centre, the function trailing off as the input vector moves away from the centre.

However, practical experience has shown that the thin plate spline actually works very well in terms of modelling accuracy [12].

Dynamics can readily be taken into account in a radial basis function network through the input vector, $u$, consisting of elements which are sampled plant or controller, input and output signals. For a system identification exercise, if $y(k-1), y(k-2), \ldots$ indicate the system output at time instant $k-1, k-2$, etc; and $u(k-1), u(k-2), \ldots$ indicate the system input at time instant $k-1, k-2$, etc, from the network input vector can be formed from

$$u = (y(k-1), y(k-2), \ldots : u(k-1), u(k-2), \ldots) \tag{2}$$

The network output, $y$, can then be made to model the system output, at time instant $k$, by an appropriate choice of basis functions, centre positions and weightings. For multiple output systems, several RBFs can be combined to form a network as shown in figure 2. All outputs share the same set of centres but have different coefficients. In a basic form of operation, one basis function type is selected and the basis function centres are fixed. The network weights, $w_i$, can then be adjusted to satisfy an error criterion such as linear least squares.

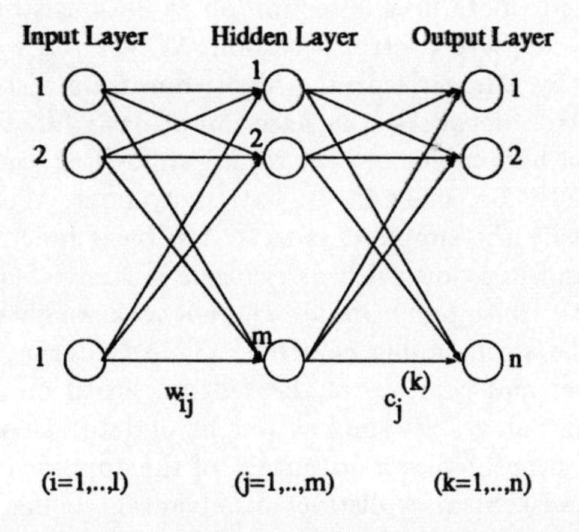


**Fig. 2.** Radial Basis Function Network Architecture

The crucial feature of radial basis function networks is the selection of basis function centres, both in terms of their number and respective positions [20]. The choice made is critical to the successful operation of the network, too few centres and the network's approximation properties will be poor, however unfortunately the number of required centres tends to increase exponentially with regard to the input space dimension, thereby making radial basis function networks unsuitable for a use on a highly complex input space [14].

It is possible to select radial basis function network centres in a number of ways, e.g. a uniform distribution across the input space, or indeed a random distribution, of a set total number of centres. More interestingly a clustering algorithm, such as the Mean-Tracking algorithm [20] can be employed, this giving a statistically best selection of centre quantity and position. A brief comparison of the different methods of centre selection, in terms of their approximation properties, can be found in [21].

Some of the activation functions require an extra parameter which determines the "width" of the basis function. Again no rigorous method exists to calculate this parameter. In the case of Gaussian basis functions, a suitable heuristic is to set the width to the mean of the distances from the centre to the nearest neighbouring centres.

Where the approximation problem is particularly complex, i.e. the input space is of high dimension, a reasonable modelling approximation can be achieved by means of a fairly low number of centres, rather along the lines of a reduced order or reduced structure model. On the understanding that the accuracy of the approximation achieved by the resultant radial basis function network will be, perhaps at best, only reasonable, for practical purposes the end product may well be sufficient. Effectively by using a lower number of centres, the nonlinearity of the input space is being linearized about the basis function centre points. The results achievable by the use of a reduced number of centres are therefore strongly dependent on the nature of the input space and just how linearizable it is.

One final point to note in this section on radial basis functions is their applicability to the adaptive control situation. Whilst the weight training characteristics offered by a linear learning algorithm are a distinct advantage over multilayer perceptron networks, the selection of basis function centres is a disadvantage. With a bad choice of centres any modelling and/or control carried out will subsequently be, most likely, extremely poor. With adaptive control, one possibility, easily the simplest, is to fix the basis function centres in terms of both number and position, with a resultant recursive least squares algorithm being applied to continually update the network weights, in line with plant changes, rather like a self-tuning controller [8]. An alternative is to continually update the number and position of the centres, based on adaptive clustering, as well as carrying out a least squares weight update. This topic is very much an ongoing research problem, an advantage of the approach being improved potential tracking and control, a distinct disadvantage being the much increased computational requirements.

## 2.1 Adaptive RBF Networks

As mentioned above, there are three sets of parameters to be determined when using RBF networks:

1. The Centre Positions
2. The Widths (where appropriate)
3. The Coefficients

Although the coefficients can be rigorously determined (assuming the other parameters are fixed), only heuristic methods are available for the centres and the widths. In addition selecting the centres and widths requires advance data for off-line pre-processing. As a result, it was felt that an adaptive RBF network which would self-tune the parameters whilst on-line would be preferable.

RBFs with non-local basis functions, e.g. thin plate spline, have been shown to be very effective in several applications [12]. For adaptive purposes thin plate splines also have the added advantage of no width parameter. Thus the number of variables to be updated is reduced. For these reasons thin plate splines were used in these trials.

For the purposes of these experiments, the centres were fixed in positions determined by data collected in advance. This enabled the remaining variables, i.e. the coefficients, to be adapted using a fast linear learning algorithm. Work is in progress to vary the other parameters for example by gradient descent methods. The fact that adjustment of the coefficients is essentially linear allows us to employ learning algorithms such as recursive least squares.

## 3 Learning Algorithms

With artificial neural networks there are a number of network parameters which have to be decided *a priori* or determined during learning. With MLPs these parameters are the number of layers, the number of nodes in each layer and the weights associated with the connections. Often the overall architecture of the MLP is fixed and training algorithm is employed to determine the weights alone. The algorithm is such that the error between network output and target output is reduced. The most popular algorithm is error back propagation (BP), which belongs to a family of gradient descent algorithms. However, with MLPs the independent variables (the weights) appear in a non-linear manner and hence the BP algorithm is very slow in determining the optimum weights. On the other hand RBF networks have the interconnections appearing linearly but have additional parameters in the form of centres, which need to be determined in a manner similar to that of the weights. However, these centres can be fixed in advance through utilisation of knowledge of the problem. If the centres are fixed then the objective of the learning algorithm is to determine the parameter vector $\theta_0$ in the following equation:

$$y(t) = \theta_0^{\mathrm{T}} x(t) \tag{3}$$

Where $y(t)$ is the output, $\theta_0$ is the vector of parameters (in this case the coefficients $a_1 \cdots a_{\mathrm{N}}$) and $x(t)$ is the input vector. In the case of adaptive RBFs, the $i$th element of $x(t)$ will be determined by:

$$x_i(t) = \phi_i(||u(t) - r_i||) \tag{4}$$

The RBF generates an estimate of the output, $\hat{y}(t)$, using the current estimate of the parameters $\hat{\theta}(t)$, according to the following formula:

$$\hat{y}(t) = \hat{\theta}_0^{\mathrm{T}} x(t) \tag{5}$$

The objective of the learning algorithm is to minimise the difference between $y(t)$ and $\hat{y}(t)$. The data is in the form of observations, each of which consists of an input/output pair. Batch learning has access to the whole of a (finite) set of observations, whereas on-line adaptive learning only has access to the current values.

Note that equation 5 is in a linear form and hence traditional regression methods can be employed to estimate $\theta_0$. Examples of such methods are Recursive Least Squares and Kaczmarz's algorithm.

### 3.1 Recursive Least Squares

The Recursive Least Squares (RLS) algorithm is a well known method of determining the parameter vector, $\hat{\theta}(t)$, such that the following error function is minimised:

$$J_{\mathrm{N}} = \frac{1}{N} \sum_{t=1}^{N} (y(t) - \hat{y}(t))^2 \tag{6}$$

The recursive least squares algorithm is then:

$$\hat{\theta}(t) = \hat{\theta}(t-1) + K(t) \left[ y(t) - \hat{\theta}(t-1)^{\mathrm{T}} x(t) \right] \tag{7}$$

where:

$$K(t) = P(t-1) \frac{x(t)}{1 + x(t)^{\mathrm{T}} P(t-1) x(t)} \tag{8}$$

and the covariance matrix $P(t)$ can be updated as follows:

$$P(t) = P(t-1) - \frac{P(t-1) x(t) x(t)^{\mathrm{T}} P(t-1)}{1 + x(t)^{\mathrm{T}} P(t-1) x(t)} \tag{9}$$

### 3.2 Kaczmarz's Algorithm

Kaczmarz's algorithm [10] simply adjusts the parameters in order to reduce the estimation error for the current input/output pair to zero. The algorithm is described in the following equation:

$$\hat{\theta}(t) = \hat{\theta}(t-1) + \frac{(y(t) - \hat{\theta}(t-1)^{\mathrm{T}} x(t))}{x(t)^{\mathrm{T}} x(t)} x(t) \tag{10}$$

The algorithm is computationally much less costly than RLS, but can suffer from poor convergence if the system of equations are ill-conditioned. This algorithm is discussed further in section 4.2.

## 4 Control Strategies

Although many different neuro-controller architectures exist [9] as yet they have not achieved widespread acceptance in industrial applications. The lack of a general stability theory prevents their use in life-critical situations, and slow training algorithms can cause problems when using adaptive systems.

### 4.1 Adaptive Control Loops using Neural Networks

A typical learning control system with two neural networks is shown in Fig. 3, [19]. One neural network or "associative memory system" (AMS) is used to build a model of the unknown process, while the second neural network is used to build up a function representing some optimised control strategy. Current literature has considered the process modelling problem extensively; whereas the modelling of the control strategy has not so far received much attention. An understanding of the behaviour of both these neural networks, and of the effects of their interconnections is necessary in formulating an overall theory of stability and performance of the complete system. Moreover, an analogy with more conventional adaptive control can be made where we note that (i) use of such adaptive controls in safety-critical situations was not possible until a reasonably complete stability theory became available and (ii) such a theory took some 30 years to develop, involving many leading research workers around the world.

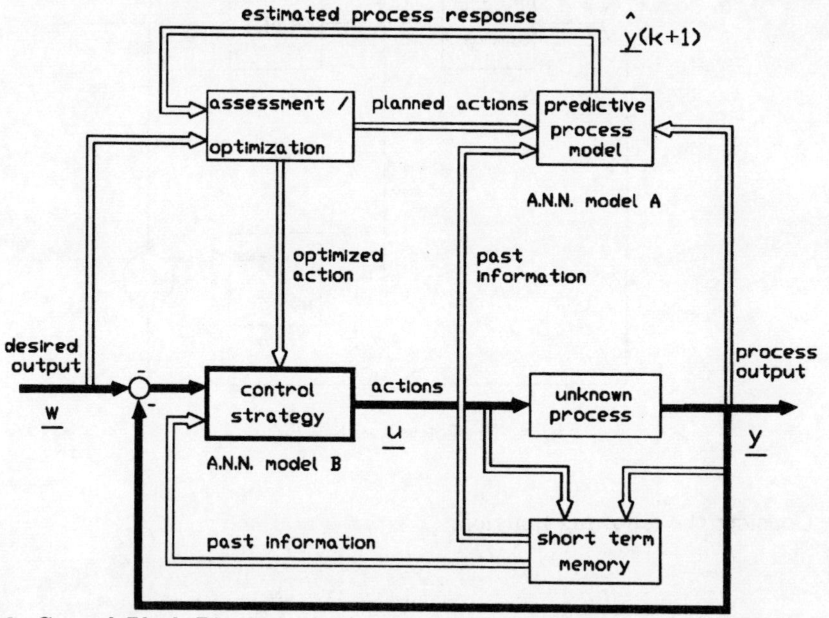


**Fig. 3.** General Block Diagram of Adaptive System With Two Neural Networks, A and B

If Radial Basis Functions (RBFs) are employed instead of the more common Multi-Layer Perceptron (MLP) paradigm, then faster training times can be achieved. If the RBF centres are fixed then training degenerates to a linear problem. In addition, the simpler form of RBF based networks allows easier mathematical analysis which should aid stability proofs.

Neural networks can be used within a control context in two ways:

1. Neural Network Based Predictive Control
2. Indirect control strategies

**Predictive Control** Predictive control schemes minimise future output deviations from set point, while taking into account the control action needed to achieve the objective. Such schemes have been proved to be very successful for linear systems [17, 6, 11]. These schemes have also been used with non-linear models. These non-linear models are limited in that they do not necessarily describe all the non-linearities and complexities of plants such as chemical processes. It is the ability of neural networks to describe and model many, if not all, the non-linearities of such plants which makes them ideal candidates for use within a predictive control scheme. The resulting controller would prove to be more robust in practical situations where the nature of the non-linearity in the process is unknown. The general schematic representation of the predictive control scheme is shown in figure 4.

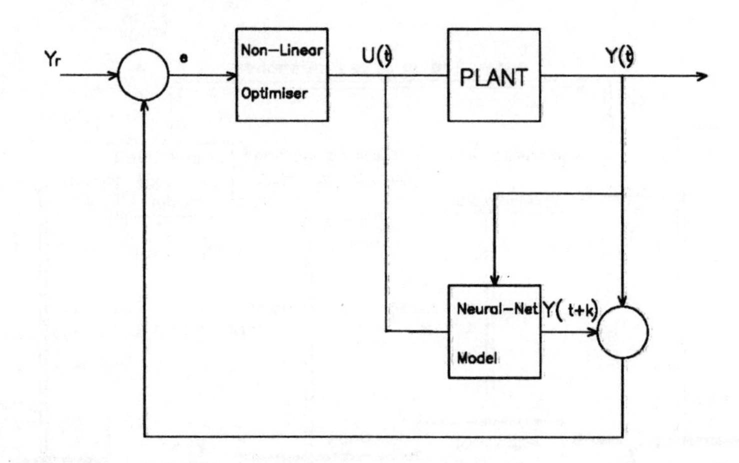


**Fig. 4.** Predictive Control Scheme

Consider the following function:

$$J(N_1, N_2, N_u) = \sum_{j=N_1}^{N_2} \{y(t+j) - y_s(t+j)\}^2 + \sum_{1}^{N_u} \rho(k) \Delta U(t+k-1)^2 \quad (11)$$

where $N_1$ is the minimum output prediction horizon, $N_2$ is the maximum output prediction horizon, and $N_u$ is the control horizon. $\rho(k)$ is the move suppression factor (control weighting sequence).

The prediction horizons specify the range of future predicted outputs to be considered, and the number of control moves required to reach the desired objective is defined by the control horizon. The attraction of predictive controllers is that excessive control effort is penalised by having a non-zero $\rho(k)$. The role of the neural network model here is to supply the future outputs $y(t+j)(j = N_1 ........., N_2)$. Thus if a known sequence of future set point values is used, then a sequence of future control moves can be found out. The number of future control moves which is calculated depends on the control horizon. To avoid the problems of extrapolation, the controls are applied in a receding horizon manner, that is, only the first of the calculated values of the future control moves is used at each sample instant. The minimisation algorithm which can be used to solve this problem can be one of the many algorithms available. For example the Conjugate direction method [5] may be used to minimise the function. The predictive control algorithm used is as follows:

1. Select the minimum prediction horizon $N_1$, the maximum prediction horizon $N_2$, the control horizon $N_u$ and the move suppression factor $\rho(k)$.
2. Set $y_s(t+j)$ (the future set point sequence)
3. minimise $J(N_1, N_2, N_u)$ using Powell's conjugate direction method to obtain the optimal sequence of controls $u(t+k)$
4. Implement $u(t)$ and go to step (ii).

Note that this assumes the neural network model is fixed. This particular scheme can be used for one-step ahead control, since as yet there is no way in which the neural network model can be made to predict many steps ahead, as can be done with linear models [6].

**Indirect Control** A possible architecture for a RBF network based controller is an indirect controller, as shown in figure 5. This employs two networks; one performs a system identification of the plant to be controlled and the other learns the desired control law using information from the first network. The controller network parameters are updated by passing the error at the output of the plant back through the system identification network to determine the error at the output of the controller network. Obviously this technique can only function correctly if a sensible inverse of the plant exists.

The usual training procedured for RBF networks are off-line batch methods. However, if the plant to be controlled is likely to change with time, then an on-line adaptive method of adjusting the parameters will be necessary.

### 4.2 Dynamics in Weight Space

The *convergence* of the weight adjustment process is now quite well understood, at least for neural networks in which a linear combination of "local" functions

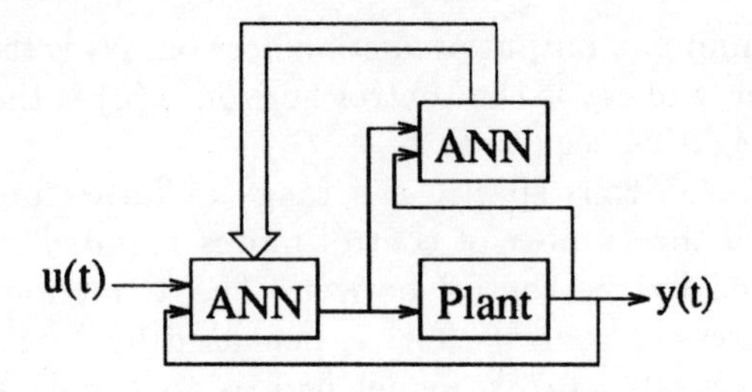


**Fig. 5.** Indirect Control Strategy

are used to approximate a given non-linear function. The "CMAC" is typical of this class of network, but the results also apply to many other networks whose weight learning is based on "local" functions such as certain types of radial basis functions or "B-splines". Radial Basis Functions are considered further in Section 2.

The weight adjustment process usually uses a gradient-descent type of algorithm tending towards a least-means-squares solution. An important consideration is the *consistency* of the set of equations, linear in the weights, that the algorithm is attempting to solve. Fig. 6 shows some simple examples of a particularly simple algorithm, Kaczmarz's, where the weights vector $w$ can tend to a fixed point, enter a point-wise limit cycle or become trapped in a "minimal capture zone", [15].

The question of consistency conditions for functions modelled by the "C-MAC" device has recently been more extensively studied by Brown and Harris [3]. As inconsistency is more likely in practice, possibly because of noisy measurements, it might appear attractive to use a stochastic approximation algorithm in which smaller and smaller weight adjustments are made as (discrete) time progresses: however, this is *not* appropriate in an adaptive time-varying situation as the weight vector will be unable to track new changes in the process being modelled. Various improvements to the simple Kaczmarz algorithm were considered in [16].

### 4.3 Coupling of the Modelling Process to the Controller

The *coupling* of the neural network modelling the process to the second neural network forming the control strategy (Fig. 7) has not yet been considered to any extent in the current literature. Figure 7 which depicts (for illustrative purposes) a well-known linear second-order optimal step response problem, based on the fact that an optimum step response is obtained with a damping ratio of 0.7, supposedly accomplished here by using two neural network models. We suppose that the step-response of the second order system is adjusted by the velocity feedback as shown, where the quantity $c_1/k$ is obtained from the neural network B (which generates the surface $\frac{c_1}{k} = \frac{1.4}{\sqrt{k}} - \frac{c}{k}$ – this is known *theoretically* in this simple case). Thus values of $c$ and $k$ are required from the process model-neural

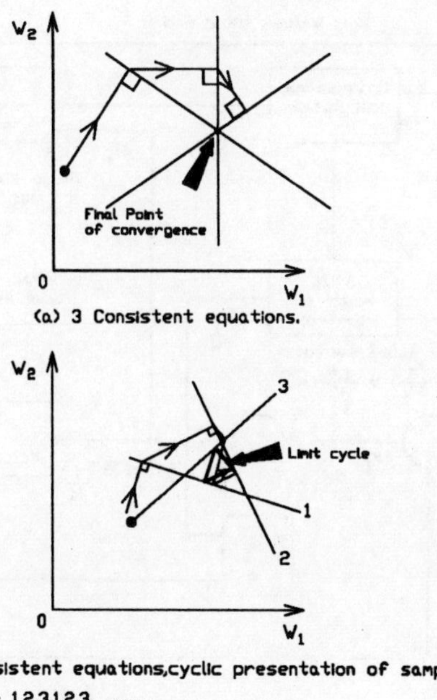


(a) 3 Consistent equations.

(b) 3 Inconsistent equations,cyclic presentation of samples in order 1,2,3,1,2,3, .........

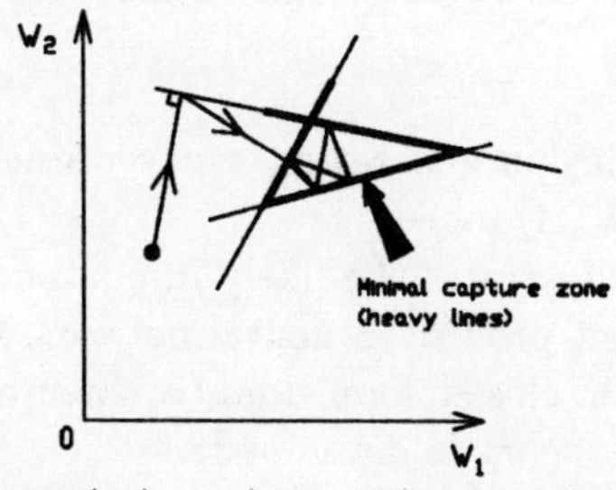


(c) 3 Inconsistent equations, random presentation of samples.

**Fig. 6.** Dynamics of the Weight Vector, using Kaczmarz's Algorithm, Simple 2-D Examples.

network A. These can be found from the (plane) surface generated by A (the linear function $\ddot{x} = ku_1 - c\dot{x}$).

Consider now the following situation in which initially the natural damping $c$ is large, and is so large that the artificial damping provided by $c_1/k$ is negative, so as to *reduce* the total effective damping coefficient $c+c_1$. Suppose $c$ is suddenly reduced to zero (e.g. by loss of oil in a mechanical damper); the new equation $\ddot{x}+c_1\dot{x}+k(x-x_D) = 0$ is unstable as $c_1 < 0$ and so an unstable transient motion builds up with an exponential factor $e^{\mu t}$, say. Of course this motion will cause neural network A to generate a new (zero) value for $c$ and then a new positive value of $\frac{c_1}{k} = \frac{1.4}{\sqrt{k}}$ will be generated by neural network B, and so stability will be restored. A criterion for a satisfactory response will be the size of the factor $e^{\mu d}$,

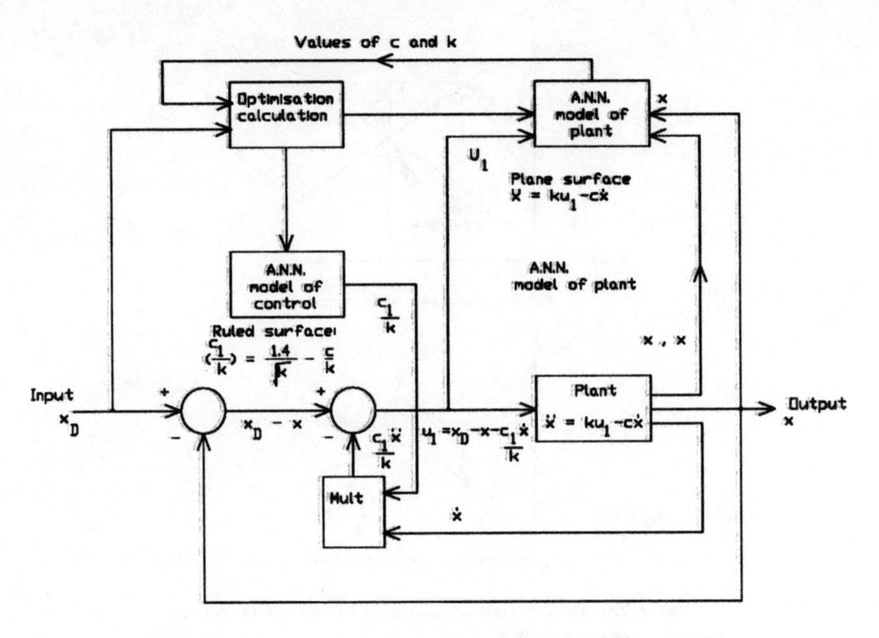


**Fig. 7.** Linear Second Order (Time Varying Example)

where $d$ is the time delay in generating a new value of $c(= 0)$ and then a new positive value for $c_1$ $(= 1.4/\sqrt{k})$.

This simple model illustrates the likely importance of the convergence speed of the weight adjustment process in neural network A, and the likely form that performance criteria for closed loop adaptive neural network controllers may take as an appropriate theory is developed.

## 5 System

To illustrate the effectiveness of adaptive RBF networks a simple example system was chosen. The application is a second order system, with reasonably high damping. At some time $t = \tau$, the damping is drastically reduced. This models a component failure, e.g. loss of oil from a damper, within the plant. For control to remain effective and stability be maintained rapid adjustment of the controller parameters must take place. Hence, fast convergence of the system identification networks is essential.

The example system is described by the following equation:

$$G(s) = \frac{144}{s^2 + 21.6s + 144} \tag{12}$$

This is a second order system with undamped natural frequency $w_n = 12$ rad s$^{-1}$ (1.91 Hz) and damping ratio $\zeta = 0.9$. At time $t = \tau$, the damping ratio is reduced to $\zeta = 0.2$. The new system equation is then:

$$G(s) = \frac{144}{s^2 + 4.8s + 144} \tag{13}$$

The system identification network was updated ten times every second, and the failure point $\tau$ was set to $t = 100$s.

## 6 Results

The system identification network used ten centres distributed over the convex hull of the input data space. A thin plate spline was used for $\phi$ and Kaczmarz algorithm used for the update. Figure 8 shows the initial convergence with a square wave input (note the 'kicks' in the error when the input signal swings from peak to peak). Figure 9 shows the transition (simulated damper failure) and reconvergence.

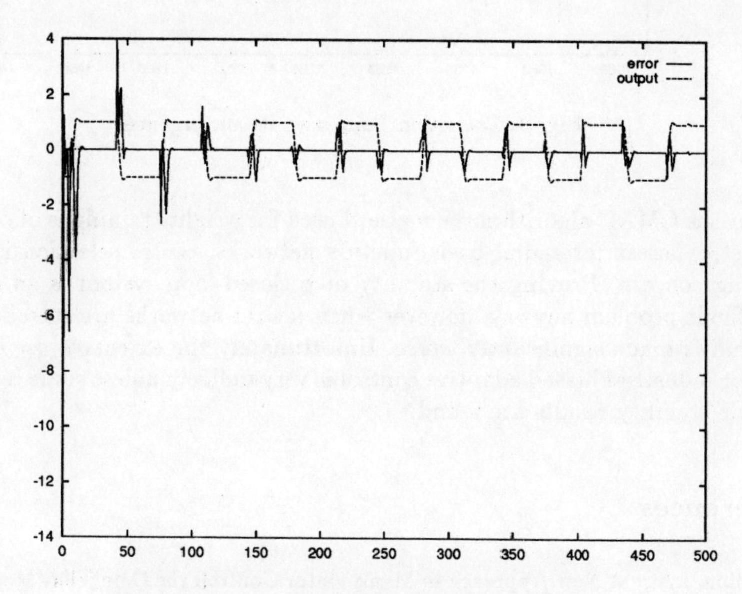


**Fig. 8.** Initial Convergence of Adaptive RBF

## 7 Conclusions

A number of different aspects of the use of neural networks, for the adaptive control of dynamic systems, have been considered here. In particular CMAC and radial basis function networks were discussed and the problems associated with them were considered.

A simple example showing the ability to adapt to track a second order system after component failure demonstrates the ability of adaptive RBF networks.

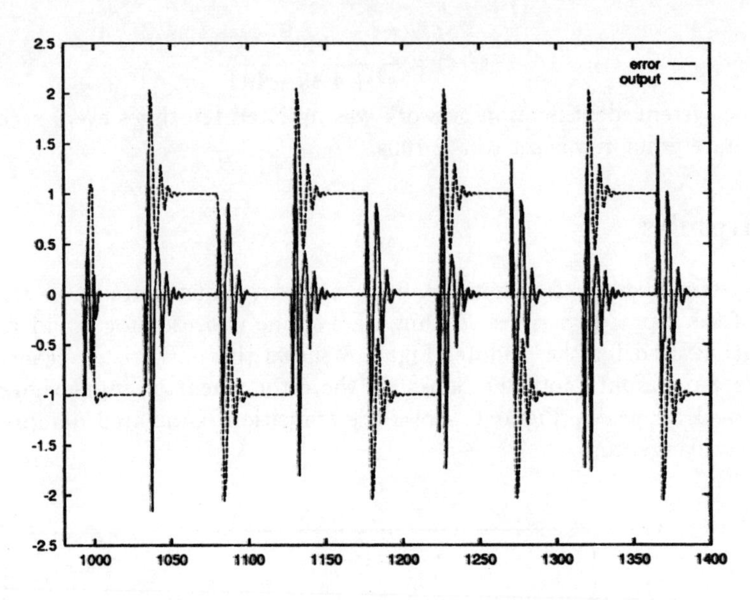


**Fig. 9.** Transition Point and Reconvergence

For the CMAC algorithm the method used for weight training is of particular interest, whereas for radial basis function networks, centre selection is a more pressing concern. Proving the stability of a closed-loop system is an extremely difficult problem anyway, however when neural networks are introduced the problem is made significantly worse. Unfortunately the extensive use of neural nets for industrial biased adaptive control is very unlikely unless some reasonably sensible stability results are found.

## References


1. Albus J.A., "A New Approach to Manipulator Control: the Cerebellar Model Articulation Controller (CMAC)", *Trans. ASME, J. Dynamic Syst. Meas. and Control*, **97**, 220-227, 1975.
2. Bishop C.M., Haynes P.S., Smith M.E.V., Todd T.N., Trotman D.L., "Real-Time Control of a High Temperature Plasma Using a Hardware Neural Network", *in "Neural Networks, Neuro-fuzzy and other Learning Systems for Engineering Applications and Research"*, (ed.) J.A. Powell, Proceedings of an EPSRC Conference, London, April 18th-19th 1994, 16-25, 1994. (Rutherford-Appleton Laboratory publication)
3. Brown M., Harris C.J., "Neurofuzzy Adaptive Modelling and Control", Prentice-Hall, 1994.
4. Cichocki A. and Unbehauen R., "Neural Networks for Optimization and Signal Processing", John Wiley and Sons, 1993.
5. Clarke D.W., Gawthrop P.J., "Self Tuning Control", *Proc IEE Part D*, **126:**6, 1979.

6. Clarke D.W., Mohtadi C., Tuffs P.S. "Generalised Predictive Control - Part 1 The Basic Algorithm", *Automatica*, 1987.
7. Girosi F. and Poggio T., "Neural Networks and the Best Approximation Property", *Biol. Cybernetics*, **63**, 169-176, 1990.
8. Harris C.J. and Billings S.A. (eds), "Self-Tuning Control: Theory and Applications", rev. 2nd ed., Peter Peregrinus Ltd., 1985.
9. Hunt K.J., Sbarbaro D., Żbikowski R., Gawthrop P.J., "Neural Networks for Control Systems - A Survey", *Automatica*, **28**:6, pp 1083-1112, 1992.
10. Kaczmarz S., "Angenäherte Auflösung von Systemen linearer Gleichungen." *Bulletin International de l'Academie Polonaise des Sciences, Cl. d. Sc. Mathém*, **A**, pp 355-357. 1937.
11. Kambhampati C., Warwick K., Berger C.S., "A Comparative Study of Multilayered and Single Layered Neural Network Based Predictive Controllers", *IEE International Conf. Intelligent Systems Engineering*, pp 293-298, 1992.
12. Lowe D., "Non Local Radial Basis Functions for Forecasting and Density Estimation", *IEEE Intern. Conf. on Neural Networks*, volume II, pp 1197-1198, 1994.
13. Mason J.D., Craddock R.J., Warwick K., Mason J.C., Parks P.C., "Towards a Stability and Approximation Theory for Neuro-Controllers", *IEE Control '94*, pp 100-103, 1994.
14. Narendra K.S., "Adaptive Control of Dynamical Systems Using Neural Networks", in *"Handbook of Intelligent Control"*, Van Nostrand, 1994.
15. Parks P.C., Militzer J., "Convergence Properties of Associative Memory Storage for Learning Control Systems", *Automation and Remote Control*, **50**, 254-286, 1989. (Plenum Press, New York).
16. Parks P.C., Militzer J., "A Comparison of Five Algorithms for the Training of CMAC Memories for Learning Control Systems", *Automatica*, **28**, 1027-1035, 1992.
17. Peterka V., "Predictor Based Self-Tuning Control", *Automatica*, **20**:1, pp 39-50, 1984.
18. Tolle H., "Neurocontrol: Learning Control Systems Inspired By Neuronal Architectures and Human Problem Solving Strategies", *Springer-Verlag, Lecture Notes in Control and Information Sciences*, **172**, 1992.
19. Tolle H., Parks P.C., Ersü E., Hormel M., Militzer J., "Learning Control with Interpolating Memories, General Ideas, Design, Layout, Theoretical Approaches and Practical Applications", *Int. J. Control*, **56**, 291-318, 1992.
20. Warwick K., Mason J.D. and Sutanto E., "Centre Selection for Radial Basis Function Networks", Proc. ICANNGA, Ales, France, 1995.
21. Warwick K., "An Overview of Neural Networks in Control Applications", in *"Neural Networks in Robotic Control Theory and Application"*, A. Zalzala and A. Morris (eds), Ellis Horwood, 1995.

# Adaptive Neurocontrol of a Certain Class of MIMO Discrete-Time Processes Based on Stability Theory

Jean-Michel Renders[1], Marco Saerens[2], and Hugues Bersini[2]

[1] Laboratoire d'automatique (cp. 165); [2] IRIDIA Laboratory (cp. 194/6)
Université Libre de Bruxelles, 50 av. F. Roosevelt, 1050 Bruxelles, BELGIUM

**Abstract.** In this chapter, we prove the stability of a certain class of nonlinear discrete MIMO (Multi-Input Multi-Output) systems controlled by a multilayer neural net with a simple weight adaptation strategy. The proof is based on the Lyapunov stability theory. However, the stability statement is only valid if the initial weight values are not too far from their optimal values which allow perfect model matching (local stability). We therefore propose to initialize the weights with values that solve the linear problem. This extends our previous work (Renders, 1993; Saerens, Renders & Bersini, 1994), where single-input single-output (SISO) systems were considered.

## 1. Introduction

There has been a renewed interest for the old cybernetic project to exploit neural networks for control applications (Psaltis, Sideris & Yamamura, 1987, 1988; Lan, 1989; Barto, 1990; Miller, Sutton & Werbos, 1990; Levin, Gewirtzman & Inbar, 1991; Hunt & Sbarbaro, 1991; White & Sofge, 1992; Narendra & Parthasarathy, 1990; Hunt et al., 1992; Narendra & Mukhopadhyay, 1994). However, most of the algorithms for adaptive control with neural networks developed up to now are gradient-based (see for instance Narendra & Parthasarathy, 1990, 1992; Gupta, Rao & Wood, 1991; Gupta, Rao & Nikiforuk, 1993; Psaltis, Sideris & Yamamura, 1987, 1988; Hunt & Sbarbaro, 1991; Puskorius & Feldkamp, 1992, 1994; Saerens & Soquet, 1991), and do not provide any information on the stability of the overall process. Only recently, researchers tried to design adaptation laws that ensure the stability of the closed-loop process (see Kawato, 1990; Renders, 1991; Gomi & Kawato, 1993 in the context of robotics applications; see Parthasarathy & Narendra, 1991; Sanner & Slotine, 1992; Narendra, 1992; Tzirkel-Hancock & Fallside, 1992; Renders, 1993; Saerens, Renders & Bersini, 1994; Jin, Pipe & Winfield, 1993; Chen & Khalil, 1991; Chen & Liu, 1992; Liu & Chen, 1993; Jin, Nikiforuk & Gupta, 1993; Rovithakis & Christodoulou, 1994; Johansen, 1994, in the context of adaptive control with neural nets), or to ensure the viability of the regulation law (Seube, 1990; Seube & Macias, 1991).

---

Contacting author is Marco Saerens, Laboratoire IRIDIA (cp 194/6). Email: saerens@ulb.ac.be.

In this chapter, we prove the stability of a certain class of nonlinear discrete-time MIMO (Multi-Input Multi-Output) systems controlled by a multilayer neural net with a simple weight adaptation strategy. The proof is based on the Lyapunov theory. However, the stability statement is only valid if the initial weight values are not too far from their optimal values that allow perfect model matching (local stability). We therefore propose to initialize the weights with values that solve the linear problem. This extends our previous work (Renders, 1993; Saerens, Renders & Bersini, 1994), where single-input single-output (SISO) systems were considered; it differs from the work of Parthasarathy & Narendra (1991), Narendra (1992), Tzirkel-Hancock & Fallside (1992), Sanner & Slotine (1992), Jin, Pipe & Winfield (1993), Rovithakis & Christodoulou (1994), Johansen (1994), Chen & Khalil (1991), Chen & Liu (1992), and Liu & Chen (1993), where affine processes were considered.

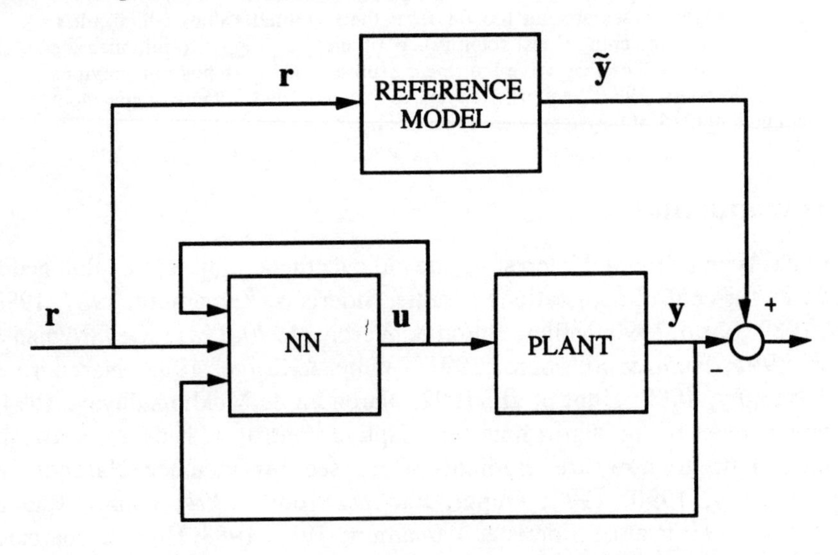


*Figure 1. General overview of the system.*

## 2. Process description

The process is described in terms of its input-output representation; it is supposed to be put in a predictor form:

$$y_i(k+d_i) = F_i[\mathbf{y}(k), ..., \mathbf{y}(k-p_y+1), \mathbf{u}(k), ..., \mathbf{u}(k-p_u+1)],\ i = 1, ..., n \quad (1a)$$

where $\mathbf{y}(k)=[y_1(k), y_2(k), ..., y_n(k)]^T$ is the output of the plant, and $\mathbf{u}(k)=[u_1(k), u_2(k), ..., u_n(k)]^T$ is the input of the plant. $d_i$ is the delay corresponding to output $y_i$. The functions $F_i$ are supposed to be continuously differentiable in terms of all the arguments. Notice that the representation (1a) can be obtained from the more familiar form $y_i(k+1) = f_i[\mathbf{y}(k), ..., \mathbf{y}(k-q_y+1),$

$\mathbf{u}(k-\Delta_i)$, ..., $\mathbf{u}(k-q_u+1)$] by recursively substituting the outputs $\mathbf{y}(l)$ posterior to the inputs $\mathbf{u}(k-\Delta_i)$ until the outputs and the inputs appear simultaneously. By definition of the delays $d_i$, we must have $\partial y_i(k+\delta_i)/\partial u_j(k) = 0$ ($\delta_i = 1, ..., d_i-1$), while $\partial y_i(k+d_i)/\partial u_j(k) \neq 0$ for at least one $j$. We assume that every input $u_j(k)$ ($j = 1, ..., n$) explicitly appears at least once in the set of equations (1a). The delays $d_i$ as well as the "orders" $p_y$, $p_u$ are fixed and known.

## 3. Assumptions About the Process and the Neural Network

The final layer of the neural network provides at each time step $k$ the $n$ inputs $u_i(k)$ of the plant. The net must supply the sequence of control actions $\mathbf{u}(k)$ that minimize an error criterion. Let us describe the basic assumptions that are made.

**(i)** The form (1a) implies that at any time $k$, we can reconstruct the state of the process by the observation of the last $p_y$ system outputs $\mathbf{y}(k-i)$ ($i = 0, ..., p_y-1$) and the last ($p_u-1$) inputs $\mathbf{u}(k-i)$ ($i = 1, ..., p_u-1$) in the operating region. Let us define the vector $\mathbf{\Phi}_k = [\mathbf{y}(k), \mathbf{y}(k-1), ..., \mathbf{y}(k-p_y+1), \mathbf{u}(k-1), ..., \mathbf{u}(k-p_u+1)]^T$. The process equation (1a) can be rewritten

$$y_i(k+d_i) = F_i[\mathbf{\Phi}_k, \mathbf{u}(k)], \qquad i = 1, ..., n \tag{1b}$$

Now, in the case where $\mathbf{x}(k)$, the state of the process, is measurable, we have $y_i(k+d_i) = F_i[\mathbf{x}(k), \mathbf{u}(k)]$ ($i = 1, ..., n$), and $\mathbf{\Phi}_k \equiv \mathbf{x}(k)$. Notice that we could also use any other information from which the state can be deduced, such as an image of the system (as in Tolat & Widrow, 1988). This is one of the major advantage of using neural nets.

**(ii)** There exists a uniquely defined control vector $\mathbf{u}(k)$, depending on the state, that allows to reach the desired targets $\tilde{y}_i(k+d_i)$ , $i = 1, ..., n$, (the outputs of the reference model, see *Figure 1*), the relationship between the $n$ inputs $u_j(k)$ and the $n$ delayed desired outputs $\tilde{y}_i(k+d_i)$ being a global diffeomorphism on the operating region. This means that, whatever the value of $\mathbf{\Phi}_k$ in the operating region, the set of equations $y_i(k+d_i) = F_i[\mathbf{\Phi}_k, \mathbf{u}(k)]$ ($i = 1, ..., n$) can be solved uniquely in terms of $\mathbf{u}(k)$, for every set of values $y_i(k+d_i) = \tilde{y}_i(k+d_i)$ ($i = 1, ..., n$) in the operating region. This implies that the Jacobian matrix is non-singular (of rank $n$) in the operating region: $\det[\partial y_i(k+d_i)/\partial u_j(k)] = \det[\partial F_i/\partial u_j(k)] \neq 0$.

**(iii)** The process is inverse stable (asymptotically minimum phase).

From assumptions **(i)** to **(iii)**, there exists a unique stable control law in terms of the desired outputs:

$$\mathbf{u}(k) = \mathcal{U}[\mathbf{Z}(z)\,\tilde{\mathbf{y}}(k), \boldsymbol{\Phi}_k] \tag{2}$$

that allows perfect tracking of the desired output $\tilde{\mathbf{y}}(k)$ in the absence of perturbations and in the absence of initial error. $\mathbf{Z}(z)$ is a delay matrix ($z$ is the forward operator) that has diagonal elements $[\mathbf{Z}]_{ij} = z^{d_i}\delta_{ij}$. We also posed $\tilde{\mathbf{y}}(k) = [\tilde{y}_1(k), \tilde{y}_2(k), ..., \tilde{y}_n(k)]^{\mathrm{T}}$.

**(iv)** The reference model is

$$\mathbf{A}_{\mathrm{m}}(z^{-1})\,\tilde{\mathbf{y}}(k) = \mathbf{B}_{\mathrm{m}}(z^{-1})\,\mathbf{r}(k) \tag{3}$$

and $\mathbf{W}_{\mathrm{m}}(z^{-1})=\mathbf{A}_{\mathrm{m}}^{-1}(z^{-1})\mathbf{B}_{\mathrm{m}}(z^{-1})$ is the transfer function matrix of the reference model. We assume that the transfer function matrix is chosen in such a way that $\mathbf{A}_{\mathrm{m}}(z^{-1})=\hat{\mathbf{A}}_{\mathrm{m}}(z^{-1})\mathbf{Z}(z)$ with $\hat{\mathbf{A}}_{\mathrm{m}}(0)=\mathbf{I}$; $\hat{\mathbf{A}}_{\mathrm{m}}^{-1}(z^{-1})$ being Strictly Positive Real (SPR) and $\mathbf{B}_{\mathrm{m}}(0)=\mathbf{B}_0$ non-singular. We further assume that the transfer matrices are designed in such a way that

$$\hat{\mathbf{A}}_{\mathrm{m}}(z^{-1}) = \mathbf{I} + \mathbf{A}_1\, z^{-1} + \mathbf{A}_2\, z^{-2} + ... + \mathbf{A}_{p_y}\, z^{-p_y} \tag{4a}$$

$$\mathbf{B}_{\mathrm{m}}(z^{-1}) = \mathbf{B}_0 + \mathbf{B}_1\, z^{-1} + \mathbf{B}_2\, z^{-2} + ... + \mathbf{B}_{p_u}\, z^{-p_u} \tag{4b}$$

The initial conditions for the outputs of the reference model are chosen as $\tilde{\mathbf{y}}(k)=\mathbf{y}(k)$ ($k = 1, ..., p_y$).

In this case, there exists a unique control law in terms of the reference signal $\mathbf{r}(k)$:

$$\mathbf{u}(k) = \mathcal{U}[\mathbf{B}_{\mathrm{m}}(z^{-1})\mathbf{r}(k), \boldsymbol{\Phi}_k] = \mathcal{U}[\boldsymbol{\rho}(k), \boldsymbol{\Phi}_k], \tag{5}$$

where we have posed $\boldsymbol{\rho}(k) = \mathbf{B}_{\mathrm{m}}(z^{-1})\mathbf{r}(k)$, that insures perfect model matching in the absence of perturbations.

The function provided by the neural net will be denoted as $\mathbf{u}(k) = \mathbf{N}[\boldsymbol{\rho}(k), \boldsymbol{\Phi}_k; \mathbf{w}_k]$, where $\mathbf{w}_k$ is the weight vector of the net at time step $k$.

**(v)** We also assume that the neural controller $\mathbf{u}(k) = \mathbf{N}[\boldsymbol{\rho}(k), \boldsymbol{\Phi}_k; \mathbf{w}]$ can approximate the control law (5) to any degree of accuracy in the region of interest, for some "perfectly tuned" weights $\mathbf{w} = \mathbf{w}^*$. This is the standard "perfect model matching" condition.

**(vi)** The speed of adaptation of the weights is low in order to be able to easily separate in the measurements of the error the effects of the parameters adjustment from the input signal variations (Landau, 1979).

## 4. Gradient-Based Algorithm

Taking account of these assumptions, we can design a learning algorithm, based on a gradient descent, that can be used to train the weights of the network (Narendra & Parthasarathy, 1990, 1992). From assumptions (i) to (iii), the neural net will be a feed-forward net which is given as input the vector $\boldsymbol{\phi}_k$ as well as the reference $\boldsymbol{\rho}(k)$. It supplies as output the control action $\mathbf{u}(k)$ at time $k$.

At each time step $k$, the neurocontroller will try to reduce the error $E(k)=1/2[(\mathbf{y}(k)-\tilde{\mathbf{y}}(k))^{\mathrm{T}}(\mathbf{y}(k)-\tilde{\mathbf{y}}(k))]$ on-line by achieving a gradient descent in the weight space. We have:

$$\begin{aligned} \mathbf{w}_k &= \mathbf{w}_{k-1} - \eta \frac{\partial E(k)}{\partial \mathbf{w}} \\ &= \mathbf{w}_{k-1} - \eta \left(\frac{\partial \mathbf{y}(k)}{\partial \mathbf{w}}\right)^{\mathrm{T}} [\mathbf{y}(k) - \tilde{\mathbf{y}}(k)] \end{aligned} \tag{6}$$

where $\partial \mathbf{y}(k)/\partial \mathbf{w}$ is defined as a matrix having elements: $[\partial \mathbf{y}(k)/\partial \mathbf{w}]_{ij} = \partial y_i(k)/\partial w_j$, and $\eta$ is the learning rate.

Now referring to Narendra & Parthasarathy's developments (Narendra & Parthasarathy, 1992), the computation of the derivative matrix $\partial \mathbf{y}(k)/\partial \mathbf{w}$ turns out to be a laborious task because of a double recurrence. Indeed, from

$$\mathbf{u}(k) = \mathbf{N}[\boldsymbol{\rho}(k), \boldsymbol{\phi}_k; \mathbf{w}] = \mathbf{N}_k \tag{7}$$

$$\mathbf{Z}(z)\, \mathbf{y}(k) = \mathbf{F}[\boldsymbol{\phi}_k, \mathbf{u}(k)] = \mathbf{F}_k \tag{8}$$

we obtain

$$\mathbf{Z}(z) \frac{\partial \mathbf{y}(k)}{\partial \mathbf{w}} = \frac{\partial \mathbf{F}_k}{\partial \mathbf{w}} = \sum_{i=0}^{p_u-1} \frac{\partial \mathbf{F}_k}{\partial \mathbf{u}(k-i)} \frac{\partial \mathbf{u}(k-i)}{\partial \mathbf{w}} + \sum_{i=0}^{p_y-1} \frac{\partial \mathbf{F}_k}{\partial \mathbf{y}(k-i)} \frac{\partial \mathbf{y}(k-i)}{\partial \mathbf{w}} \tag{9a}$$

with

$$\begin{aligned} \frac{\partial \mathbf{u}(k-i)}{\partial \mathbf{w}} &= \left(\frac{\partial \mathbf{N}}{\partial \mathbf{w}}\right)_{k-i} + \sum_{j=1}^{p_u-1} \frac{\partial \mathbf{N}_{k-i}}{\partial \mathbf{u}(k-i-j)} \frac{\partial \mathbf{u}(k-i-j)}{\partial \mathbf{w}} \\ &\quad + \sum_{j=0}^{p_y-1} \frac{\partial \mathbf{N}_{k-i}}{\partial \mathbf{y}(k-i-j)} \frac{\partial \mathbf{y}(k-i-j)}{\partial \mathbf{w}} \end{aligned} \tag{9b}$$

where $\partial \mathbf{y}(k)/\partial \mathbf{w}$ is the total derivative introduced by Narendra & Parthasarathy (1991). This algorithm is similar to the one that has been designed for the on-line computation of the gradient for recurrent networks (see for instance Williams & Zipser, 1989). Notice that nothing is said about the stability of the closed loop system.

We will suppose that either a process model or a neural identifier of the process is available (and this can be done efficiently by using orthogonal least square methods; see for instance Chen, Billings, Cowan & Grant, 1990; Chen, Billings & Luo, 1989). Both Narendra & Parthasarathy (1990), Nguyen & Widrow (1989, 1990) and Jordan (1989) have proposed to incorporate the process model in a second neural network, the neural emulator, and then to link the neurocontroller with this neural emulator. Consequently, it corresponds to the classical indirect control approach in which the parameters of the model are estimated either on-line or in a preliminary phase, and the parameters of the controller (in this case the weights of the neural controller) are adjusted assuming that the estimated parameters of the process represent the real values (the certainty equivalence principle). Provided such a process model (or neural emulator) is available, the derivatives $\partial \mathbf{F}_k/\partial \mathbf{y}(k-i)$ and $\partial \mathbf{F}_k/\partial \mathbf{u}(k-i)$ are easily estimated either directly from the model or through a "backpropagation-like" algorithm applied to the respective inputs of the neural emulator. The derivatives $\partial \mathbf{y}(k-i)/\partial \mathbf{w}$ are computed in the same recurrent way as $\partial \mathbf{y}(k)/\partial \mathbf{w}$, but at time $(k-i)$, with $\partial \mathbf{y}(0)/\partial \mathbf{w} = 0$.

In the same vein, the derivatives $\partial \mathbf{N}_{k-i}/\partial \mathbf{y}(k-i-j)$ and $\partial \mathbf{N}_{k-i}/\partial \mathbf{u}(k-i-j)$ can be obtained through backpropagation applied now to the respective inputs of the neurocontroller. Finally, the first derivative of expression (9b), $\partial \mathbf{N}/\partial \mathbf{w}$, is simply computed by using the backpropagation algorithm applied, in the usual way, to the synaptic weights of the neurocontroller.

The problems raised by the recurrent calculations just described relate first to the quantity of items to store in memory so as to compute derivatives at time $(k+1)$ as a function of derivatives obtained at time $k$, then to the computer time required to achieve all the operations, and, finally, to the amount of knowledge about the process (the derivatives $\partial \mathbf{F}_k/\partial \mathbf{y}(k-i)$ and $\partial \mathbf{F}_k/\partial \mathbf{u}(k-i)$) that must be available in order to compute the gradient.

Our purpose has been to try to *reduce the complexity* of this gradient-based algorithm. We will provide some heuristic arguments that allow us to simplify this training algorithm (section 5). Thereafter, we give some theoretical support to this simplified algorithm (section 6).

## 5. Simplification of the Gradient-Based Algorithm

We can simplify the learning algorithm by considering another point of view on the problem. Instead of computing the gradient in terms of the *weights*, we will compute the gradient in terms of the *output of the controller*; that is, we search for the control actions that allow us to reach the desired value. Since for each time step $k$, given the current value of the state, there exists a unique control action vector, $\mathbf{u}(k)$, that allows to reach the delayed desired output, $[\mathbf{Z}(z)\tilde{\mathbf{y}}(k)]$, we just keep the first term for the computation of the gradient in the right-hand side of (9a); therefore considering the previous control actions and observations (the state) as fixed. From assumption (ii), the relationship is a global diffeomorphism, so that there is only one global minimum. Now, since the control is provided by the neural network, we just propagate this gradient on the weights; therefore correcting the weights in order to provide a more accurate control action when the same situation is encountered (same reference to reach, same current state of the process). Considering the state, $\boldsymbol{\Phi}_k$, as fixed in (7) and (8) yields

$$\mathbf{Z}(z)\frac{\partial \mathbf{y}(k)}{\partial \mathbf{w}} = \frac{\partial \mathbf{F}_k}{\partial \mathbf{w}} = \frac{\partial \mathbf{F}_k}{\partial \mathbf{u}(k)}\frac{\partial \mathbf{u}(k)}{\partial \mathbf{w}}$$

$$\frac{\partial \mathbf{u}(k)}{\partial \mathbf{w}} = \left(\frac{\partial \mathbf{N}}{\partial \mathbf{w}}\right)_k$$

This results in the following weight adaptation rule:

$$\mathbf{w}_k = \mathbf{w}_{k-1} - \eta\left[\frac{\partial \mathbf{y}(k)}{\partial[\mathbf{Z}^{-1}\mathbf{u}(k)]}\frac{\partial[\mathbf{Z}^{-1}\mathbf{N}_k]}{\partial \mathbf{w}}\right]^{\mathrm{T}}[\mathbf{y}(k) - \tilde{\mathbf{y}}(k)]$$

$$= \mathbf{w}_{k-1} - \eta\left[\mathbf{Z}^{-1}\frac{\partial \mathbf{F}_k}{\partial \mathbf{u}(k)}\frac{\partial \mathbf{N}_k}{\partial \mathbf{w}}\right]^{\mathrm{T}}[\mathbf{y}(k) - \tilde{\mathbf{y}}(k)] \quad (10)$$

which is a very simple adaptation law. We observe that the adaptation of the weights is performed with a delay (given by the matrix $\mathbf{Z}$), when the error is available. However, if we use a predictor of the process, we do not have to wait in order to adjust the weights. Indeed, a predictor can be used in order to predict the unknown future values of $\mathbf{y}(k)$ in (10), which allows us to compute the gradient (10) immediately (Renders, Bersini and Saerens, 1993). We will now provide some theoretical support to the simplified algorithm (10).

## 6. Stability of the Simplified Algorithm

We will show that if we start in the neighborhood of the perfect weight tuning, and if the speed of adaptation is low, the convergence can be guaranteed. Indeed, with the restrictions mentioned above (section 3), we will prove the following statement:

*If* $\mathbf{v}_k = \frac{\partial \mathbf{F}_k}{\partial \mathbf{u}(k)} \frac{\partial \mathbf{N}}{\partial \mathbf{w}}[\boldsymbol{\rho}(k), \boldsymbol{\phi}_k; \mathbf{w}_k]$*, supposed measurable, is globally bounded, the adaptation rule*

$$\mathbf{w}_k = \mathbf{w}_{k-1} - \eta\, (\mathbf{Z}^{-1}\mathbf{v}_k)^{\mathrm{T}}\, \mathbf{e}(k) \tag{11}$$

*with* $\eta$ *being a small positive constant, will drive the error* $\mathbf{e}(k) = [\mathbf{y}(k) - \tilde{\mathbf{y}}(k)]$ *asymptotically to zero*: $\mathbf{e}(k) \to 0$.

Let us suppose that at a given time step we have $\mathbf{u}(k) = \mathbf{N}[\boldsymbol{\rho}(k), \boldsymbol{\phi}_k; \mathbf{w}_k]$. Now, we define $\tilde{\mathbf{w}}_k = (\mathbf{w}_k - \mathbf{w}^*)$ where $\mathbf{w}^*$ is an optimal weight vector that allows perfect model matching with $\mathbf{u}(k) = \mathbf{N}[\boldsymbol{\rho}(k), \boldsymbol{\phi}_k; \mathbf{w}^*]$. We suppose that $\mathbf{w}^*$ is sufficiently close to $\mathbf{w}_k$ to permit a first order expansion around $\mathbf{w}_k$:

$$\mathbf{N}[\boldsymbol{\rho}(k), \boldsymbol{\phi}_k; \mathbf{w}^*] \cong \mathbf{N}[\boldsymbol{\rho}(k), \boldsymbol{\phi}_k; \mathbf{w}_k] - [\frac{\partial \mathbf{N}}{\partial \mathbf{w}}[\boldsymbol{\rho}(k), \boldsymbol{\phi}_k; \mathbf{w}_k]]\, \tilde{\mathbf{w}}_k \tag{12}$$

The symbol $\cong$ signifies that the equality is verified at the first order. Notice that if the output of the net is linear in its parameters (for instance, a radial basis function network where the basis functions are not modified), this development is strictly valid. We now define a modified reference $\boldsymbol{\rho}'(k)$ such that

$$\begin{aligned} \mathbf{N}[\boldsymbol{\rho}'(k), \boldsymbol{\phi}_k; \mathbf{w}^*] &= \mathbf{N}[\boldsymbol{\rho}(k), \boldsymbol{\phi}_k; \mathbf{w}^*] + [\frac{\partial \mathbf{N}}{\partial \mathbf{w}}[\boldsymbol{\rho}(k), \boldsymbol{\phi}_k; \mathbf{w}_k]]\, \tilde{\mathbf{w}}_k \\ &\cong \mathbf{N}[\boldsymbol{\rho}(k), \boldsymbol{\phi}_k; \mathbf{w}_k] \end{aligned} \tag{13}$$

Now, $\boldsymbol{\rho}'(k)$ will be sufficiently close to $\boldsymbol{\rho}(k)$ in order to make a first-order expansion of $\mathbf{N}[\boldsymbol{\rho}'(k), \boldsymbol{\phi}_k; \mathbf{w}^*]$ at $\mathbf{N}[\boldsymbol{\rho}(k), \boldsymbol{\phi}_k; \mathbf{w}^*]$. We have

$$\begin{aligned} \mathbf{N}[\boldsymbol{\rho}'(k), \boldsymbol{\phi}_k; \mathbf{w}^*] \cong\; &\mathbf{N}[\boldsymbol{\rho}(k), \boldsymbol{\phi}_k; \mathbf{w}^*] \\ &+ \frac{\partial \mathbf{N}}{\partial \boldsymbol{\rho}(k)}[\boldsymbol{\rho}(k), \boldsymbol{\phi}_k; \mathbf{w}^*]\, [\boldsymbol{\rho}'(k) - \boldsymbol{\rho}(k)] \end{aligned} \tag{14}$$

According to the definition of $\boldsymbol{\rho}'(k)$, we find

$$\boldsymbol{\rho}'(k) \cong \boldsymbol{\rho}(k) + [\frac{\partial \mathbf{N}}{\partial \boldsymbol{\rho}(k)}[\boldsymbol{\rho}(k), \boldsymbol{\phi}_k; \mathbf{w}^*]]^{-1}\, [\frac{\partial \mathbf{N}}{\partial \mathbf{w}}[\boldsymbol{\rho}(k), \boldsymbol{\phi}_k; \mathbf{w}_k]]\, \tilde{\mathbf{w}}_k \tag{15}$$

Now, since we have $\tilde{\mathbf{y}}(k) = \mathbf{A}_{\mathrm{m}}^{-1}(z^{-1})\boldsymbol{\rho}(k)$, from the definition of $\boldsymbol{\rho}'(k)$, we also have $\mathbf{y}(k) \cong \mathbf{A}_{\mathrm{m}}^{-1}(z^{-1})\boldsymbol{\rho}'(k)$, so that

$$\mathbf{e}(k) = [\mathbf{y}(k) - \tilde{\mathbf{y}}(k)] \cong \mathbf{A}_{\mathrm{m}}^{-1}(z^{-1})\, [\boldsymbol{\rho}'(k) - \boldsymbol{\rho}(k)]$$

or

$$\mathbf{Z}(z)\, \mathbf{e}(k) \cong \hat{\mathbf{A}}_{\mathrm{m}}^{-1}(z^{-1})\, [\boldsymbol{\rho}'(k) - \boldsymbol{\rho}(k)] \tag{16}$$

Therefore we obtain the error equation relating the error at the output of the process and the error in the weight values:

$$\mathbf{Z}(z)\,\mathbf{e}(k) \cong \hat{\mathbf{A}}_m^{-1}(z^{-1})\ [\frac{\partial \mathbf{N}}{\partial \boldsymbol{\rho}(k)}[\boldsymbol{\rho}(k), \boldsymbol{\phi}_k; \mathbf{w}^*]]^{-1}\ [\frac{\partial \mathbf{N}}{\partial \mathbf{w}}[\boldsymbol{\rho}(k), \boldsymbol{\phi}_k; \mathbf{w}_k]]\ \tilde{\mathbf{w}}_k$$

$$\cong \hat{\mathbf{A}}_m^{-1}(z^{-1})\ \mathbf{v}_k\,\tilde{\mathbf{w}}_k \qquad (17)$$

where we defined

$$\mathbf{v}_k = [\frac{\partial \mathbf{N}}{\partial \boldsymbol{\rho}(k)}[\boldsymbol{\rho}(k), \boldsymbol{\phi}_k; \mathbf{w}^*]]^{-1}\ [\frac{\partial \mathbf{N}}{\partial \mathbf{w}}[\boldsymbol{\rho}(k), \boldsymbol{\phi}_k; \mathbf{w}_k]].$$

Now, let us show that

$$[\frac{\partial \mathbf{N}}{\partial \boldsymbol{\rho}(k)}[\boldsymbol{\rho}(k), \boldsymbol{\phi}_k; \mathbf{w}^*]]^{-1} = \frac{\partial(\mathbf{Z}(z)\mathbf{y}(k))}{\partial \mathbf{u}(k)}\ , \qquad (18)$$

the derivative in (18) being computed at $\mathbf{w}^*$.

Indeed, for $\mathbf{w}=\mathbf{w}^*$ (perfect tuning), the whole system behaves like the reference model:

$$\mathbf{Z}(z)\,\mathbf{y}(k) = \mathbf{Z}(z)\,\mathbf{A}_m^{-1}(z^{-1})\,\mathbf{B}_m(z^{-1})\,\mathbf{r}(k) = \hat{\mathbf{A}}_m^{-1}(z^{-1})\,\boldsymbol{\rho}(k)$$

and from (4a), we obtain

$$\frac{\partial(\mathbf{Z}(z)\mathbf{y}(k))}{\partial \boldsymbol{\rho}(k)} = \mathbf{I}$$

On the other hand, we have

$$\frac{\partial(\mathbf{Z}(z)\mathbf{y}(k))}{\partial \boldsymbol{\rho}(k)} = \frac{\partial(\mathbf{Z}(z)\mathbf{y}(k))}{\partial \mathbf{u}(k)}\ \frac{\partial \mathbf{N}}{\partial \boldsymbol{\rho}(k)}[\boldsymbol{\rho}(k), \boldsymbol{\phi}_k; \mathbf{w}^*]$$

and the result (18) follows. Therefore, we have

$$\mathbf{v}_k = \frac{\partial(\mathbf{Z}(z)\mathbf{y}(k))}{\partial \mathbf{u}(k)}\ \frac{\partial \mathbf{N}}{\partial \mathbf{w}}[\boldsymbol{\rho}(k), \boldsymbol{\phi}_k; \mathbf{w}_k] = \frac{\partial \mathbf{F}_k}{\partial \mathbf{u}(k)}\ \frac{\partial \mathbf{N}}{\partial \mathbf{w}}[\boldsymbol{\rho}(k), \boldsymbol{\phi}_k; \mathbf{w}_k] \qquad (19)$$

Now, the transfer function matrix $[\mathbf{Z}^{-1}\,\hat{\mathbf{A}}_m^{-1}(z^{-1})]$ appearing in the error equation (17) is not SPR, while $\hat{\mathbf{A}}_m^{-1}(z^{-1})$ is SPR. The standard technique to cope with this problem is to introduce an augmented error, which is computable from the past values (see, for instance, Goodwin & Sin, 1984; Astrom & Wittenmark, 1989; Narendra & Annaswamy, 1989). The augmented error is defined as

$$\mathbf{e}'(k) = \mathbf{e}(k) - \hat{\mathbf{A}}_m^{-1}(z^{-1})\,\mathbf{Z}^{-1}\,[\mathbf{v}_k\,\mathbf{w}_k] + \hat{\mathbf{A}}_m^{-1}(z^{-1})\ [\mathbf{Z}^{-1}\,\mathbf{v}_k]\ \mathbf{w}_{k-1} \qquad (20)$$

$$= \mathbf{e}(k) + \hat{\mathbf{A}}_{\mathrm{m}}^{-1}(z^{-1}) [(\mathbf{Z}^{-1}\mathbf{v}_k) \mathbf{w}_{k-1} - \mathbf{Z}^{-1}(\mathbf{v}_k \mathbf{w}_k)] \tag{21}$$

and the error augmentation is proportional to the variation in weight values. By expanding (20) tanks to equation (17), we obtain a new error equation:

$$\mathbf{e}'(k) \cong \hat{\mathbf{A}}_{\mathrm{m}}^{-1}(z^{-1}) (\mathbf{Z}^{-1}\mathbf{v}_k) \tilde{\mathbf{w}}_{k-1} \tag{22}$$

The augmented error is therefore an estimation of the error that would have been produced by using the weights $\mathbf{w}_{k-1}$ instead of the delayed weights. Now, the transfer function matrix is SPR, so that we are in the conditions of the stability lemma (see appendix) with

$$\mathbf{h}(k) = \mathbf{e}'(k) \tag{23a}$$

$$\mathbf{H}(z) = \hat{\mathbf{A}}_{\mathrm{m}}^{-1}(z^{-1}) \tag{23b}$$

$$\boldsymbol{\nu}_k = \mathbf{Z}^{-1} \frac{\partial \mathbf{F}_k}{\partial \mathbf{u}(k)} \frac{\partial \mathbf{N}}{\partial \mathbf{w}}[\boldsymbol{\rho}(k), \boldsymbol{\phi}_k; \mathbf{w}_k] = \mathbf{Z}^{-1} \mathbf{v}_k \tag{23c}$$

$$\boldsymbol{\psi}_k = \tilde{\mathbf{w}}_{k-1} \tag{23d}$$

and the fact that $\mathbf{e}'(k) \to 0$ when adjusting the weights by

$$\mathbf{w}_k = \mathbf{w}_{k-1} - \eta (\mathbf{Z}^{-1}\mathbf{v}_k)^{\mathrm{T}} \mathbf{e}'(k) \tag{24}$$

follows immediately.

Moreover, since $\mathbf{e}'(k)$ converges to zero, from the adaptation law (24) and the boundedness of $\mathbf{v}_k$, $(\mathbf{w}_k - \mathbf{w}_{k-1})$ also tends towards zero. It follows that $[(\mathbf{Z}^{-1}\mathbf{v}_k) \mathbf{w}_{k-1} - \mathbf{Z}^{-1}(\mathbf{v}_k \mathbf{w}_k)]$ tends towards zero, and since $\hat{\mathbf{A}}_{\mathrm{m}}^{-1}(z^{-1})$ is SPR, from equation (21), $\mathbf{e}(k) \to 0$ asymptotically .

We therefore obtain the following stability statement:

*If* $\mathbf{v}_k = \frac{\partial \mathbf{F}_k}{\partial \mathbf{u}(k)} \frac{\partial \mathbf{N}}{\partial \mathbf{w}}[\boldsymbol{\rho}(k), \boldsymbol{\phi}_k; \mathbf{w}_k]$, *supposed measurable, is globally bounded, the adaptation rule*

$$\mathbf{w}_k = \mathbf{w}_{k-1} - \eta (\mathbf{Z}^{-1}\mathbf{v}_k)^{\mathrm{T}} \mathbf{e}'(k) \tag{24}$$

*with* $\eta$ *being a small positive constant, will drive the error* $\mathbf{e}(k) = [\mathbf{y}(k) - \tilde{\mathbf{y}}(k)]$ *asymptotically to zero*: $\mathbf{e}(k) \to 0$.

The adaptation law (24) can be simplified further by noticing that the contribution of the error augmentation to the weight adaptation is of order $O(\eta^2)$ in equation (24), and can therefore be ignored (we already ignored the second-order term for the stability proof, see equation (A6)). We finally obtain the following adaptation law:

$$\mathbf{w}_k = \mathbf{w}_{k-1} - \eta\, (\mathbf{Z}^{-1}\mathbf{v}_k)^{\mathrm{T}}\, \mathbf{e}(k) \tag{25}$$

which is the same adjustment law as proposed in section 5 (equation (10)). Of course, (24) can be used as well.

Notice that the term $\partial\mathbf{N}/\partial\mathbf{w}[\boldsymbol{\rho}(k), \boldsymbol{\Phi}_k; \mathbf{w}_k]$ appearing in (25) can be computed thanks to backpropagation algorithm, while $\partial\mathbf{F}_k/\partial\mathbf{u}(k)$ is the Jacobian matrix of the process defined by (1). As proposed by Jordan (1989), Nguyen & Widrow (1989, 1990) and Narendra & Parthasarathy (1990), this Jacobian matrix can be deduced from a model of the process; possibly in the form of a neural identifier. As mentioned above, the adjustment strategy is indirect in that it uses a model of the plant (in our case, we use a neural network that identifies the plant) in order to compute the term $\partial\mathbf{F}_k/\partial\mathbf{u}(k)$ required in the adaptation law. Notice that, since the adaptation law relies on the availability of a perfect model of the process, the global stability of the overall system *cannot be guaranteed* (in practice, a perfect model is, of course, not available; moreover, the process can change in time so that the model is no more appropriate). However, during our experiments, we observed empirically that the convergence rate does not seem to depend crucially on the precision of the term $\partial\mathbf{F}_k/\partial\mathbf{u}(k)$. In Renders & Saerens (1994), we derived a robust algorithm that takes the perturbations and the modeling errors into account. Various simulations have been carried out both for the SISO and the MIMO case (Renders, 1993; Renders, Bersini & Saerens, 1993; Saerens, Renders & Bersini, 1994; Saerens et al., 1992).

## 7. Weight Initialization Procedure

Now, the result is valid only when the weights are not too far from their optimal value. We do not expect that this hypothesis could be easily removed because there is always a risk of falling in a local minimum. This suggests a preliminary initialization of the weight values in order to be in the basin of attraction of the optimal values. One way to initialize the weights is to consider direct linear connections from the inputs to the outputs of the net. The weights of these connections can be trained by linear techniques, or initialized to values that solve the linear problem (Kawato, 1990; Gomi & Kawato, 1993; Scott, Shavlik & Ray, 1992). The other weight values should be set near zero in order to have negligible effect on the control law. Tuning of these weights can then be started, while maintaining the direct weights from input to output constant. In other words, we have

$$\mathbf{u}(k) = \mathbf{u}_1(k) + \mathbf{u}_2(k) \tag{26}$$

with $\mathbf{u}_1(k)$ being the result of the linear transform from the input to the output, and $\mathbf{u}_2(k)$ being the output of the multi-layer net, without the direct connections.

Several simulation results for this algorithm can be found in (Renders, 1993, Renders, Saerens & Bersini, 1994; Saerens, Renders & Bersini, 1994; Renders & Saerens, 1994).

## 8. Conclusion

In this chapter, we proved the input-output stability of a nonlinear MIMO (Multi-Input Multi-Output) system controlled by a multilayer neural net with a simple weight adaptation strategy. This extends our previous work (Renders, 1993; Saerens, Renders & Bersini, 1994), where single-input single-output (SISO) systems were considered. The stability proof is based on the Lyapunov stability theory. In the SISO case, for radial basis function nets and when the reference signal is the delayed desired output, the stability result is strictly valid: the weights do not have to be initialized around the perfectly tuned values (Renders, 1993; Saerens, Renders & Bersini, 1994). We were not able to prove a similar result in the MIMO case: the stability statement is only valid if the initial weight values are not too far from their optimal values which allow perfect model matching (local stability). For this reason, following Kawato (1990), we propose to introduce direct linear connections from the inputs to the outputs of the net. The weights of these connections can be trained by linear techniques, or initialized to values that solve the linear problem. The other connection weights should be initialized around zero in order to have negligible effect on the control. Notice that continuous-time SISO processes were considered in (Saerens, Renders & Bersini, 1994).

Further work will be devoted to the design of algorithms that do not require the a priori knowledge of process delays and orders. Furthermore, in this work, the hypothesis that the process is linearizable and decouplable by static state-feedback has been implicitly assumed. The use of recurrent neural networks may enable to overcome this limitation by automatically constructing a general dynamic state-feedback law if necessary.

Finally, let us mention that we also developed another weight adaptation law, based on the $e_1$-modification scheme (Narendra & Annaswamy, 1987), for robust adaptive control in the presence of perturbations and modeling errors

(Renders & Saerens, 1994). Finally, we showed that the same arguments can be applied to adaptive fuzzy control (Renders, Saerens & Bersini, 1993).

## Appendix: Stability Lemma

This lemma (see, for instance, Landau, 1979; Slotine & Li, 1991; Vidyasagar, 1993) will allow us to design a weight adaptation law that is asymptotically stable, with the restriction that the initial values of the weights are not too far from their optimal value. Let us consider two signals related by

$$\mathbf{h}(k) = \mathbf{H}(z)\, \boldsymbol{\nu}_k\, \boldsymbol{\psi}_k \tag{A1}$$

where $\mathbf{h}(k)$ is an output vector signal, $\mathbf{H}(z)$ is a Strictly Positive Real transfer matrix, $\boldsymbol{\psi}_k$ is a vector function of time, and $\boldsymbol{\nu}_k$ is a measurable matrix. If the vector $\boldsymbol{\psi}_k$ varies according to

$$\boldsymbol{\psi}_{k+1} = \boldsymbol{\psi}_k - \eta\, (\boldsymbol{\nu}_k)^{\mathrm{T}}\, \mathbf{h}(k) \tag{A2}$$

with $\eta$ being a positive small constant value, then, if $\boldsymbol{\nu}_k$ is bounded, $\mathbf{h}(k) \to 0$ asymptotically.

***Proof of the stability lemma.*** Let us define a positive definite function $V_k$ of the form

$$V_k = (\mathbf{x}(k))^{\mathrm{T}}\, \mathbf{P}\, \mathbf{x}(k) + \frac{1}{\eta}\, (\boldsymbol{\psi}_k)^{\mathrm{T}}\, \boldsymbol{\psi}_k \tag{A3}$$

where $\mathbf{P}$ is a symmetric positive definite matrix and $\mathbf{x}(k)$ is a state vector of the system described by (A1). Since $\mathbf{H}(z)$ is Strictly Positive Real (SPR), we know that there exists two symmetric positive definite matrices $\mathbf{P}$ and $\mathbf{Q}$, and two matrices $\mathbf{K}$ and $\mathbf{L}$ such that

$$\mathbf{x}(k+1) = \mathbf{A}\, \mathbf{x}(k) + \mathbf{B}\, \boldsymbol{\xi}(k) \tag{A4a}$$

$$\mathbf{h}(k) = \mathbf{C}\, \mathbf{x}(k) + \mathbf{D}\, \boldsymbol{\xi}(k) \tag{A4b}$$

which is the state representation of $\mathbf{H}(z)$, where we have posed $\boldsymbol{\xi}(k) = \boldsymbol{\nu}_k\, \boldsymbol{\psi}_k$, and

$$\mathbf{A}^{\mathrm{T}}\mathbf{P}\mathbf{A} - \mathbf{P} = -\mathbf{L}\mathbf{L}^{\mathrm{T}} - \mathbf{Q} \tag{A5a}$$

$$\mathbf{B}^{\mathrm{T}}\mathbf{P}\mathbf{A} + \mathbf{K}^{\mathrm{T}}\mathbf{L}^{\mathrm{T}} = \mathbf{C} \tag{A5b}$$

$$\mathbf{K}^{\mathrm{T}}\mathbf{K} = \mathbf{D} + \mathbf{D}^{\mathrm{T}} - \mathbf{B}^{\mathrm{T}}\mathbf{P}\mathbf{B} \tag{A5c}$$

Notice that the fact that the speed of adaptation must be low (assumption vi) implies

$$[(\boldsymbol{\psi}_{k+1})^{\mathrm{T}} \boldsymbol{\psi}_{k+1} - (\boldsymbol{\psi}_k)^{\mathrm{T}} \boldsymbol{\psi}_k] \cong 2\,(\boldsymbol{\psi}_k)^{\mathrm{T}} \Delta\boldsymbol{\psi}_k \tag{A6}$$

where we neglected the term in $O(\eta^2)$.

We have to demonstrate that $V_k$ is a Lyapunov function of the global system, by showing that $\Delta V_k$ is negative. By using (A4a) to compute $\mathbf{x}(k+1)$, and by using the relationships (A5abc) and (A4b), we obtain

$$\begin{aligned}
V_{k+1} - V_k &= (\mathbf{x}(k+1))^{\mathrm{T}} \mathbf{P}\, \mathbf{x}(k+1) - (\mathbf{x}(k))^{\mathrm{T}} \mathbf{P}\, \mathbf{x}(k) \\
&+ \frac{1}{\eta} [(\boldsymbol{\psi}_{k+1})^{\mathrm{T}} \boldsymbol{\psi}_{k+1} - (\boldsymbol{\psi}_k)^{\mathrm{T}} \boldsymbol{\psi}_k] \\
&\cong (\mathbf{x}(k+1))^{\mathrm{T}} \mathbf{P}\, \mathbf{x}(k+1) - (\mathbf{x}(k))^{\mathrm{T}} \mathbf{P}\, \mathbf{x}(k) \\
&+ \frac{1}{\eta} [2\,(\boldsymbol{\psi}_k)^{\mathrm{T}} \Delta\boldsymbol{\psi}_k] \\
&= -\,(\mathbf{x}(k))^{\mathrm{T}} \mathbf{Q}\, \mathbf{x}(k) \\
&- [\mathbf{L}^{\mathrm{T}} \mathbf{x}(k) + \mathbf{K}\, \boldsymbol{\xi}(k)]^{\mathrm{T}} [\mathbf{L}^{\mathrm{T}} \mathbf{x}(k) + \mathbf{K}\, \boldsymbol{\xi}(k)] \\
&\leq 0
\end{aligned}$$

Now, since $V_k$ is bounded from below and $\boldsymbol{\nu}_k$ is bounded, $\Delta V_k$ must decrease to zero, which implies $\mathbf{x}(k) \rightarrow 0$ asymptotically, and therefore $\mathbf{h}(k) \rightarrow 0$ asymptotically.

## Acknowledgments

This work was partially supported by the ARC 92/97-160 (BELON) project from the "Communauté Française de Belgique", and the FALCON (6017) Basic Research ESPRIT project from the European Communities.

## References

ÅSTRÖM K.J. & WITTENMARK B., 1989, "Adaptive control". Addison-Wesley Publishing Company.

BARTO A., 1990, "Connectionist learning for control: an overview". In Neural networks for control, W Thomas Miller, R. Sutton & P. Werbos (editors), The MIT Press.

CHEN F.-C. & KHALIL H., 1991, "Adaptive control of nonlinear systems using neural networks – A dead-zone approach". Proceedings of the American Control Conference, pp. 667-672.

CHEN F.-C. & LIU C.-C., 1992, "Adaptively controlling nonlinear continuous-time systems using neural networks". Proceedings of the American Control Conference, pp. 46-50.

CHEN S., BILLINGS S. & LUO W., 1989, "Orthogonal least squares methods and their application to non-linear system identification". International Journal of Control, 50 (5), pp. 1873-1896.

CHEN S., BILLINGS S., COWAN C. & GRANT P., 1990, "Practical identification of NARMAX models using radial basis functions". International Journal of Control, 52 (6), pp. 1327-1350.

GOMI H. & KAWATO M, 1993, "Neural network control for a closed-loop system using feedback-error-learning". Neural Networks, 6, pp. 933-946.

GOODWIN G. & SIN K. S., 1984, "Adaptive filtering, prediction and control". Prentice-Hall.

GUPTA M., RAO D. & WOOD H., 1991, "Learning adaptive neural controller". Proceedings of the IEEE International Joint Conference on Neural Networks, Singapore, pp. 2380-2384.

GUPTA M., RAO D. & NIKIFORUK P., 1993, "Neuro-controller with dynamic learning and adaptation". Journal of Intelligent and Robotic Systems, 7, pp. 151-173.

HUNT K. & SBARBARO D., 1991, "Neural networks for nonlinear internal model control". IEE Proceedings-D, 138 (5), pp. 431-438.

HUNT K., SBARBARO D., ZBIKOWSKI R. & GAWTHROP P., 1992, "Neural networks for control systems – A survey". Automatica, 28 (6), pp. 1083-1112.

JIN L., NIKIFORUK P. & GUPTA M., 1993, "Direct adaptive output tracking control using multilayer neural networks". IEE Proceedings-D, 140 (6), pp. 393-398.

JIN Y., PIPE T. & WINFIELD A., 1993, "Stable neural adaptive control for discrete systems". Proceedings of the World Congres on Neural Networks, Portlend, Vol III, pp. 277-280.

JOHANSEN T.A., 1994, "Robust adaptive control of slowly-varying discrete-time non-linear systems". Technical report of the Norwegian Institute of Technology, Department of Engineering Cybernetics.

JORDAN M.I., 1989, "Generic constraints on underspecified target trajectories". Proceedings of International Joint Conference on Neural Networks, Washington, Vol I, pp. 217-225.

KAWATO M., 1990, "Feedback-error-learning neural network for supervised learning". In "Advanced neural computers", Eckmiller R. (editor), Elsevier Publishers, pp. 365-372.

LAN M-S., 1989, "Adaptive control of unknown dynamical systems via neural network approach". Proceedings of the 1989 American Control Conference, Pittsburgh, pp. 910-915.

LANDAU Y., 1979, "Adaptive control. The model reference approach". Marcel Dekker.

LEVIN E., GEWIRTZMAN R. & INBAR G., 1991, "Neural network architecture for adaptive system modeling and control". Neural Networks, 4, pp. 185-191.

LIU C.-C. & CHEN F.-C., 1993, "Adaptive control of nonlinear continuous-time systems using neural networks - general relative degree and MIMO cases". International Journal of Control, 58 (2), pp. 317-335.

MILLER T., R. SUTTON and P. WERBOS (eds.) (1990). Neural networks for control. MIT Press, Cambridge (MA).

NARENDRA K., 1992, "Adaptive control of dynamic systems using neural networks". In "Handbook of intelligent control: Neural, fuzzy and adaptive approaches", White D. A. & Sofge D. A. (editors). Van Nostrand Reinhold.

NARENDRA K.S. & ANNASWAMY A., 1987, "A new adaptive law for robust adaptation without persistent excitation". IEEE Transactions on Automatic Control, 32 (2), pp. 134-145.

NARENDRA K.S. & ANNASWAMY A., 1989, "Stable adaptive systems". Prentice-Hall.

NARENDRA K.S. & MUKHOPADHYAY S., 1994, "Adaptive control of nonlinear multivariate systems using neural networks". Neural Networks, 7 (5), pp. 737-752.

NARENDRA K.S. & PARTHASARATHY K., 1990, "Identification and control of dynamic systems using neural networks". IEEE Transactions on Neural Networks, 1 (1), pp. 4-27. Reprinted in "Artificial Neural Networks: Concepts and Control Applications" by Rao Vemuri (editor), IEEE Computer Society Press, 1992.

NARENDRA K.S. & PARTHASARATHY K., 1991, "Gradient methods for the optimization of dynamical systems containing neural networks". IEEE Transactions on Neural Networks, 2 (2), pp. 252-262. Reprinted in "Artificial Neural Networks: Concepts and Control Applications" by Rao Vemuri (editor), IEEE Computer Society Press, 1992.

NARENDRA K.S. & PARTHASARATHY K., 1992, "Neural networks and dynamical systems". International Journal of Approximate Reasoning, 6, pp. 109-131.

NGUYEN D. & WIDROW B., 1989, "The truck backer-upper: an example of self-learning in neural networks". Proceedings of International Joint Conference on Neural Networks, Washington, Vol II, pp. 357-353.

NGUYEN D. & WIDROW B., 1990, "Neural networks for self-learning control systems". IEEE Control Systems Magazine, 10 (3), pp. 18-23.

PARTHASARATHY K. & NARENDRA K., 1991, "Stable adaptive control of a class of discrete-time nonlinear systems using radial basis networks". Report No. 9103, Yale University, Center for Systems Science, New Haven, Connecticut.

PSALTIS D., SIDERIS A. & YAMAMURA A., 1987, "Neural controllers". Proceedings of IEEE first International Conference on Neural Networks, San Diego, Vol 4, pp. 551-558.

PSALTIS D., SIDERIS A. & YAMAMURA A., 1988, "A multilayered neural network controller". IEEE Control Systems Magazine, 8 (2), pp. 17-21. Reprinted in "Artificial Neural Networks: Concepts and Control Applications" by Rao Vemuri (editor), IEEE Computer Society Press, 1992.

PUSKORIUS G. & FELDKAMP L., 1992, "Model reference adaptive control with recurrent networks trained by the dynamic DEKF algorithm". Proceedings of the International Joint Conference on Neural Networks, Baltimore, pp. 106-113.

PUSKORIUS G. & FELDKAMP L., 1994, "Neurocontrol of nonlinear dynamical systems with Kalman filter trained recurrent networks". IEEE Transactions on Neural Networks, 5 (2), pp. 279-297.

RENDERS J.-M., 1991, "A new approach of adaptive neural controller design with application to robotics control". In IMACS Annals on Computing and Applied Mathematics, 10: Mathematical and Intelligent Models in System Simulation, J.C. Baltzer AG, Scientific Publishing Co., pp. 361-366.

RENDERS J.-M., 1993, "Biological metaphors for process control" (in french). Ph.D Thesis, Université Libre de Bruxelles, Faculté Polytechnique, Belgium.

RENDERS J.-M., BERSINI H. & SAERENS M., 1993, "Adaptive neurocontrol: How black-box and simple can it be ?". Proceedings of the tenth International Workshop on Machine Learning, Amherst, pp. 260-267.

RENDERS J.-M., SAERENS M. & BERSINI H., 1994, "On the stability of a certain class of adaptive fuzzy controllers". Proceedings of the European Conference on Fuzzy and Intelligent Technologies (EUFIT), Aachen, pp. 27-34.

RENDERS J.-M., SAERENS M. & BERSINI H., 1994, "Adaptive neurocontrol of MIMO systems based on stability theory". Proceedings of the IEEE International Conference on Neural Network, Orlando, pp. 2476-2481

RENDERS J-M. & SAERENS M., 1994, "Robust adaptive neurocontrol of MIMO processes". Submitted for publication.

ROVITHAKIS G. & CHRISTODOULOU M., 1994, "Adaptive control of unknown plants using dynamical neural networks". IEEE Transactions on Systems, Man, and Cybernetics, 24 (3), pp. 400-412.

SAERENS M. & SOQUET A., 1991, "Neural Controller Based on Back-Propagation Algorithm". IEE Proceedings–F, 138 (1), pp. 55-62.

SAERENS M., SOQUET A., RENDERS J-M. & BERSINI H., 1992, "Some preliminary comparisons between a neural adaptive controller and a model reference adaptive controller". In "Neural Networks in Robotics" by Bekey G. & Goldberg K. (editors), pp. 131-146. Kluwer Academic Press.

SAERENS M., RENDERS J.-M. and BERSINI H., 1994 (in press), "Neural controllers based on backpropagation algorithm". Chapter to appear in the "IEEE Press Book on Intelligent Control", M. Gupta and N. Sinha (editors). IEEE Press.

SANNER R. & SLOTINE J.-J., 1992, "Gaussian networks for direct adaptive control". IEEE Transactions on Neural Networks, 3 (6), pp. 837-863.

SCOTT G., SHAVLIK J. & HARMON RAY W., 1992, "Refining PID controllers using neural networks". Neural Computation, 4, pp. 746-757.

SEUBE N., 1990, "Construction of learning rules in neural networks that can find viable regulation laws to control problems by self-organization". Proceedings of the International Neural Networks Conference, Paris, pp. 209-212.

SEUBE N. & MACIAS J.-C., 1991, "Design of neural network learning rules for viable feedback laws". Proceedings of the European Control Conference, Grenoble, pp. 1241-1246.

SLOTINE J.-J. & LI W., 1991, "Applied nonlinear control". Prentice-Hall.

TOLAT V. & WIDROW B., 1988, "An adaptive broom balancer with visual inputs". Proceedings of the International Joint Conference on Neural Networks, San Diego, pp. 641-647.

TZIRKEL-HANCOCK E. & FALLSIDE F., 1992, "Stable control of nonlinear systems using neural networks". International Journal of Robust and Nonlinear Control, 2, pp. 63-86.

VIDYASAGAR M., 1993, "Nonlinear systems analysis, second edition". Prentice-Hall.

WHITE D. A. & SOFGE D. (editors), 1992, "Handbook of intelligent control. Neural, fuzzy, and adaptive approaches". Van Nostrand Reinhold.

WILLIAMS R.J. & ZIPSER D., 1989, "A learning algorithm for continually running fully recurrent neural networks". Neural Computation, 1 (2), pp. 270-280.

# Local Model Architectures for Nonlinear Modelling and Control

Roderick Murray-Smith and Kenneth Hunt

Daimler-Benz AG, Alt-Moabit 91 b, D-10559 Berlin, Germany.
E-mail: murray or hunt @DBresearch-berlin.de

**Abstract.** Local Model Networks are learning systems which are able to model and control unknown nonlinear dynamic processes from their observed input-output behaviour. Simple, locally accurate models are used to represent a globally complex process. The framework supports the modelling process in real applications better than most artificial neural network architectures. This paper shows how their structure also allows them to more easily integrate knowledge, methods and *a priori* models from other paradigms such as fuzzy logic, system identification and statistics. Algorithms for automatic parameter estimation and model structure identification are given.
Local Models intuitively lend themselves to the use of Local Controllers, where the global controller is composed of a combination of simple locally accurate control laws. A Local Controller Network (LCN) for controlling the lateral deviation of a car on a straight road is demonstrated.

## 1 Introduction

This paper presents methods for the automatic construction of local model networks. These are architectures capable of representing nonlinear dynamic systems by smoothly integrating a number of simpler locally accurate models. Local methods of controlling systems, such as gain scheduling, are very useful methods for nonlinear control, but determining the gain schedule becomes increasingly difficult as the process complexity increases. The *Local Controller Networks* presented here can be viewed as a general formulation of the gain scheduling idea, where the 'schedule' can be automatically determined by the *Local Model* structure constructed to represent the process. The effect of the choice of local or global parameter estimation methods on the resulting model structure is discussed.

### 1.1 Model Structures

We consider discrete-time nonlinear systems having the general form

$$y(t) = f(y(t-1), \ldots y(t-n_y), u(t-k), \ldots u(t-k-n_u), e(t-1), \ldots e(t-n_e)) + e(t). \tag{1}$$

Here, $y(t)$ is the system output, $u(t)$ the input and $e(t)$ is a zero-mean disturbance term. We restrict attention to single-input single-output systems so that $y(t) \in$

$Y \subset \mathcal{R}$, $u(t) \in U \subset \mathcal{R}$ and $e(t) \in E \subset \mathcal{R}$. $k$ represents a time delay. This type of model is known as the NARMAX (Nonlinear ARMAX) model [1] and has been studied widely in nonlinear systems identification.

When we define the information vector as

$$\psi(t-1) = [y(t-1), \ldots y(t-n_y), u(t-k), \ldots u(t-k-n_u), e(t-1), \ldots e(t-n_e)]^T \tag{2}$$

then the system (1) can be written as

$$y(t) = f(\psi(t-1)) + e(t). \tag{3}$$

The $^T$ in (2) denotes the transpose operator. The aim in empirical modelling is to find a parameterised structure which emulates the nonlinear function $f$.

### 1.2 Linear Models

Standard linear approaches to modelling consider a linearisation of (1) about a nominal operating point (e.g. by taking a first-order Taylor series expansion) which results in the linear ARMAX model

$$y(t) = \phi_{ar}^T(t-1)\theta + e(t) \tag{4}$$

where

$$\phi_{ar}(t-1) = [-y(t-1), \ldots -y(t-n_y), u(t-k), \ldots u(t-k-n_u), e(t-1), \ldots e(t-n_e)]^T, \tag{5}$$

and

$$\theta = [a_1, \ldots a_{n_y}, b_0, \ldots b_{n_u}, c_1, \ldots c_{n_e}]^T \tag{6}$$

is a constant parameter vector. Note that in (4)–(5) all signals now strictly represent deviations from the nominal operating point. With the definitions (5) and (6) the linear system (4) can be written in the familiar transfer-function form

$$y(t) = \frac{q^{-k}B(q^{-1})}{A(q^{-1})}u(t) + \frac{C(q^{-1})}{A(q^{-1})}e(t). \tag{7}$$

Here, $q^{-1}$ is the unit delay operator and the polynomials $A, B$ and $C$ are

$$\begin{aligned} A(q^{-1}) &= 1 + a_1 q^{-1} + \ldots + a_{n_y} q^{-n_y}, \\ B(q^{-1}) &= b_0 + b_1 q^{-1} + \ldots + b_{n_u} q^{-n_u}, \\ C(q^{-1}) &= 1 + c_1 q^{-1} + \ldots + c_{n_e} q^{-n_e}. \end{aligned} \tag{8}$$

### 1.3 Local Model Networks

Linear, or other simple models, can be used in a more general way, so that a more general class of systems can be represented. A number of such simple models, each weighted by a basis function associated with a different area of the input space, can be combined to create a globally complex model

$$\hat{y} = \hat{f}(\psi) = \sum_{i=1}^{n_{\mathcal{M}}} \hat{f}_i(\psi)\rho_i(\tilde{\phi}), \tag{9}$$

where $\tilde{\phi}$ defines the *operating point* of the system. The network form of equation (9) is shown in Fig. 1. The trained network structure can be viewed as a decomposition of the complex, nonlinear system into a set of $n_{\mathcal{M}}$ locally accurate sub-models, which are then smoothly integrated by their associated basis functions. To illustrate the workings of a local model network, a one dimensional

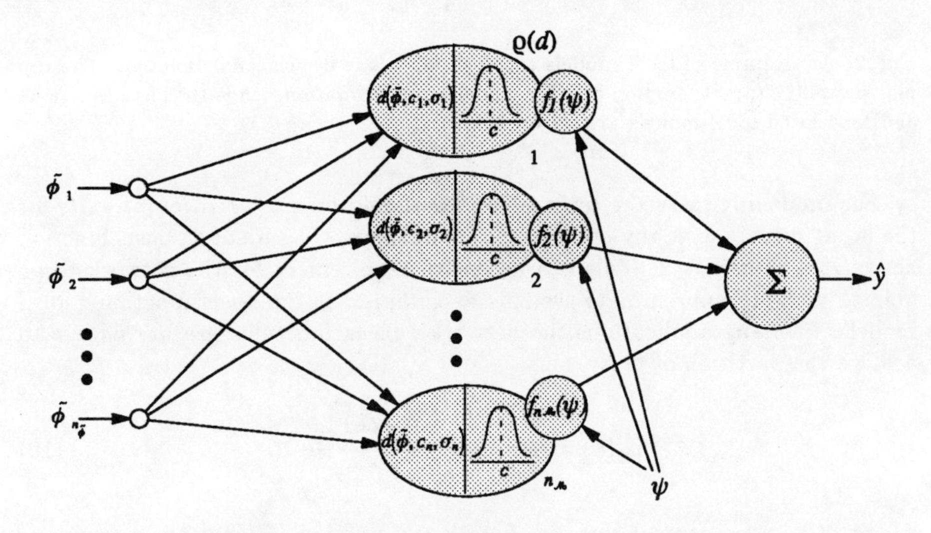


**Fig. 1.** Local Model Basis Function network

function is mapped using local models in Fig. 2.

The basis, or *model validity functions* used in this work are radial, i.e. they use a *distance metric* $d(\tilde{\phi}; c_i, \sigma_i)$ which measures the distance of the current operating point $\psi$ from the basis function's centre $c_i$, relative to the width variable $\sigma_i$. See Fig. 3 for a simple representation of operating regimes in a two dimensional operating space. The overlapping operating regimes allow the basis functions to smooth the transfer from one region of the model structure to the next. As there is a limited amount of overlap, however, there is a limited number of local models associated with any point in the input space, making the full model more interpretable.

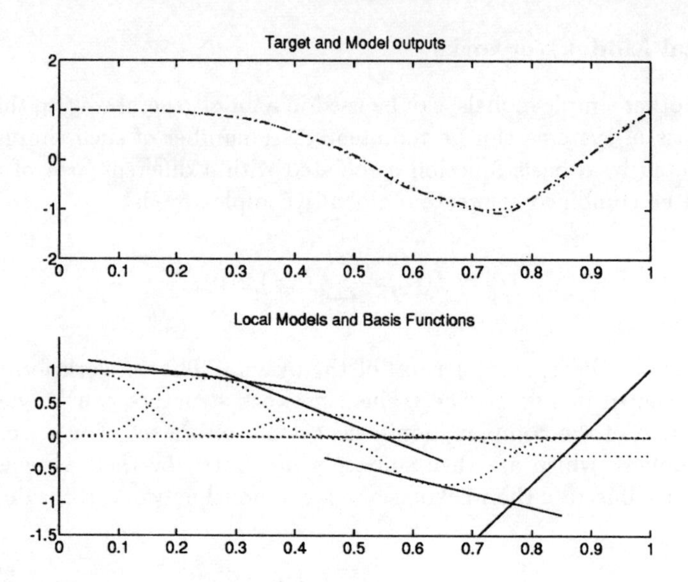


**Fig. 2.** An example of local models representing a one dimensional function. The top plot shows the target function and the model's approximation, while the basis functions and associated local models are shown below.

For modelling tasks the basis functions should form a *partition of unity* for the input space, i.e. at any point in the input space, the sum of all basis function activations should be 1. This is a necessary requirement for the network to be able to globally approximate systems as complex as the basis functions' local models. In many applications the network's basis functions are *normalised* to achieve the partition of unity, i.e.

$$\rho_k(\tilde{\phi}) = \frac{\rho\left(d\left(\tilde{\phi}, c_k, \sigma_k\right)\right)}{\sum_{i=1}^{n_{\mathcal{M}}} \rho\left(d\left(\tilde{\phi}, c_i, \sigma_i\right)\right)} \tag{10}$$

where $\rho(\cdot)$ is the general *unnormalised* basis function, so that the *normalised* basis functions $\rho_k(\cdot)$ sum to unity,

$$\sum_{i=1}^{n_{\mathcal{M}}} \rho_i\left(d\left(\tilde{\phi}, c_i, \sigma_i\right)\right) = 1. \tag{11}$$

[2] discusses the advantages of normalisation in RBF nets, promoting somewhat simplistically the advantages of a partition of unity produced by normalisation. Normalisation can be important for basis function nets, often making the model less sensitive to poor choice of basis functions, but it also has a number of side-effects which we discussed in detail in [3]. These side-effects make the argument for or against the use of normalisation more complicated than is often assumed.

The use of local basis functions means that the structure has the advantages inherent to the local nature of the basis functions while, because of the more

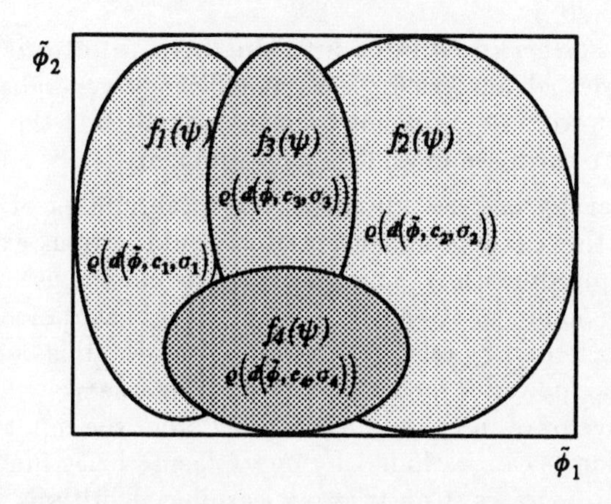


**Fig. 3.** Local Model Operating Regimes. Each local model is associated with an operating regime. These regimes overlap, and the gradual decay of 'validity' provides interpolation between models.

powerful local models associated with the basis functions, not requiring as many basis functions as a simple RBF net (e.g. as used in [4]) to achieve the desired accuracy. This improvement is more significant in higher dimensional and dynamic problems.

### 1.4 Pre-structuring Local Model Nets with *a priori* knowledge

A major advantage of local model nets is that they are not only useful architectures for general learning tasks, but that it is relatively easy to introduce *a priori* knowledge about a particular problem. This leads to more interpretable models which can be more reliably identified from a limited amount of observed data.

**Incorporating local models based on *a priori* knowledge** If the knowledge is available, local models could be physically motivated models, possibly requiring only a subset of their variables to be identified, thus allowing the engineer to easily create *grey-box* models. The most general form of information is the expected dynamic order of the process, and the form of model to be identified (e.g. simple linear ARX models, where $\theta_i$ is a local parameter vector corresponding to linearisation about a local nominal operating point). A generalised form outlined later allows the designer to specify a *pool* of feasible local models, which could be locally tested for suitability in the various operating regimes defined by the basis functions.

In many cases, there will not be sufficient data to train the model throughout the input space. This is especially true in areas outside normal desired operation, where the model may have to be very robust, and well understood. These situations can be covered by fixing *a priori* models in the given areas, and applying learning techniques only where the data is available and reliable.

**Incorporating *a priori* knowledge into the basis functions** Locality of representation provides advantages for learning efficiency, generalisation and transparency. It is, however, very difficult to automatically find the 'correct' level of locality for a given subspace of an arbitrary problem.

For a fixed density of basis functions, the total number of local models required rises – as the 'curse of dimensionality' would have us expect – exponentially with the input dimension. This does not rule out the use of local methods, however, as many nonlinear systems in the real world have smooth nonlinearities which can be represented by relatively few local models (this becomes even more relevant with more powerful local models). Also, as the system being modelled is often only active or of interest in a small region of the input space, a further saving of redundancy can be found by only placing basis functions in regions where the system operates. Constructive learning algorithms which automatically attempt this are described in section 2.3, but it is still important to use the available knowledge to limit the size of the search space faced by the learning algorithm.

The problems of dimensionality can be reduced in many systems with a large number of inputs, as there are often combinations of input dimensions which are of no interest, or which are additively or linearly related. The problem can then be decomposed, if the user already has the necessary *a priori* knowledge, so that the system can be approximated over an additive combination of lower dimensional sub-models (e.g. the theory of additive modelling techniques [5, 6] was developed to support such decompositions).

The Local Model net can thus be generalised to cater for possible redundancy in the operating point for certain operating regimes; some operating regimes may be specified by only a sub-vector of the operating vector. To allow this the argument to each $\rho_i(\cdot)$, which we will now call $\phi_i^{\mathrm{op}}$, is defined as belonging to a set $\Phi_i^{\mathrm{op}}$ defined on a subspace of $\mathcal{R}^{n_{\tilde{\phi}}}$, i.e. $\phi_i^{\mathrm{op}} \in \Phi_i^{\mathrm{op}} \subset \mathcal{R}^{n_{\phi_i^{\mathrm{op}}}}$, with $n_{\phi_i^{\mathrm{op}}} \leq n_{\tilde{\phi}}$. The *generalised local model network*, or GLMN, is defined as

$$\hat{y}(t) = \hat{f}(\psi) = \sum_{i=1}^{n_{\mathcal{M}}} \hat{f}_i(\psi)\rho_i(\phi_i^{\mathrm{op}}). \tag{12}$$

Using the above framework, the strong links between Fuzzy membership functions and Basis Functions (see [7, 8]) become even more apparent. *A priori* knowledge of how best to decompose the problem could therefore be expressed as fuzzy rules with accompanying basis/membership functions.

In summary, for many high-dimensional nonlinear dynamic processes, although the system may be globally strongly nonlinearly dependent on the inputs, it may now be possible to use the most important subset of the inputs to partition the input space into subspaces which are more manageable for the automated learning algorithm.

## 2 Learning in Local Model Nets

### 2.1 Parameter estimation in Local Model Nets

The problem of parameter estimation for systems which are linear in the parameters is reasonably well understood, with a variety of efficient optimisation algorithms existing to optimise the parameters $\theta$ of local models $\hat{f}_i(\cdot)$ in equation (9) to minimise the cost functional $J(\theta, \mathcal{M}, \mathcal{D})$ for a given local model structure $\mathcal{M}$, where $\mathcal{M} = (c, \sigma, n_{\mathcal{M}}, \mathcal{M}_{1..n_{\mathcal{M}}})$ (i.e. the basis functions' centre locations and basis function sizes, as well as local model types) and training set $\mathcal{D} = (\psi(t-1), y(t)), t = 1..N$. Parameter optimisation for a given model structure finds the optimal cost $J^*$

$$J^*(\mathcal{M}, \mathcal{D}) = \min_{\theta} J(\theta, \mathcal{M}, \mathcal{D}). \tag{13}$$

The methods described in the literature are usually *global* optimisation methods based on the assumption that all of the parameters $\theta$ can be optimised simultaneously with a single linear regression operation,

$$\mathbf{Y} = \Phi\theta \tag{14}$$

where $\Phi$ is the design matrix, where the rows are defined by

$$\phi_i = \left[\rho_1(\tilde{\phi}_i)[1\ \psi_{i_1} \ldots \psi_{i_{n_\psi}}] \ldots \rho_{n_{\mathcal{M}}}(\tilde{\phi}_i)[1\ \psi_{i_1} \ldots \psi_{i_{n_\psi}}]\right], \tag{15}$$

so that the design matrix $\Phi$, and vector of output measurements $\mathbf{Y}$ are

$$\Phi = \begin{pmatrix} \phi_1 \\ \cdot \\ \cdot \\ \phi_N \end{pmatrix}, \mathbf{Y} = \begin{pmatrix} y_1 \\ \cdot \\ \cdot \\ y_N \end{pmatrix} \tag{16}$$

The Moore-Penrose pseudoinverse of $\Phi$, $\Phi^+$ is then used to estimate the weights,

$$\hat{\theta} = \Phi^+\mathbf{Y} = (\Phi^T\Phi)^{-1}\Phi^T\mathbf{Y} \tag{17}$$

but this is not always computationally feasible for larger problems. A further problem is that the global nature of the observation can lead to the trained network being less transparent, as the parameters of the local models cannot be interpreted independently of neighbouring nodes. Also, even with robust identification algorithms, ill-conditioning in the design matrix can lead to the 'optimal' network parameters consisting of delicately balancing large positive and negative weights which minimise the output error on the training set, but which are not robust when confronted with new examples – i.e. the model generalises poorly.

The lack of independence in globally trained local models is significant for later use in *Local Controller Networks* (described in section 3.1).

**Local learning** An alternative to global learning is to locally estimate the parameters of each of the local models independently[1] using the basis functions as local weighting criteria. This results in a set of local estimation criteria for the $i$-th local model (where $i = 1..n_{\mathcal{M}}$) of

$$J_i(\theta_i) = \frac{1}{N_i} \sum_{k=1}^{N_i} \rho_i(\tilde{\phi}_{i_k})(y_{i_k} - \hat{y}_{i_k})^2, \tag{18}$$

where $N_i$ is the number of examples in the local training set $\mathcal{D}_i$ limited to the receptive field of local model $i$, and $\hat{y}_{i_k}$ is the output from local model $i$, for data point $k$ from $\mathcal{D}_i$. In this case the estimate of the local model parameter vector $\theta_i$ is given by $\hat{\theta}_i = \arg\min J_i(\theta_i)$. In matrix terms the operation is now

$$\hat{\theta}_i = (\Phi_i^T Q_i \Phi_i)^{-1} \Phi_i^T Q_i \mathbf{Y}, \tag{19}$$

where $\Phi_i$ is an $N_i \times (n_\psi + 1)$ vector,

$$\Phi_i = \begin{pmatrix} \phi_{i1} \\ \cdot \\ \cdot \\ \phi_{iN_i} \end{pmatrix}, \phi_{ik} = [1 \ \psi_k] \tag{20}$$

where the $k$ refers to the $k$th example in local training set $\mathcal{D}_i$. $Q_i$ is an $N_i \times N_i$ diagonal matrix, where the diagonal elements are the activations of the basis function of the $i$th model over the training set $\mathcal{D}_i$,

$$Q_i = \begin{pmatrix} \rho_i(\tilde{\phi}_{i_1}) & 0 & 0 & 0 \\ 0 & \rho_i(\tilde{\phi}_{i_2}) & 0 & 0 \\ 0 & 0 & \ddots & 0 \\ 0 & 0 & 0 & \rho_i(\tilde{\phi}_{iN_i}) \end{pmatrix}. \tag{21}$$

The local learning method involves the computation of $n_{\mathcal{M}}$ locally weighted least squares regressions, one for each local model, using only the training data $\mathcal{D}_i$ within the model's receptive field, and with only the bases related to the given local model's parameters.

### 2.2 Structure Identification in Local Model Nets

The use of existing *a priori* knowledge, as described in Section 1.4, to define the model structure is important, but as many problems are not well enough understood for the model structure to be fully specified in advance, it will often be necessary to adapt the structure for a given problem, based on information in the training data. The optimisation of the network structure $\mathcal{M}$ is an important but difficult non-convex optimisation problem. For a review of structure identification algorithms see [9] or the paper by Johansen and Foss in this volume.

[1] This assumes that the basis functions achieve a partition of unity.

The goal of the structure identification procedure is to provide a problem-adaptive learning scheme which automatically relates the complexity of the local models, and the density and overlap of basis functions to the local complexity and importance of the system being modelled. The desirable features of a structure identification algorithm are:

- *Consistency* – as the number of training points increases the algorithm should produce models which approximate the real process more accurately.
- *Parsimony* – the model structure produced by the algorithm should be the simplest possible which can represent the process to the required accuracy.
- *Robustness* – the model structures produced should be as robust as possible with regard to noisy data or missing data.
- *Interpretability* – the model structure produced should ideally be as interpretable as possible, given the available data, local models and basis functions.

The aim is therefore to find a model structure $\mathcal{M}$ which allows the network to best minimise the given cost function in a robust manner, taking the above points into consideration. Minimising $J^*(\mathcal{M}, \mathcal{D})$, from equation (13), over the range of possible model structures leads to the 'super-optimal' cost, using *a priori* knowledge about the process structure $\mathcal{K}_\mathcal{S}$,

$$J^{**}(\mathcal{D}, \mathcal{K}_\mathcal{S}) = \min_{\mathcal{M}} J^*(\mathcal{M}, \mathcal{D}). \tag{22}$$

The robustness is an important aspect, as constructive structure identification algorithms can obviously be very powerful, enabling the network to represent the training data very accurately by using a large number of parameters, but usually then leading to a high variance. The choice of model structure plays a major role in the *bias-variance trade-off* (see [10] for details about the trade-off), and this should be reflected in the cost functions $J$and $J^*$ (from equation (13)) in the form of regularisation terms for $J$ and terms which penalise over-parameterisation in the structure functional $J^*$.

## 2.3 The constructive approach

The constructive method is to start off with a simple model, to estimate its parameters, determine where the representation is still unsatisfactory and to dynamically add new models to the network. This leads to a sequence of model structures $\mathcal{M}_1 \to \mathcal{M}_2 \to \ldots \to \mathcal{M}_{n_\mathcal{M}}$, where $\mathcal{M}_i \to \mathcal{M}_{i+1}$ indicates an increase in the representational ability (more degrees of freedom) in the model structure followed by a parameter identification and confidence estimation stage. Constructive techniques which gradually enhance the model representation in this manner have a number of advantages. They automate the learning process by letting the network grow to fit the complexity of the target system, but they do this robustly, by forcing growth to be guided by the availability of data and the complexity of the local models. This automatically determines the size of the network needed to approximate the function adequately, while preventing overfitting. Two such constructive algorithms are described in [11] and [12].

The construction becomes a multi-step process: At each step various options are constructed, the model parameters optimised, and the structure with the best cost-complexity value chosen. The procedure is then repeated at the next stage of construction. Unfortunately the search space is usually very large and such algorithms therefore suffer from the 'curse of dimensionality' and scale up badly to larger problems.

**The Multi-Resolution Constructive Algorithm** To produce efficient, practical algorithms the following observations about the modelling problem should be noted:

1. Although highly desirable, the distribution of training data will probably not be directly related to the complexity of the observed process.
2. The process will probably have varying levels of complexity throughout different areas of the input space.
3. The training data will not be uniformly distributed, and there will be areas of the input space which *cannot* be filled with data.

The first point implies that we should consider the local complexity of the process output when altering model structure, as opposed to unsupervised learning techniques, which only consider the density of the input data, regardless of the output response. The second point implies the need for a multi-resolution technique which will find model structure representing varying volumes of the input space (varying levels of 'locality'). The third point lets us reduce the volume of the input space we consider for new local models to only points covered by the available training data.

The MRC algorithm differs from other constructive algorithms in a number of ways. The conventional methods tend to basically apply search techniques to find a model structure which minimises the cost-complexity functional. The method used here is to use a 'model mismatch' or 'complexity heuristic' to indicate the areas of the input space where the current model differs most from the underlying process, i.e. where the greatest complexity is. The model structure is then developed in these areas. The options for model structure extension are also drastically reduced by restricting the possible positions of basis function centres to be on input points in the training set. This leads to a relatively simple method for extending the model structure which expends its effort in predicting where new structure would be useful, rather than trying out the options and selecting the best of them. The process is described below:

1. The model starts off with a minimal representation (perhaps only one linear model, depending on the state of the *a priori* knowledge) and searches for 'coarse' complexity. It refines the model structure at ever increasing levels of resolution until the desired accuracy has been achieved, or the training data has been exhausted.

2. To determine where to add extra representation the 'complexity' heuristic is needed. This decides where new models should be placed, based on a weighted local statistic of the training data, or from measured model residuals. To enforce the gradual nature of the approximation the new centres must be a minimum distance $d_{min}$ from existing centres.[2]
3. Given the suggested location of the new model centre the desired overlap with neighbouring regions is determined, thus completing the basis function optimisation for this stage of the model construction.
4. If a pool of local model structures has been defined, the best fitting local model for each basis function can be chosen by estimating the local model parameters for the new model structure and running cross-validation runs. If the receptive field of any given basis function has too few units to reliably estimate the associated local model parameters it can be removed (and Step 3 is repeated), or the local model structure simplified.
5. If the model is still not accurate enough, the search for the next most 'complex' area of the input space is then restarted. This is repeated until either no further local models can be added, or the added models do not bring any improvement, whereupon the scale of the 'complexity window' is reduced, and the search is restarted at the finer resolution.

The constructive procedure is illustrated in Fig. 4 where the model complexity is increased gradually for a two-dimensional function approximation problem.

**Complexity detection – where are extra units needed?** The 'complexity detection' heuristic used to place new local models was inspired by the *Vector Field Approach to Cluster Analysis* [13]. Observation 3 above indicates that we should simplify the search task by assuming that the training data covers the significant areas of the input space adequately for the initial search (an assumption which must be true for any degree of learning to take place). Initially, all training points which are a minimum distance $d_{min} = \gamma\sigma_{win}$ from existing centres are viewed as possible centres $c_{new}$ for the new basis function. The centre $c_{new} \in \mathcal{R}^{n_{\tilde{\phi}}}$, and the 'complexity' of the mapping in a windowed area, where $\rho(\cdot)$ is the windowing function[3] around this point is measured using

$$F_{total}(c, \sigma_{win}) = \frac{1}{N} \sum_{i=1}^{N} \rho(d(\tilde{\phi}_i, c, \sigma_{win}))e_i, \tag{23}$$

where $N$ is the number of neighbouring data points used, $e_i$ can be a general error statistic, but which is often simply

$$e_i = |y_i - \hat{y}_i| \,. \tag{24}$$

[2] $d_{min}$ is related to the current resolution of search $\sigma_{win}$.

[3] In this work a Gaussian bell was used.

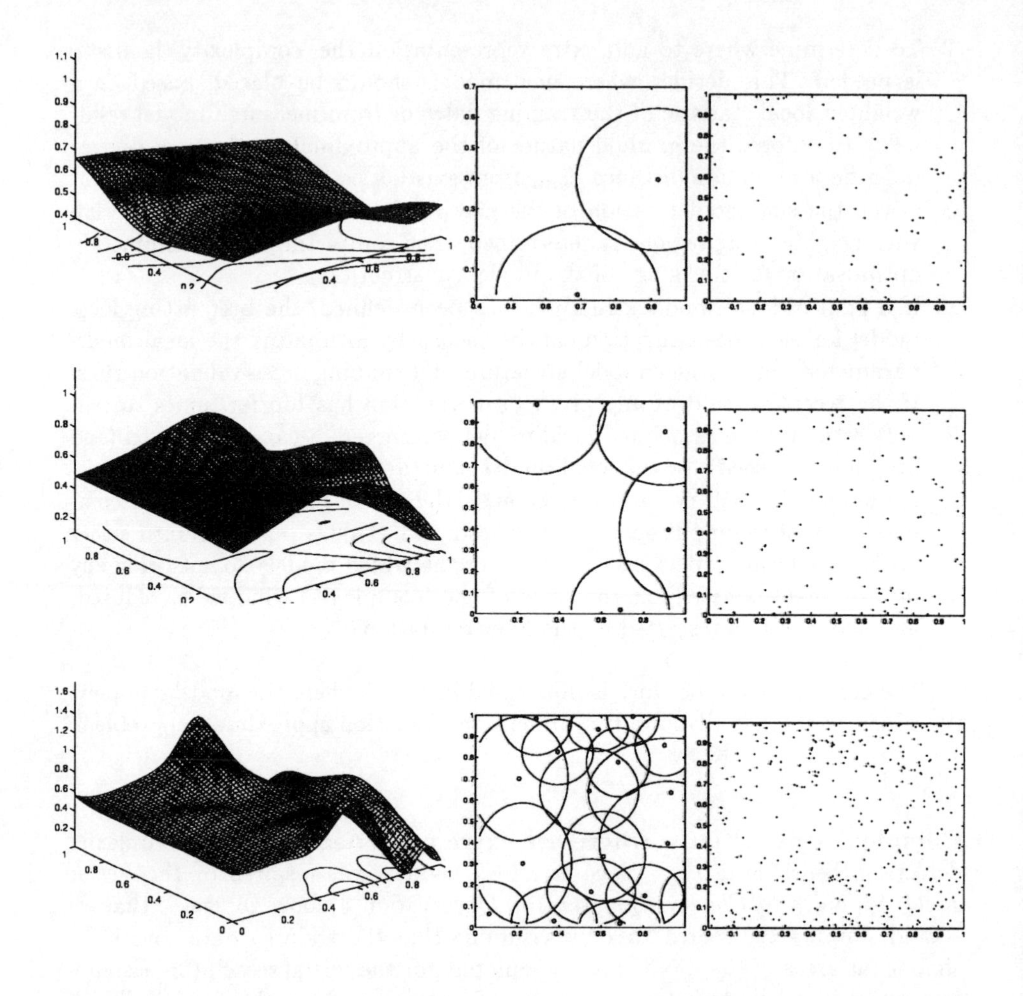

**Fig. 4.** A gradual approach to constructing a model. The model responses are shown for a series of stages in model development, shown by the contour plots of the basis functions. The right hand column shows the training data used – note that the set is expanded with the model structure.

The function $d(\cdot)$ is a distance measure. The complexity is estimated by an analogy to the concepts of forces acting on a mass in physics. The weighting of the forces depends on their associated error statistic $(e_i)$. The windowing function focusses the heuristic's attention on the level of locality currently being examined. The larger the level of $F_{total}$, the larger the estimated complexity. The windowing function operates at different scales $\sigma_{win}$, starting off by searching for coarse complexity and refining the search as learning progresses.

As new models must be a certain minimum distance away from previous

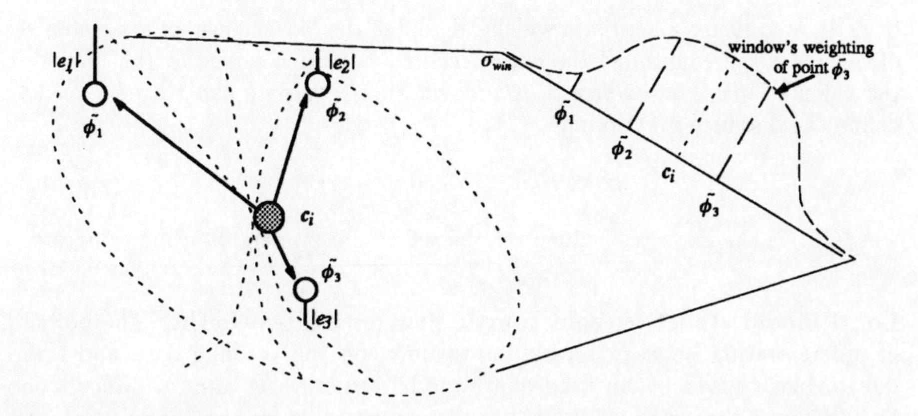


**Fig. 5.** Windowed Complexity Estimate. The weighting function $\rho(\cdot)$ weights the error measures $e_i$ at points $\tilde{\phi}_i$ around the prototype centre. The window function is usually chosen such that points further from the centre have less effect on the outcome of the complexity estimate.

models, the input space will gradually be filled with basis functions, the density being determined by the current resolution. Once no more basis functions can be inserted at this resolution the algorithm moves on to the next, finer search stage [4].

The major disadvantage of the complexity heuristic is its computational load. It can require a maximum of $N^2$ calculations of the weighting function and the associated offset for each prototype centre. However, as extra basis functions are added for any given resolution, the number of potential centres sinks, due to the distance constraint $d_{min}$ covering a higher percentage of the training points. This means that the search for complexity gets faster as the model grows. The computational effort can also be reduced by limiting the search by using only a subset of the training data as potential with *active selection* of the training data.

**Overlap determination** The conventional method for overlap determination is to set the radius $\sigma_j$ proportional to the average distance of the $k$ nearest neighbours from the centre $c_j$. In many cases this will be unsatisfactory, as the resulting level of overlap with the neighbouring basis functions would vary greatly[5]. One alternative method is to observe that a covariance matrix $\boldsymbol{\sigma}_i$, can

[4] To prevent non-complex areas of the input space being unnecessarily filled with local models, the search at a given resolution is abandoned if over a window of $n_{cutoff}$ successive insertions no improvement in mean cross-validation error is made ($n_{cutoff} = 4$ was used for the experiments in this work).

[5] Too much overlap leads to problems with poor estimation and singularities in the regression process. With too little overlap and normalisation the mapping loses smooth interpolation between local models, and the behaviour further away from the centre becomes unpredictable, because of the interaction with other basis functions. In [9]

be estimated using a heuristic which calculates the 'covariance' of a selected set of centres $c_n$ surrounding[6] the chosen centre $c_i$ (which serves as the 'mean' in the calculation). The inverse of the 'covariance' estimate can then be used to define the distance function:

$$\boldsymbol{\sigma}_i \propto E_n \left[ (c_n - c_i)(c_n - c_i)^T \right], \quad (25)$$

where $E_n$ is the expected value over the set of chosen neighbouring centres $\mathbf{c}_n$.

**Local model structure selection & Preventing overfitting** The problem of differentiating between errors due to noise on the training data and errors due to bias caused by an inadequate model structure is often a difficult one. Overfitting is the result of extending the structure so far that noise is learned, instead of system structure. Earlier we discussed ways of reducing the variance in the parameter estimation phase. It is also possible to reduce variance by limiting the model structure, which can be done by *stopping growth*, *pruning structure*, or careful *selection of local model structures.*

Local Model Structures need not be homogeneous – the user can define a pool of possible local models which can then be inserted into a given operating regime, with the 'best' one being chosen. This would allow a more robust fitting of models to operating regimes, taking the amount of data and the local process complexity into account[7]. Such methods encourage the algorithm to choose simpler models when data is sparse, and *stopping growth* can be viewed as the extension of this local structure selection to the case of preferring no increase in model structure. The most basic constraint on the structure identification algorithm is to require a minimum number of data points within the receptive field of a given local model for it to be considered viable, i.e. only if $\rho_{total_l} > N_{min}$, where $N_{min}$ is the minimum number of training points needed and $\rho_{total_l} = \sum_{i=1}^{N} \rho_l(\tilde{\phi}_i)$ is a heuristic 'count' of the local data points for smoothly overlapping basis functions. $N_{min}$ is dependent on the level of noise on the data and the complexity of the local model.

A further method is to use *pruning techniques*, which merge neighbouring local models if their parameters are similar enough[8], have also been successfully

it was shown that the condition of the design matrix is highly dependent on the level of overlap between basis functions.

[6] It is important to choose the neighbours in as wide a range of directions as possible, see [9] for details.

[7] The resulting local model net will be a heterogeneous structure, where each local model could be different. If all local models are linear in the parameters, the standard global optimisation techniques remain valid. This will be true in the most straightforward example of heterogeneous LMN's, where the local models are linear, but with varying dynamic order. If some local models require nonlinear optimisation techniques, a variety of local learning methods can be used.

[8] The distance between the local model parameter vectors is then calculated $\delta_{ik} = |\theta_i - \theta_k|$ , and the most similar min $\delta_{ij}, i, j = 1..n_{\mathcal{M}}$ local models were merged into one, and the new basis function centred between the old ones.

applied to simplify the networks. These rely on the use of local learning techniques, as described earlier, to ensure that the parameters have a strictly local interpretation.

# 3 Control Based on Local and Time-varying Models

We turn our attention now to control methods based upon the local model network. In Section 3.1 a direct analogue of the local model network, the *local controller network*, is considered. In Section 3.2 on-line adaptive control is considered. This is based upon a linear time-varying interpretation of the local model network, coupled with control redesign and perhaps on-line parameter update. Further details of the control designs discussed here can be found in Żbikowski *et al* [14].

## 3.1 Local Controller Networks

We postulate a general nonlinear controller for the system (1):

$$u(t) = C(u(t-1), \ldots u(t-n_u^C), y(t), \ldots y(t-n_y^C), r(t), \ldots r(t-n_r)). \quad (26)$$

The signal $r(t)$ is a reference value while $C$ is the nonlinear function characterising the controller. The controller can be written more compactly as

$$u(t) = C(\psi^C(t)), \quad (27)$$

where

$$\psi^C(t) = [u(t-1), \ldots u(t-n_u^C), y(t), \ldots y(t-n_y^C), r(t), \ldots r(t-n_r)]^T. \quad (28)$$

Note that when the controller is designed on the basis of a model which is valid around some operating point the signals in the preceding equations represent deviations from the operating point.

By directly exploiting the structure of the local model network a *local controller network* (LCN) can be defined. For each operating regime $\Phi_i$ the system is characterised by the local model $\hat{f}_i$, to a degree given by the validity function $\rho_i$. It is natural to exploit the given partition of the operating set and for each operating regime to design a local controller $C_i$ whose validity is also given by the same $\rho_i$. The overall controller is then defined analogously to the local model network by using the validity functions as smooth interpolators. The local controller network is described by

$$u(t) = C(\psi^C(t)) = \sum_{i=1}^{n_{\mathcal{M}}} C_i(\psi^C(t-1))\rho_i(\phi_i^{\mathrm{op}}) \quad (29)$$

and this should be compared to equation (12). For linear local models of the form $\hat{f}_i(\psi(t-1)) = \psi^T(t-1)\theta_i$, linear local controllers are natural;

$$C_i(\psi^C(t)) = {\psi^C}^T \theta_i^C. \quad (30)$$

The $\theta_i^C$ are the local controller parameters which can be directly obtained from some linear control design algorithm mapping the local model parameters;

$$\theta_i^C = \Xi(\theta_i, J_i^C). \tag{31}$$

Here, $\Xi$ denotes the mapping of the linear control design algorithm, while $J_i^C$ is a set of local control design specifications.

The LCN can be rearranged by combining equations (29) and (30) to give

$$u(t) = C(\psi^C(t)) = \psi^{C^T} \sum_{i=1}^{n_{\mathcal{M}}} \theta_i^C \rho_i(\phi_i^{\mathrm{op}}). \tag{32}$$

This can be interpreted as a linear controller with operating-point-dependent parameters:

$$u(t) = \psi^{C^T} \tilde{\theta}^C(\tilde{\phi}(t)), \tag{33}$$

with $\tilde{\theta}^C(\tilde{\phi}(t)) = \sum_{i=1}^{n_{\mathcal{M}}} \theta_i^C \rho_i(\phi_i^{\mathrm{op}})$. Here the controller parameters $\tilde{\theta}^C$ are obtained by interpolating the local controller parameters $\theta_i^C$ using the validity functions $\rho_i(\phi_i^{\mathrm{op}})$. This is clearly very similar to gain scheduling control [15, 16], whereby the LCN structure provides a more general framework and a systematic way to do the actual scheduling. Further details of the link to gain scheduling are discussed in [14]. Other special cases of the LCN structure include the modular control architecture of Jacobs and Jordan [17], and the heterogeneous control approach of Kuipers and Åström [18]. As mentioned earlier, there is also a direct link with fuzzy control and this is discussed in depth in [19, 20, 21, 22].

In the case of linear local models it is natural to propose a network of linear local controllers. A local two-degrees-of-freedom control structure for each local model is defined by

$$u(t) = \frac{1}{H(q^{-1})}(S(q^{-1})r(t) - G(q^{-1})y(t)), \tag{34}$$

where the polynomials $G, H$ and $S$ have the form

$$S(q^{-1}) = s_0 + s_1 q^{-1} + \ldots + s_{n_r} q^{-n_r}, \tag{35}$$

$$G(q^{-1}) = g_0 + g_1 q^{-1} + \ldots + g_{n_y^C} q^{-n_y^C}, \tag{36}$$

$$H(q^{-1}) = 1 + h_1 q^{-1} + \ldots + h_{n_u^C} q^{-n_u^C}. \tag{37}$$

The controller (34) can be expressed as

$$u(t) = -H'(q^{-1})u(t) - G(q^{-1})y(t) + S(q^{-1})r(t) \tag{38}$$

where $H'(q^{-1}) = H(q^{-1}) - 1$. This structure can furthermore be expressed in the vector form of equation (30) as

$$u(t) = \psi^{C^T} \theta^C \tag{39}$$

where the controller parameter vector $\theta^C$ has the form

$$\theta^C = [-h_1, \ldots - h_{n_u^C}, -g_0, \ldots - g_{n_y^C}, s_0, \ldots s_{n_r}]^T. \qquad (40)$$

When we have a local model network representation of the nonlinear system this controller structure is repeated $n_{\mathcal{M}}$ times and there exist $n_{\mathcal{M}}$ local controller parameter vectors $\theta_i^C$. Each local controller is designed for the local linear model. The model parameter vectors $\theta_i$ can be decoded to give a local polynomial representation $A_i, B_i, C_i$ of the form (7). The above control structure then defines a notional characteristic equation for each model/controller pair. This has the form

$$A_i(q^{-1})H_i(q^{-1}) + q^{-k}B_i(q^{-1})G_i(q^{-1}) = T_i(q^{-1}). \qquad (41)$$

The local controller polynomials $G_i, H_i$ whose coefficients form part of the $\theta_i^C$ are solutions to this linear equation. The desired characteristic polynomial $T_i(q^{-1})$ can be set in a number of ways, e.g. by directly choosing closed-loop poles or by minimising a cost function ($T_i$ are then given by spectral factorisation). The remaining controller polynomials $S_i$ (which act on the reference signal $r$) can be designed to achieve desirable command tracking properties.

### 3.2 Adaptive Control Interpretation

An alternative approach is to interpret a nonlinear local model network as a linear time-varying system [23, 24]. Standard linear control design, of the form outlined above, can then be repeatedly applied on-line to the time varying linear model. This can also be done in conjunction with on-line parameter estimation for update of the LMN parameters. This scheme then becomes strongly reminiscent of standard self-tuning control. To see how this works consider the LMN based upon linear local models:

$$y(t) = \psi^T(t-1)\sum_{i=1}^{n_{\mathcal{M}}} \theta_i \rho_i(\phi_i^{\text{op}}) + e(t) \qquad (42)$$

with

$$\psi(t-1) = [y(t-1), \ldots y(t-n_y), u(t-k), \ldots u(t-k-n_u)]^T. \qquad (43)$$

The system can be written in a linear form with time-varying parameters,

$$y(t) = \psi^T(t-1)\tilde{\theta}(\tilde{\phi}(t)) + e(t). \qquad (44)$$

The time-varying parameter vector $\tilde{\theta}$ is defined as

$$\tilde{\theta} = [a_1(\tilde{\phi}(t)), \ldots a_{n_y}(\tilde{\phi}(t)), b_0(\tilde{\phi}(t)), \ldots b_{n_u}(\tilde{\phi}(t))]^T. \qquad (45)$$

The individual coefficients have the specific parameterisation

$$a_1(\tilde{\phi}(t)) = \sum_{i=1}^{n_{\mathcal{M}}} \theta_{i,1}\rho_i(\phi_i^{\text{op}}), \qquad (46)$$

$$\vdots$$

$$a_{n_y}(\tilde{\phi}(t)) = \sum_{i=1}^{n_{\mathcal{M}}} \theta_{i,n_y} \rho_i(\phi_i^{\text{op}}), \tag{47}$$

$$b_0(\tilde{\phi}(t)) = \sum_{i=1}^{n_{\mathcal{M}}} \theta_{i,n_y+1} \rho_i(\phi_i^{\text{op}}), \tag{48}$$

$$\vdots$$

$$b_{n_u}(\tilde{\phi}(t)) = \sum_{i=1}^{n_{\mathcal{M}}} \theta_{i,n_y+n_u+1} \rho_i(\phi_i^{\text{op}}). \tag{49}$$

When the time-delay is $k = 1$ the predictive form of the system equation can be expressed as

$$\begin{aligned} y(t) = {} & a_1(\tilde{\phi}(t))y(t-1) + \ldots + a_{n_y}(\tilde{\phi}(t))y(t-n_y) + \\ & b_0(\tilde{\phi}(t))u(t-1) + \ldots + b_{n_u}(\tilde{\phi}(t))u(t-1-n_u) + e(t). \end{aligned} \tag{50}$$

Note that the procedure can be easily generalised to arbitrary $k$, but for ease of notation we restrict attention to the case $k = 1$. From this model the time-varying $A$ and $B$ polynomials of a linear ARX-type model can be identified:

$$\begin{aligned} A(q^{-1}) &= 1 - a_1(\tilde{\phi}(t))q^{-1} - \ldots - a_{n_y}(\tilde{\phi}(t))q^{-n_y}, \\ B(q^{-1}) &= b_0(\tilde{\phi}(t)) + \ldots + b_{n_u}(\tilde{\phi}(t))q^{-n_u}. \end{aligned} \tag{51}$$

These $A$ and $B$ polynomials then form the basis for redesign.

### 3.3 Example

We now illustrate the LCN of Section 3.1 with a simple example. The problem of controlling the lateral deviation of a car on a straight road is considered [25, 26]. The model is given by:

$$\begin{aligned} \dot{\beta} &= \left(-2\,\frac{k}{mv}\right)\,\beta + \left(\frac{v}{a} - \frac{k}{mv}\right)\,\lambda \\ \dot{y} &= v(\psi + \beta) \\ \dot{\psi} &= \left(\frac{v}{a}\right)\,\lambda \\ \dot{\lambda} &= u \end{aligned}$$

The variables in this model are:

- $y$ – lane centre offset (to be controlled),
- $\psi$ – vehicle yaw angle and road course angle difference,
- $\beta$ – sideslip angle,
- $\lambda$ – steering angle (control input),

– $v$ – vehicle speed,

and the constants are: $m = 5800\ kg$ - vehicle mass, $a = 4.25\ m$ - wheel base and $k = 150\ kN/rad$ - lateral friction.

The task is to regulate the offset to zero by appropriate manipulation of the steering angle. It can be seen from the above equations that the model (the transfer-function between steering angle and offset) is a linear time-varying system; the model parameters depend on the vehicle speed $v$. This is therefore a prime candidate for control with the LCN structure using speed as the scheduling variable. Note that this system is unstable (3 poles at $s = 0$).

First, we illustrate the difficulty of attempting to control this system with a fixed linear controller. A linear controller was designed for an operating speed of 20m/s. A simple pole-assignment regulator based on a pair of dominant closed-loop poles giving a rise-time of 5s was designed. The response of this controller at a speed of 20m/s to initial offsets of 1m and 0.4m is shown in Fig. 6. With this

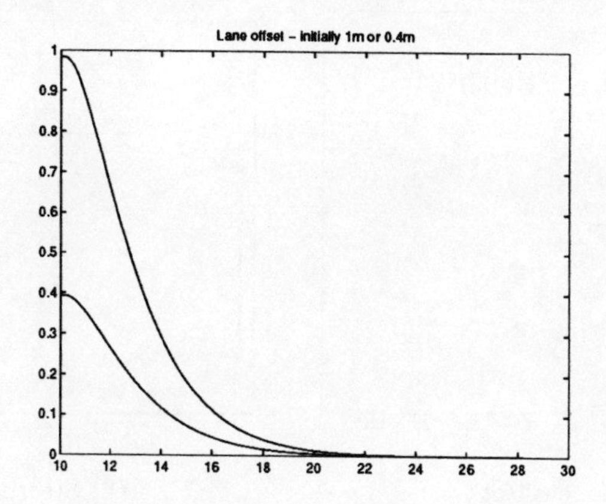


**Fig. 6.** Linear controller, $v = 20\text{m/s}$

same controller the car was controlled at a speed of 15m/s. Although reasonable offset correction was still achieved, the control input was unacceptably active. This is illustrated in Fig. 7 where the steering angle for an initial offset of 1m is plotted. Note that for speeds lower than 15m/s or higher than 24m/s the controller became unstable.

A local controller network was then designed based upon linear models for a number of operating speeds. For each operating point the control specification was the same, i.e. the system should have a rise-time of 5s in reponse to step-like offsets. The controller response for this LCN system was consistent over all speeds. An example is illustrated in Fig. 8(a) for $v = 10\text{m/s}$ (the single linear controller designed at 20m/s was unstable for this condition). The system responds to the initial offset with the desired speed, and the response is visually indistinguishable from that shown in Fig. 6 even though the vehicle speed is very much lower.

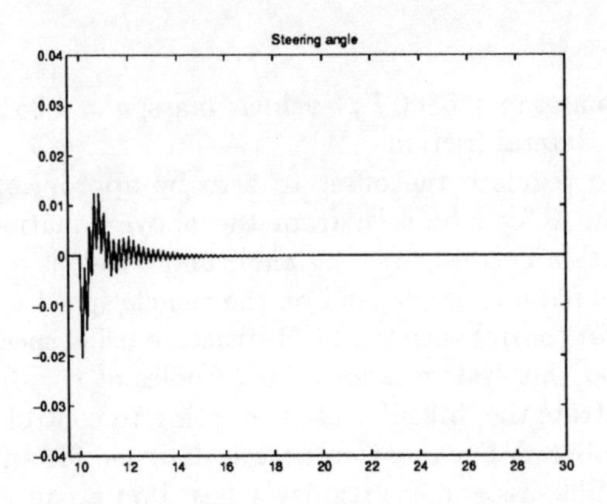


**Fig. 7.** Linear controller, $v = 15\text{m/s}$

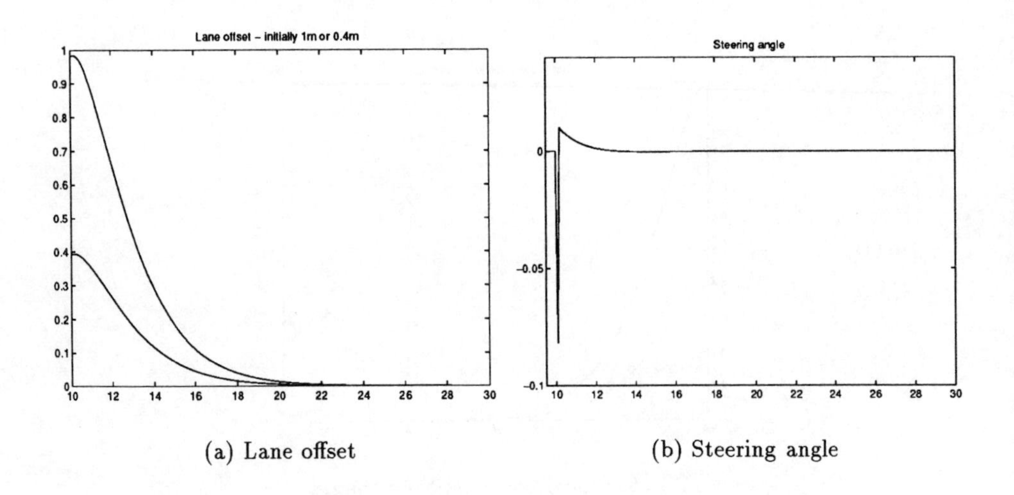


(a) Lane offset (b) Steering angle

**Fig. 8.** LCN controller, $v = 10\text{m/s}$

## 4 Conclusions

**Local Model Nets** Local Model networks are practical learning architectures for use in real problems. The structure is general enough to be applicable to a wide range of processes, while being relatively simple to understand, and lending itself well to the interpretation of results and introduction of *a priori* knowledge to simplify training. The use of constructive learning algorithms, as described in this paper, provides the user a way of automatically creating a model structure capable, given the training data, of representing the target process.

**Constructive Learning Algorithms** Constructive techniques which gradually enhance the model representation in this manner have a number of advantages. The network first allocates representation where most needed, according to the complexity heuristic. The main features of a process are captured first, then the details. This is an implicit style of regularisation, as the model construction process can now be seen as a gradual increase in variance and decrease in bias. Learning continues until the desired level of bias-variance trade-off is achieved. Modelling accuracy and generalisation ability tend to be improved, as the model structure is extended as a far as possible to fit the data, while the overfitting protection inherent to the constructive algorithm limits overtraining. A further important point is that *a priori* knowledge can be introduced in the form of a pool of local model structures, so that the local model structure best suited to a local area of the input space is chosen. This automatically creates a heterogeneous model structure.

**Local Controller Networks** Once the *Local Model Net (LMN)* structure has been estimated during learning it is relatively straightforward to produce a matching *Local Controller Net (LCN)* using standard linear control design methods. The method requires, however, that the linear local model be a local approximation of the system in question, so a local learning method should be used during construction of the Local Model Net. This forces the algorithm to extend the model structure so that the local models can become independent local approximations of the target system, rather than the smaller model structures produced by global learning which cannot be interpreted individually.

The results were applied to a simple simulation of a car steering problem to illustrate the easy applicability of the ideas, but it should be remembered that the problems associated with conventional gain scheduling [15] are also relevant to LCNs.

## References


1. I. J. Leontaritis and S. A. Billings, "Input-output parametric models for non-linear systems," *Int. J. Control*, vol. 41, no. 2, pp. 303–344, 1985.
2. H. W. Werntges, "Partitions of unity improve neural function approximation," in *Proc. IEEE Int. Conf. Neural Networks*, (San Francisco, CA), pp. 914–918, 1993. Vol. 2.
3. R. Shorten and R. Murray-Smith, "On Normalising Basis Function networks," in *4th Irish Neural Networks Conf., Univ. College Dublin*, Sept. 1994.
4. J. Moody and C. Darken, "Fast-learning in networks of locally-tuned processing units," *Neural Computation*, vol. 1, pp. 281–294, 1989.
5. T. J. Hastie and R. J. Tibshirani, *Generalized Additive Models.* Monographs on Statistics and Applied Probability 43, London: Chapman and Hall, 1990.
6. J. H. Friedman, "Multivariate Adaptive Regression Splines," *Annals of Statistics*, vol. 19, pp. 1–141, 1991.
7. C. Harris, C. G. Moore, and M. Brown, *Intelligent Control: Aspects of Fuzzy Logic and Neural Nets.* World Scientific, 1993.
8. M. Brown and C. Harris, *Neurofuzzy Adaptive Modelling and Control.* Hemel-Hempstead, UK: Prentice Hall, 1994.

9. R. Murray-Smith, *A Local Model Network Approach to Nonlinear Modelling.* Ph.D. Thesis, Department of Computer Science, University of Strathclyde, Glasgow, Scotland, Nov. 1994. E-mail:murray@DBresearch-berlin.de.
10. S. Geman, E. Bienenstock, and R. Doursat, "Neural networks and the Bias/Variance Dilemma," *Neural Computation*, vol. 4, no. 1, pp. 1–58, 1992.
11. R. Murray-Smith and H. Gollee, "A constructive learning algorithm for local model networks," in *Proc. IEEE Workshop on Computer-intensive methods in control and signal processing, Prague, Czech Republic*, pp. 21–29, 1994. E-mail:murray@DBresearch-berlin.de.
12. R. Murray-Smith, "A Fractal Radial Basis Function network for modelling," in *Inter. Conf. on Automation, Robotics and Computer Vision, Singapore*, vol. 1, pp. NW–2.6.1–NW–2.6.5, 1992. E-mail:murray@DBresearch-berlin.de.
13. H. C. Andrews, *Introduction to Mathematical Techniques in Pattern Recognition.* Robert E. Krieger, 1983.
14. R. Żbikowski, K. J. Hunt, A. Dzieliński, R. Murray-Smith, and P. J. Gawthrop, "A review of advances in neural adaptive control systems," Technical Report of the ESPRIT NACT Project TP-1, Glasgow University and Daimler-Benz Research, 1994.
15. J. S. Shamma and M. Athans, "Gain scheduling: Potential hazards and possible remedies," *IEEE Control Systems Magazine*, pp. 101–107, June 1992.
16. W. J. Rugh, "Analytical framework for gain scheduling," *IEEE Control Systems Magazine*, vol. 11, pp. 79–84, 1991.
17. R. A. Jacobs and M. I. Jordan, "Learning piecewise control strategies in a modular neural network," *IEEE Transactions on Systems, Man, and Cybernetics*, vol. 23, no. 3, pp. 337–345, 1993.
18. B. Kuipers and K. Åström, "The composition and validation of heterogeneous control laws," *Automatica*, vol. 30, no. 2, pp. 233–249, 1994.
19. K. J. Hunt, R. Haas, and M. Brown, "On the functional equivalence of fuzzy inference systems and spline-based networks," *International Journal of Neural Systems*, 1995. To appear - March issue.
20. M. Brown and C. Harris, "A nonlinear adaptive controller: A comparison between fuzzy logic control and neurocontrol," *IMA J. Math. Control and Info.*, vol. 8, no. 3, pp. 239–265, 1991.
21. M. Brown and C. Harris, *Neurofuzzy Adaptive Modelling and Control.* Hemel-Hempstead, UK: Prentice Hall, 1994.
22. T. A. Johansen, "Fuzzy model based control: stability, robustness, and performance issues," *IEEE Trans. on Fuzzy Systems*, vol. 2, no. 3, pp. 221–233, 1994.
23. T. A. Johansen, "Adaptive control of MIMO non-linear systems using local ARX models and interpolation," in *IFAC ADCHEM 94*, (Kyoto, Japan), May 1994.
24. H. Wang, M. Brown, and C. Harris, "Modelling and control of nonlinear, operating point dependent systems via associative memory networks," *J. of Dynamics and Control*, accepted for publication 1994.
25. J. Ackermann, *Robuste Regelung.* Berlin: Springer-Verlag, 1993.
26. K. Mecklenburg, T. Hrycej, U. Franke, and H. Fritz, "Neural control of autonomous vehicles," in *Proc. IEEE Vehicular Technology Conference, Denver, USA*, 1992.

# On ASMOD – An Algorithm for Empirical Modelling using Spline Functions

Tom Kavli[1] and Erik Weyer[2]

[1] SINTEF Box 124 Blindern, 0314 Oslo, Norway Email: Tom.Kavli@si.sintef.no
[2] Department of Chemical Engineering, The University of Queensland
Brisbane QLD 4072, Australia Email: erikw@cheque.uq.oz.au

**Abstract.** Empirical modelling algorithms build mathematical models of systems based on observed data. This chapter describes the theoretical foundation and principles of the ASMOD algorithm, including some improvements on the original algorithm. The ASMOD algorithm uses B-splines for representing general nonlinear models of several variables. The internal structure of the model is, through an incremental refinement procedure, automatically adapted to the dependencies observed in the data. Only input variables which are found of relevance are included in the model, and the dependency of different variables are decoupled when possible. This makes the model more parsimonious and also more transparent to the user. Case studies are included which confirm the usefulness of the algorithm.

## 1 Introduction

Mathematical models of systems are commonly used in science and engineering. Application areas of such models include control, optimisation, prediction, filtering and fault detection to mention a few. Mathematical models are usually obtained either by physical modelling or empirical modelling. In physical modelling one starts with first principles like conservation of energy or mass and builds a model based on physical insight in to the system. Empirical modelling on the other hand does not require detailed physical knowledge of the system. As the name suggest, models are now built based upon observed (empirical) data from the system. In recent years there has also been a growing interest in combining physical and empirical modelling, see e.g. [3].

The ASMOD (Adaptive Spline Modelling of Observation Data) algorithm is an algorithm for empirical modelling. It uses B-splines to represent general nonlinear and coupled dependencies in multivariable observation data [18, 4]. B-splines are commonly used in computer graphics and CAD systems for representing three dimensional curves and surfaces with high accuracy. The B-spline theory is also applicable for higher dimensional spaces and is therefore a suitable starting point for nonlinear and multivariable empirical modelling. However, by direct expansion of the standard spline representation to higher dimensions the number of parameters grows rapidly to impractical levels. Methods must thus be found to reduce the complexity of the models. The ASMOD algorithm represents one attempt to solve this problem by adapting the model structure to the dependencies (coupled or decoupled) that are observed in the data. The model complexity can then be dramatically reduced compared to using models of a general structure, and the model accuracy can be correspondingly improved since more accurate estimates can be obtained for the fewer parameters.

The theoretical approximation properties of the ASMOD models rely on the approximation properties of the B-splines, and hence existing theory is available for analysing their capabilities and properties [29]. Data interpolation and extrapolation with B-splines have proven to be very well behaved, with no oscillatory behaviour as commonly observed e.g. with polynomial fitting [20].

In the ASMOD algorithm the output variable is modelled as a sum of several low dimensional submodels, where each submodel only depends on a small subset of the input variables. The decomposition of the high dimensional input space into low dimensional additive subspaces makes the model more transparent to the user, at the same time as the the complexity (number of parameters) of the model is dramatically reduced.

The use of B-splines in empirical modelling is not new. Moody and Darken [22] use B-splines for modelling multivariable data. To attain good interpolation and also local flexibility they use a hierarchy of B-spline models with increasing resolution. There is no adaption of the grid resolutions, and these have to be manually set prior to training. Sanger [28] describes an algorithm for incrementally constructing models using B-spline basis functions.

The MARS algorithm, developed by Friedman [9], uses a nonuniform partitioning of the input space by recursively splitting hyperrectangular subdomains by a plane perpendicular to a selected input variable. "Natural spline" basis functions are then formed based upon the partitioning tree. After a fine partitioning of the input space has been obtained, the model is pruned back by removal of basis functions to get a model structure which gives good generalization. A similar approach has been attempted by Kavli with the ABBMOD algorithm using B-spline basis functions [19].

The use of radial basis functions for interpolation or approximation of observation data can be found in early statistical approximation literature [25], in the spline literature [31, 8], and later in neural network literature [24, 23]. Artificial neural networks have been proposed as very flexible and general methods for nonlinear data modelling [14], and the relationship between spline approximation with radial basis functions and the radial basis artificial neural networks has been pointed out by Poggio and Girosi [24].

Finally it should be noted that the B-spline basis functions have many similarities with the membership functions used in fuzzy logic, and spline models such as the ASMOD can be mapped into a special form of fuzzy rules as proposed by Brown and colleagues [5, 6].

This chapter is organised as follows: In the following section a formal description of B-spline functions and the ASMOD model structure is given. Some basic properties of this model representation and methods for model optimisation are also briefly discussed. Then the ASMOD algorithm for model construction through an incremental model refinement procedure is described in section 3, including a set of optimisation criteria that can be used for selecting the best sequence of refinements. Section 4 briefly outlines some theoretical results concerning the properties of the ASMOD algorithm, and some experimental results are given in section 5. Finally some concluding remarks are given in section 6.

## 2 ASMOD models

An ASMOD model is a mapping between an input variable $x \in \Omega \subset R^n$ and an output variable $y \in R$. B-spline basis functions are used for constructing this mapping, and an ASMOD model is a linear combination of B-spline basis functions. Hence we can write the model as

$$m(x) = \sum_{i=1}^{K} c_i b_i(x) = c^T b(x) \tag{1}$$

where $b(x)$ is a vector of B-spline basis functions, and $c$ is a coefficient vector. From (1) it is clear that an ASMOD model is an element in a linear space spanned by the B-spline basis functions.

A univariate B-spline basis function is defined in terms of a polynomial degree $p$ and $p+2$ knots as follows

**Definition 1.** *Let $p$ be a non-negative integer and let $\tau_0, \tau_1, \ldots, \tau_{p+1}$ be $p+2$ real numbers with $\tau_{i-1} \leq \tau_i$ for $i = 1, \ldots, p+1$. The B-spline with knots $\tau_0, \tau_1, \ldots, \tau_{p+1}$ is defined recursively by*

$$b(x|\tau_0, \ldots, \tau_{p+1}) = (x - \tau_0)q(x|\tau_0, \ldots, \tau_p) + (\tau_{p+1} - x)q(x|\tau_1, \ldots, \tau_{p+1})$$

*where*

$$q(x|\tau_0, \ldots, \tau_{p+1}) = \begin{cases} b(x|\tau_0, \ldots, \tau_{p+1})/(\tau_{p+1} - \tau_0), & \text{if } \tau_0 < \tau_{p+1} \\ 0 & \text{otherwise} \end{cases}$$

*and*

$$b(x, \tau_0, \tau_1) = \begin{cases} 1 & \text{if } \tau_0 \leq x < \tau_1 \\ 0 & \text{otherwise} \end{cases}$$

For model building we need more than one single basis function and hence more knots. A knot vector and the associated B-spline are defined as follows

**Definition 2.** *A knot vector $\tau$ is a non-decreasing sequence of real numbers $(\tau_i)_{i=1}^{k+p+1}$ where $p$ is a given integer. A B-spline basis function can be associated with each set of $p+2$ consecutive knots*

$$b_i(x) = b_{i,p}(x) = b_{i,p,\tau}(x) = b(x|\tau_i, \ldots, \tau_{i+p+1})$$

*giving a total of $k$ basis functions. The linear space of all linear combinations of these functions is*

$$S_{p,\tau} = \left\{ \sum_{i=1}^{k} c_i b_{i,p,\tau}(x) | \; c_i \in R, \; i = 1, \ldots k \right\}$$

*An element of $S_{p,\tau}$ is called a spline function or just a spline. $S_{p,\tau}$ itself is called a spline space. The input domain of a spline in $S_{p,\tau}$ is $[\tau_{p+1}, \tau_{k+1})$.*

From this definition it follows that a B-spline is a piecewise polynomial, and the polynomial segments are given by the knot vector. Provided the knots are distinct, the splines and their derivatives, up to the $p-1$-th derivative, are continuous. (Discontinuities can be introduced if desired by having multiple knots at the same location.) It also follows that the B-spline basis functions have local support, being non-zero only over $p+1$ knot intervals. Moreover the spline basis functions form a partitioning of unity (sum to one) at all points in the input domain. Finally it can be shown that any polynomial of degree $p$ can be written as a B-spline of degree $p$. These are all desirable properties that are utilised in the ASMOD algorithm. Examples of basis functions of degree one and two are given in figure 1. For a more comprehensive presentation of spline theory see for instance [29].

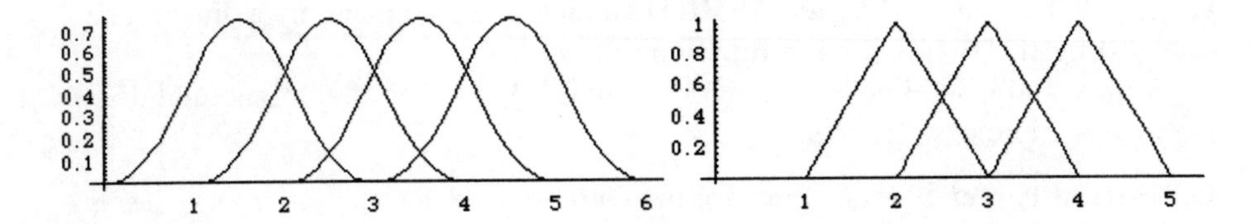


**Fig. 1.** Examples of univariate B-spline bases of degree 1 (left) and 2 (right). The knot vectors are $\tau = (1, 2, 3, 4, 5)$ and $\tau = (0, 1, 2, 3, 4, 5, 6)$ respectively. The input domain is $[2, 4)$ in both cases.

In order to model coupled dependencies of two or more input variables, ASMOD uses tensor products of two or more univariate spline spaces.

**Definition 3.** *The tensor product of two univariate spline spaces*

$$S_{p_1,\tau_1} = \left\{ \sum_{i=1}^{k_1} c_i b_{i,p_1,\tau_1}(x_1) | \; c_i \in R, \; i = 1, \ldots k_1 \right\}$$

*and*

$$S_{p_2,\tau_2} = \left\{ \sum_{i=1}^{k_2} c_i b_{i,p_2,\tau_2}(x_2) | \; c_i \in R, \; i = 1, \ldots k_2 \right\}$$

*is denoted by $S_{p_1,\tau_1} \otimes S_{p_2,\tau_2}$ and given by*

$$S_{p_1,\tau_1} \otimes S_{p_2,\tau_2} = \left\{ \sum_{i=1}^{k_1} \sum_{j=1}^{k_2} c_{i,j} b_{i,p_1,\tau_1}(x_1) b_{j,p_2,\tau_2}(x_2), \right.$$
$$\left. c_{i,j} \in R, \; i = 1, \ldots, k_1, \; j = 1, \ldots, k_2 \right\}$$

*Higher order tensor products are similarly defined.*

Obviously a spline function in $S_{p_1,\tau_1} \otimes S_{p_2,\tau_2}$ can be written as

$$\sum_{k=1}^{k_{1,2}} c_k b_{k,p_1,p_2,\tau_1,\tau_2}(x_1, x_2), \; c_k \in R, \; i = 1, \ldots, k_{1,2}$$

where $k_{1,2} = k_1 k_2$ and $b_{k,p_1,p_2,\tau_1,\tau_2}(x_1, x_2) = b_{i,p_1,\tau_1}(x_1) b_{j,p_2,\tau_2}(x_2)$ for some suitable numbering. Figure 2 shows an example of a bivariate basis function of degree two.

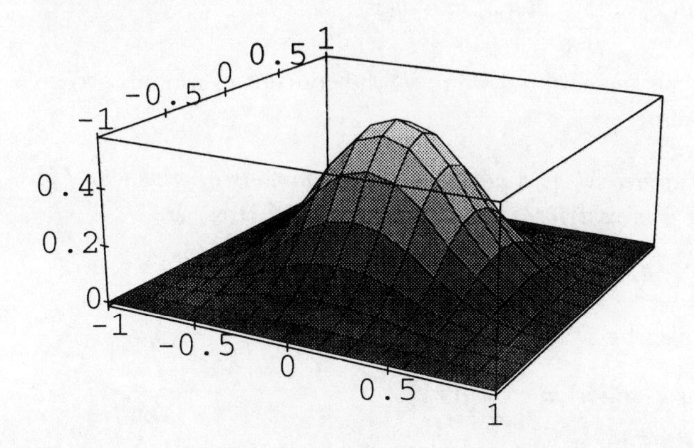


**Fig. 2.** A bivariate B-spline basis function of degree 2 formed by the knot vector $\tau = (-1, -1/3, 1/3, 1)$ for both variables.

The number of basis functions grows exponentially with the number of input variables to a tensor spline function. Thus, in most cases it is not practical to use tensor splines of more than a few input variables. To overcome this so called *curse of dimensionality* we have chosen to decompose high dimensional input spaces into several low dimensional subspaces and to model the output variable as a sum of functions, each acting on a low dimensional subspace.

More formally an ASMOD model $m(x)$ is decomposed into a sum of $U$ lower dimensional B-spline functions called submodels and denoted by $s_u(x), u = 1, \ldots, U$. The sets of input variables to the submodels are mutually disjoint so

$$m(x) = \sum_{u=1}^{U} s_u(x_u) \tag{2}$$

where $\{x_u\} \subseteq \{x\}$ and $\{x_i\} \cap \{x_j\} = \emptyset$ for $i \neq j$. ($\{x_u\}$ is the set of input variables present in the $x_u$ vector).

Hence an ASMOD model is an element of a linear space

$$L = L_1 + \cdots + L_U \tag{3}$$

where $L_u$, $u = 1, \ldots, U$ is a spline space or a tensor product of spline spaces and $s_u \in L_u$, $u = 1, \ldots, U$. Notice that $L$ is not a direct sum since there is no unique representation of a constant function.

By renaming the basis functions in $L_1, \ldots, L_U$ with a common index variable $i$ and extending their domain of definition to all variables of $x$ they can be written as $b_1(x), \ldots, b_K(x)$, and the ASMOD model (2) can be written as the inner product

$$m(x) = \sum_{i=1}^{K} c_i b_i(x) = c^T b(x) \tag{4}$$

For the theoretical analysis in section 4 we need the concept of models structures, which we will now define.

**Definition 4.** *A model structure $M$ is a set of ASMOD models as given by (4) where the coefficient vector $c$ is restricted to a compact set $\mathcal{A}$, i.e.*

$$M = \{m(x) |\ m(x) = c^T b(x),\ c \in \mathcal{A}\}$$

A submodel $s_u(x)$ can also be written as

$$s_u(x) = c^{uT} b^u(x)$$

and a submodel structure is similarly defined as a model structure.

### 2.1 Determination of the coefficients

After we have chosen a model structure $M$, we want to determine the parameter vector $c^*$ in (4) which minimise the expected value of a chosen non-negative criterion function $f(y - c^T b(x))$, i.e.

$$c^* = \arg\min_{c \in \mathcal{A}} E f(y - c^T b(x))$$

where $E$ is the expectation operator. The set $\mathcal{A}$ is chosen as a set to which we a priori know that $c^*$ belongs. Notice that there is not necessarily a unique solution to the minimisation problem.

A common choice of criterion function is the quadratic criterion given by

$$I(c, M) = E(y - c^T b(x))^2. \tag{5}$$

This criterion is known to have optimal properties when $y - c^T b(x)$ is normally distributed, but other criteria exist which are more robust with respect to unknown distributions and outliers in the data. For a more detailed discussion see e.g. [13]. Here we will only consider the quadratic criterion.

Since the underlying probability distribution is unknown, we can not evaluate the expected value of the criterion directly. Instead we use the empirical value

$$I_{\text{emp}}(c, M, l) = \frac{1}{l} \sum_{i=1}^{l} (y^i - c^T b(x^i))^2 \tag{6}$$

computed from a set of empirical data $(x^1, y^1), \ldots, (x^l, y^l)$ as an estimate for the expected value. $x^i$ and $y^i$ are pairs of input vectors and output variables observed at the same time instants $i$. In section 3.4 we present different modifications to the criterion (6) which can be used for improving the estimated value of (5). None of these modifications will change the vector $c^*$ minimising the criterion, and (6) can thus always be used for determining the optimal parameter vector within a model structure. This is a linear regression problem, and the solution is commonly known to be defined by the Moore-Penrose pseudo-inverse as [26]

$$c^* = (B^T B)^{-1} B^T y.$$

Here $B$ is the matrix of basis functions evaluated at the input vectors, i.e. $B_{i,j} = b_j(x^i)$, and $y$ is the vector of observed output variables. Singular value decomposition or another numerically stable algorithm should be used for the matrix inversion since the matrix easily becomes badly conditioned or singular. The backfitting algorithm can be used to decompose the fitting problem into several smaller problems of fitting individual submodels[12].

# 3 Model structure identification and algorithms

## 3.1 Introduction

In most identification and modelling schemes one considers a given model structure or a fixed set of given model structures and estimates parameters for these structures. However, due to the curse of dimensionality problem this is not feasible when the input space is high dimensional. Either we end up with a model structure containing far too many parameters compared to the number of available data points or we have an impractically large set of model structures. The ASMOD algorithm tries to overcome this problem by adapting the model structure to the observed data. This is done in an iterative model refinement procedure in which, for each step, the model is either *grown* to include new modelling capability or *pruned* to produce a more parsimonious model.

## 3.2 The general ASMOD algorithm

Here we describe a general version of the ASMOD algorithm. Usually ASMOD starts with a very simple model structure, containing only a few prespecified submodel structures, or none at all. In the algorithm below we call the most promising (the "best") model structure at a time for the current model structure. The linear space associated with it is called the current linear space. After an initial model structure has been specified, models and model structures are identified according to the following algorithm:

*The ASMOD algorithm:*

1. Let the initial model structure be the current model structure.
2. Let $i = 0$, and let the *stop refinement* criterion be FALSE.
3. WHILE the *stop refinement* criterion is FALSE, do:
   (a) Let $i = i + 1$
   (b) From the current model structure, generate a set $\mathcal{M}_i = \{M_1, M_2, \ldots, M_{N_i}\}$ of candidate model structures (grown and/or pruned).
   (c) Estimate the parameters in each model structure in $\mathcal{M}_i$. Denote the estimated parameter vectors by $\hat{c}_{ij}$, $j = 1, \ldots, N_i$, i.e.

$$\hat{c}_{ij} = \arg \min_{c \in \mathcal{A}_{M_j}} I_{\text{emp}}(c, M_j, l)$$

   (d) Compute a criterion function $g(M)$ for all candidate model structures.
   (e) Select the model structure with the smallest value of the criterion function $g(M)$ as the new current model structure. Denote this model structure by $\hat{M}_i$ and denote the corresponding parameter vector from 3c by $\hat{c}_i$. That is

$$\hat{M}_i = \arg \min_{M \in \mathcal{M}_i} g(M)$$

$$\hat{c}_i = \arg \min_{c \in \mathcal{A}_{\hat{M}_i}} I_{\text{emp}}(c, \hat{M}_i, l)$$

   (f) Compute the *stop refinement* criterion.
4. The identified model structure $\hat{M}$ is the one which gives the minimum value of $g(\hat{M}_j), j = 1, \ldots, i$, and the identified model within this structure is given by the corresponding parameter vector $\hat{c}$, i.e.

$$\hat{M} = \arg \min_{M \in \{\hat{M}_1, \ldots, \hat{M}_i\}} g(M)$$

$$\hat{c} = \arg \min_{c \in \mathcal{A}_{\hat{M}}} I_{\text{emp}}(c, \hat{M}, l)$$

Here $i$ represents the iteration number in the refinement procedure. The last step in the algorithm selects among the model structures, which at some stage have been the current model structure, the one that gives the minimum value of $g$. This will often be the second last one, since we will continue to refine the model structure until an increase in the value of $g$ occur.

Model growing and pruning can be done concurrently at each refinement step by making competing candidates, or the model can first be refined and then pruned in a repeated fashion.

In the above algorithm the following needs to be specified: the set of candidate model structures $\mathcal{M}_i$ to be generated for a current model structure, the criterion function $g$, and the stop refinement criterion. These items will be discussed in the following subsections.

### 3.3 Generation of candidate model structures

For a given model structure there will always be an unlimited number of possible refinements. We will thus need a set of rules for how candidate refinements can be constructed which limits the number of candidates. In the original ASMOD algorithm [18, 19] three basic methods for model growing are given. We will here also propose three methods for model pruning (model pruning was suggested as a possibility, but not used in the original algorithm).

Suppose we are given a linear space of the form

$$L = \sum_{i=1}^{U} L_i$$

where $L_i$ is a spline space or a tensor product of spline spaces. Let $\{x_i\}$ denote the set of input variables present in submodel structure $i$ and $\{x\}$ the set of all input variables.

*Model growing.* The three methods for growing a space are as follows

**M1** A new univariate submodel structure involving a new input variable is added to the model structure. That is, let $x_j \in \{x\} \setminus (\cup_{i=1}^{U}\{x_i\})$ and let $p_j$ be a given polynomial degree and $\tau_j$ a given knot vector. Then

$$L_{U+1} = S^{x_j}_{p_j,\tau_j}$$

and

$$L = \sum_{i=1}^{U+1} L_i$$

**M2** Two existing submodel structures are combined and replaced by their tensor product, allowing coupled dependencies of the combined input variables to be modelled, i.e.

$$L = L_j \otimes L_k + \sum_{i=1,\ i\neq j,k}^{U} L_i$$

**M3** A new knot is inserted in the knot vector of one of the input variables, thus locally introducing increased modelling capability in the variable. The location of new knots is restricted to the mid point between two adjacent knots in the input domain. A subspace $L_i$ is given by

$$L_i = S^{x_{i_1}}_{p_{i_1},\tau_{i_1}} \otimes S^{x_{i_2}}_{p_{i_2},\tau_{i_2}} \otimes \cdots \otimes S^{x_{i_n}}_{p_{i_n},\tau_{i_n}}$$

The new knot vector is

$$\tau'_{i_j} = \left(\tau_{i_j,1}, \ldots, \tau_{i_j,k}, \frac{\tau_{i_j,k} + \tau_{i_j,k+1}}{2}, \tau_{i_j,k+1}, \ldots, \tau_{i_j,n_{i_j}+p_{i_j}+1}\right)$$

and the new subspace is

$$L_i = S^{x_{i_1}}_{p_{i_1},\tau_{i_1}} \otimes \cdots \otimes S^{x_{i_j}}_{p_{i_j},\tau'_{i_j}} \otimes \cdots \otimes S^{x_{i_n}}_{p_{i_n},\tau_{i_n}}$$

These methods are applied to a linear space creating new linear spaces, and they can then be applied again to the newly created linear spaces in an iterative way. To limit the number of candidate models we have chosen to apply these methods only to the current linear space. For the current linear space these methods are applied to every possible combination of input variables, subspaces and knot vectors. To be more precise

**G1** For every variable in the input vector which has not already been included in any of the current submodel structures, create a candidate model structure where this variable is included in a new univariate submodel structure with a prespecified knot vector and polynomial degree.

**G2** For every combination of two submodel structures, create a candidate model structure where the two submodel structures are replaced by the tensor product of the two.

**G3** Form candidate model structures where one of the submodel structures is replaced by a submodel structure with a refined knot vector. Do this for all submodel structures, all knot vectors of the submodel structures and all adjacent knots in the input domain of the knot vectors.

*Model pruning.* The three methods for pruning a space are similar to the growing methods, but used backwards. These are defined as follows

**M4** A univariate submodel structure with no knots in the interior of the input domain and polynomial degree $p_j$ is replaced by a spline of degree $p_j - 1$, also with no interior knots. If $p_j - 1 = 1$ the spline becomes a constant and the submodel structure can be removed. Let $L_k$ be a univariate spline space of variable $x_j$, and let $p_j$ and $\tau_j = \{\tau_{j,1}, \ldots, \tau_{j,2p_j+2}\}$ be the corresponding spline degree and knot vector. Then the pruned submodel structure is

$$L_k = S^{x_j}_{p_j-1,\{\tau_{j,2},\ldots,\tau_{j,2p_j+1}\}}$$

**M5** A multivariable submodel structure is split into two submodel structures whose tensor product is the original submodel structure. Let $L_k$ be a tensor product of spline spaces which can be decomposed as $L_k = L_{k1} \otimes L_{k2}$. Then after pruning we have

$$L = L_{k1} + L_{k2} + \sum_{i=1,\ i\neq k}^{U} L_i$$

**M6** One knot which is interior to the input domain of a spline function is removed from a knot vector. A subspace $L_i$ is given by

$$L_i = S^{x_{i_1}}_{p_{i_1},\tau_{i_1}} \otimes S^{x_{i_2}}_{p_{i_2},\tau_{i_2}} \otimes \cdots \otimes S^{x_{i_n}}_{p_{i_n},\tau_{i_n}}$$

The new knot vector is

$$\tau'_{i_j} = (\tau_{i_j,1}, \ldots, \tau_{i_j,k}, \tau_{i_j,k+2}, \ldots, \tau_{i_j,n_{i_j}+p_{i_j}+1})$$

and the new subspace is

$$L_i = S^{x_{i_1}}_{p_{i_1},\tau_{i_1}} \otimes \cdots \otimes S^{x_{i_j}}_{p_{i_j},\tau'_{i_j}} \otimes \cdots \otimes S^{x_{i_n}}_{p_{i_n},\tau_{i_n}}$$

As for model growing the pruning methods can then be applied repeatedly to the newly created linear spaces in an iterative way. To limit the number of candidate models we also in this case limit ourselves to the current linear space. For the current linear space these methods are applied to every possible combination of input variables, subspaces and knot vectors, i.e.

**G4** Apply method **M4** to every subspace $L_i$ that satisfies the requirements for this method.

**G5** For every subspace $L_i$ which is a tensor product of spline spaces, form candidate model structures for every possible splitting of $L_i$.

**G6** Form candidate model structures where one of the submodel structures is replaced by a submodel structure with a pruned knot vector. Do this for all submodel structures, all knot vectors of the submodel structures and all knots interior to the input domain of the knot vectors.

### 3.4 Model structure selection and stop refinement criteria

We next turn our attention to the choice of the criterion function $g$ and the stop refinement criterion in the ASMOD algorithm. These will usually be closely related since an obvious approach is to continue refining until an increase in $g$ is observed. The choice of $g$ will of course depend on the intended use of the model, but in the case of a prediction model, a natural guideline is to choose the model structure which gives the largest improvement in accuracy and continue to refine until no further improvement can be seen.

If enough data is available the most straight forward method is to evaluate the models on independent data sets. However, often the available data are so sparse that we want to use them all for parameter estimation, and other methods must be used to assess the obtained models.

It is well known that the expected value of the prediction error

$$E(y - \hat{c}^T b(x))^2$$

can be split into a bias term and a variance term, see e.g. [10] for a discussion. The bias term is due to the model structure not being sufficiently general to model the underlying system, while the variance term is due to the variability in the estimated parameters, i.e. the optimal parameters for the structure is not identified. It is clear that with a more general model structure the bias term will be reduced. However, a more general model structure involves more parameters and hence the variability of the parameter estimate will increase, and so will the variance term. The problem is thus to balance the two terms against each other.

In the system identification literature several criteria for model order selection have been suggested. Among these criteria we find the FPE, AIC and MDL (see below). From a pragmatic point of view one can say that they aim at minimising a criterion containing two terms, one involving the prediction errors computed on the data set used for parameter estimation (bias) and a term involving the number of parameters in the model (variance). These criteria can also be used with ASMOD although we must keep in mind that the theoretical assumptions under which they were derived, may not hold.

In the following we briefly describe some model selection criteria that can be used in the ASMOD algorithm.

### Model structure selection criteria

*Cross validation* The method of cross validation has similarities with the principle of using an independent data set for testing. The data are divided into $m \leq l$ disjoint subsets, where $l$ is the total number of observations. $m$ models are then estimated in parallel on the data where one of the subsets are taken out for testing. The model structure which on average gives the smallest error on the validation set is chosen for further refinement. If $m = l$ we say that we have a full cross validation of the model.

The three next criteria which are commonly used in system identification weight the improved model accuracy when evaluated on the data against a penalty for the increased number of parameters. Let $I_{\text{emp}}(M, l)$ denote the minimum value of the criterion (6) for a model in the model structure $M$, i.e.

$$I_{\text{emp}}(M, l) = \min_{c \in \mathcal{A}_M} I_{\text{emp}}(c, M, l)$$

As before $l$ denotes the number of data points and $K_M$ the number of independent parameters in the model structure. For details on how the criteria are derived, see the references below.

*The Final Prediction Error Criterion (FPE)* ([1]) The $g$ function is given by

$$g_{\text{FPE}}(M) = I_{\text{emp}}(M, l)\frac{1 + K_M/l}{1 - K_M/l} \tag{7}$$

*Akaike's Information Theoretic Criterion (AIC)* ([2]) Now $g$ is given by

$$g_{\text{AIC}}(M) = \ln(I_{\text{emp}}(M, l)) + \frac{2K_M}{l} \tag{8}$$

where ln denotes the natural logarithm.

*The Minimum Description Length (MDL) principle* ([27]) Here

$$g_{\text{MDL}}(M) = \ln(I_{\text{emp}}(M, l)) + \frac{K_M \ln l}{l} \tag{9}$$

*Remark.* In [21] the relationship between FPE, AIC and statistical hypothesis testing is investigated. It is shown that AIC and FPE considered as hypothesis testing criteria correspond to a large significance level (the probability of rejecting a true null hypothesis is large). To reduce the significance level it has been suggested to use $4K_M/l$ (AIC) and $(1 + 2K_M/l)/(1 - 2K_M/l)$ (FPE) as correction factors.

In the above expressions for AIC and MDL we have made a tacit assumption that, for the estimation problem at hand, the used least square estimator coincides with the maximum likelihood estimator. In the general versions of AIC and MDL, $I_{\text{emp}}(M, l)$ is replaced by the likelihood function.

*Structural Risk Minimisation (SRM)* ([30]) This criterion is based on risk minimisation theory ([30]). This theory gives upper bounds on the expected value $I(c, M)$ in terms of the empirical value $I_{\text{emp}}(c, M, l)$. The SRM criterion minimises this upper bound and is given by

$$g_{\text{SRM}}(M) = \left[\frac{I_{\text{emp}}(M, l)}{1 - T(h, l)}\right]_{\infty} \tag{10}$$

where

$$T(h, l) = C_1 \sqrt{\frac{h \ln(2l) - \ln(h!) + C_2}{l}} \tag{11}$$

and

$$[z]_{\infty} = \begin{cases} z & \text{if } z \geq 0 \\ \infty & \text{if } z < 0 \end{cases} \tag{12}$$

Here $h$ is the Vapnik-Chervonenkis (VC) dimension, and since ASMOD models are linear in the parameters it is roughly the number of parameters (see section 4.1). There do exist analytical expression for $C_1$ and $C_2$, but they are derived through a worst case analysis, and give far too conservative results (again see section 4.1). This means that in practice we must set $C_1$ and $C_2$ to some suitable values chosen on the basis of prior knowledge and experience.

**The stop refinement criterion** The obvious stop refinement criterion is to keep on refining until all candidate model structures considered in a refinement step show an increase in the value of the chosen $g$ function. This means that the current model structure from which the candidate structures were generated, gave the best result in terms of the criterion specified by $g$, and this model structure is therefore taken as $\hat{M}$, the identified model structure (see step 4 in the ASMOD algorithm).

An alternative to stopping refinement as soon as all candidate model structures give an increase in $g$ is to continue on with a few more refinement steps and terminate if none of the candidate model structures show any improved performance. This alternative is appealing when the models are grown and pruned in a repeated fashion.

# 4 Theoretical aspects

In this section we very briefly examine some of the theoretical properties of the ASMOD algorithm. In subsection 4.1 we examine the finite sample properties while the asymptotic properties are investigated in subsection 4.2. A more detailed analysis of ASMOD can be found in [33].

## 4.1 Finite sample properties

To analyse the finite sample properties we make use of the theory put forward in [30] with some extensions from [32] and [34]. This theory allows us to derive upper and lower bounds on the criterion function $I(c, M)$ in terms of the empirical function $I_{\text{emp}}(c, M, l)$ for all models in a model structure $M$.

Quite naturally the bounds are dependent on the complexity of the model structure, the more complex a model structure is, the weaker are the bounds. One way to measure the complexity of a model structure is the Vapnik Chervonenkis (VC) dimension. The formal definition of the VC dimension is rather technical, see [30], but for our purposes we only need the following result.

**Lemma 5.** *Given an ASMOD model structure*

$$m(x, c) = \sum_{i=1}^{K} c_i b_i(x) \quad c \in \mathcal{A} \tag{13}$$

*and a quadratic optimisation criterion*

$$(y - m(x, c))^2 \tag{14}$$

*Let $U$ be the number of submodel structures of (13) (see (2)). Then the VC dimension $h$, of (14) is*

$$h = K - U + 2$$

*where $K$ is the number of elements in the parameter vector $c$.*

In order to derive the upper bound we make the following assumptions on the data set

1. The data $(x^1, y^1), \ldots, (x^l, y^l)$ are mutually independent.
2. For all models in the model structure $M$ we assume that the ratio of the $p$th order mean to the first order mean is bounded by a constant $\tau$ for some $p > 2$, that is

$$\frac{\left(E(y - c^T b(x))^{2p}\right)^{1/p}}{E(y - c^T b(x))^2} \leq \tau \;\; \forall c \in \mathcal{A} \tag{15}$$

The upper bound is given in the following theorem which is a special case of theorem 7.5 in [30].

**Theorem 6 Upper bound.** *Suppose the above assumptions are satisfied. Furthermore assume that the number of independent observations $l$ is larger than the VC dimension $h$ of the model structure $M$. Then with probability* $1-\eta$

$$I(c) \leq \left[\frac{I_{\mathrm{emp}}(c,l)}{1-T(h,\eta,l)}\right]_{\infty} \tag{16}$$

*is valid simultaneously for all c. Here*

$$T(h,\eta,l) = 2\tau a(p)\sqrt{\frac{h\ln(2l)-\ln(h!)-\ln(\eta/12)}{l}}$$

*where p and $\tau$ are given in (15), $[\cdot]_{\infty}$ is explained in (12) and*

$$a(p) = \left(\frac{(p-1)^{p-1}}{2(p-2)^{p-1}}\right)^{\frac{1}{p}}$$

The SRM criterion (10) is based on the upper bound (16). If we choose $C_1 = 2\tau a(p)$ and $C_2 = -\ln(\eta/12)$ in (10)-(11) we get (16).

The obtained bound is potentially very conservative. In particular the required number of independent observations we need for achieving a given upper bound on the expected value is impractically large. However, we must keep in mind the very weak assumptions we have made. In [30] it is suggested to set $2\tau a(p)$ equal to 1 when we use the upper bound in practical applications.

The important variables in this bound, apart from the empirical value, are the VC dimension $h$, and the sample size $l$. The functional dependence of the upper bound on $h$ and $l$ is quite natural in the sense that the upper bound increases with $h$ and decreases with $l$. We expect the empirical and expected values to come close when the number of observations increases, and it should not come as a surprise that the bound (remember it is uniform) gets more conservative when we consider an increasing number of parameters, i.e. we consider an increasingly complex estimation task. The bound makes a connection between the sample size, the complexity of the estimation problem and the desired reliability of the estimate.

A similar uniform lower bound on $I(c)$ is given in [32] and [34]. Since both the upper and lower bounds are uniform in $c$ they can be used to derive a confidence for the optimal parameter vector on the basis of a finite number of samples. This is however mainly of theoretical interest since the bounds are very conservative.

### 4.2 Asymptotic properties

Now we consider the asymptotic properties, that is what happens when the number of observations $l$ tends to infinity.

Under natural conditions it can be shown that for a single model structure the empirical value converges uniformly to the expected value with probability one, i.e.

$$\sup_{c\in\mathcal{A}_M} |I(c,M) - I_{\mathrm{emp}}(c,M,l)| \to 0 \text{ w.p. 1 as } l\to\infty \tag{17}$$

Obviously the above result extends to a finite number of model structures.

Assume that we base our model structure selection criterion on the upper bound in theorem 6, perhaps with some modified constants. That is we let the estimate of the parameter vector $\hat{c}(l)$ and the model structure $\hat{M}(l)$ be given by

$$(\hat{c}(l), \hat{M}(l)) = \arg \min_{M \in \mathcal{M},\ c \in \mathcal{A}_M} \left[ \frac{I_{\text{emp}}(c, M, l)}{1 - T(h_M, \eta, l)} \right]_\infty$$

where $\mathcal{M}$ is the set of model structures we consider in a refinement step, and $h_M$ is the VC dimension of the model structure $M$. Then, still under natural conditions, it can be shown that $(\hat{c}(l), \hat{M}(l))$ converges to a parameter vector and a model structure which minimise the expected value $I(c, M)$, or in mathematical terms

$$(\hat{c}(l), \hat{M}(l)) \to \mathcal{C} \text{ w.p. } 1 \text{ as } l \to \infty$$

where the set $\mathcal{C}$ is given by

$$\begin{aligned} \mathcal{C} &= \arg \min_{M \in \mathcal{M},\ c \in \mathcal{A}_M} I(c, M) \\ &= \left\{ (c, M) |\ M \in \mathcal{M},\ c \in \mathcal{A}_M,\ I(c, M) = \min_{\bar{M} \in \mathcal{M},\ \bar{c} \in \mathcal{A}_{\bar{M}}} I(\bar{c}, \bar{M}) \right\} \end{aligned} \quad (18)$$

Let

$$I(M) = \min_{c \in \mathcal{A}_M} I(c, M)$$

The above results mean that in each refinement step in the ASMOD algorithm we will asymptotically choose the best model structure in terms of the criterion $I(M)$, which is the criterion we want to minimise.

Notice that this does not guarantee that we end up in an overall optimal model structure. The reason is that in each refinement step we only consider a few model structures for computational reasons, and the optimal model structure may never be considered.

## 5 Results and applications

Different implementations of the ASMOD algorithm have in earlier work been tested on different synthetic and real world problems [7, 4]. Here we report results obtained for two synthetic problems, a 10-dimensional function and state space modelling of a biological reactor. The models were fitted with the pseudo-inverse computed by singular value decomposition. All models were initialised as empty (containing no submodels), and were then repeatedly refined until no improvement was observed in the optimisation criterion. The models were then pruned, again until no improvement was observed. The results reported here may differ from earlier reported results with the same data sets, since we are using different optimisation criteria and pruning is now used.

### 5.1 A 10-dimensional test function

The problem consisted in identifying a 10 dimensional function from small samples of noisy data. The function was given by

$$f(x) = 10\sin(\pi x_1 x_2) + 20(x_3 - 0.5)^2 + 10x_4 + 5x_5 + 0x_6 + \cdots + 0x_{10}.$$

Note that the output did not depend on $x_6, \ldots, x_{10}$, and one of the difficulties was to identify which of the input variables were relevant. The algorithm should also identify the correct decoupled and coupled dependencies as well as which dependencies were linear and which were not. This problem was one of several problems constructed by Friedman for evaluating the MARS algorithm [9].

The input variables were picked independently and randomly from a uniform distribution in the interval $[0, 1]$. The output variable in the training data sets was added to normally distributed noise, $N(0, 1)$, corresponding to approximately 20% noise. The training was done on 20 independent data sets consisting of 50 input/output pairs each. Models using different refinement criteria were trained on the same data sets, and all models were tested on the same independent test set consisting of 1000 noise free observations. The normalised mean square errors listed in Table 1 are computed for the test set as

$$NMSE = \sum_i \frac{(y_i - \hat{y}_i)^2}{(y_i - \bar{y})^2}.$$

$y_i$ and $\hat{y}_i$ are the output and estimated output values, and $\bar{y}$ is the mean of the output values $y_i$.

The models were initialised as empty. New variables were inserted with quadratic B-splines and no internal knots. An optimal identification of the model structure should end up with one bivariate quadratic submodel for $x_1$ and $x_2$ with at least 9 parameters, one univariate quadratic submodel for $x_3$ (2 additional independent parameters) and two univariate linear submodel for $x_4$ and $x_5$ (1 independent parameter each). The optimal model should thus have 13 or a few more independent parameters.

Table 1 lists the number of parameters in the final models when different optimisation criteria were used and the corresponding mean square errors. The numbers in parentheses are the standard deviations for the 20 data sets. For comparison, results obtained by Friedman with the MARS algorithm are included. These values are averages of 100 training data sets and with a test set of 5000 points [9].

By inspecting the models the following observations were made. The Akaike's information criterion (AIC), the modified final prediction error (FPE-2) and the structural risk minimisation (SRM) criterion all gave very accurate identification of the model structure. The coupled dependency of $x_1$ and $x_2$ was always found, and the initially quadratic splines were, most of the time, reduced to linear splines for variables $x_4$ and $x_5$. We see that the models obtained by these criteria have the correct number of parameters and gave very good accuracy on the test data.

**Table 1.** The number of independent parameters and scaled mean squared errors for the 10-dimensional test function using different optimisation criteria. The values are averages (and standard deviations in parenthesis) for 20 independent samples, each of 50 observations.

| Criterion | Remarks | # param. | NMSE |
|---|---|---|---|
| CV | Full cross validation | 19.0 (4.8) | 0.071 (0.06) |
| AIC | equation (8) | 14.8 (1.6) | 0.027 (0.02) |
| AIC-2 | AIC with $4K_M/l$ | 6.8 (4.7) | 0.420 (0.41) |
| FPE | equation (7) | 21.2 (3.4) | 0.180 (0.42) |
| FPE-2 | FPE with $(1+2K_M/l)/(1-2K_M/l)$ | 13.5 (0.8) | 0.019 (0.01) |
| MDL | equation (9) | 8.1 (5.0) | 0.350 (0.40) |
| SRM | equation (10) | 14.0 (1.3) | 0.025 (0.02) |
| MARS | average of 100 samples | 32.0 | 0.200 (0.12) |

For the SRM criterion we used the parameter values $C_1 = 0.85$ and $C_2 = 4.8$. We originally started with $C_1 = 1.0$ as suggested in [30], but this gave on the average too simple model structures. We also observe that the AIC-2 and the FPE gave respectively too simple and too complex model structures. The sensitivity to the tuning of parameters is a weakness of these criteria, since we do not know if the tuned parameters used here will work well for new problems. More experience is needed before we can say if the best criteria for this problem will work well for other problems.

The cross validation criterion (CV) gave on the average slightly too complex models, and thus had a tendency to overfit the data. Large variance in the cross validation estimates is a known problem [12], and since we select the "best" candidate for every step in the refinement procedure we will have a tendency to always find one which is good. The result is that the refinement will continue a few steps after the optimal structure is found, and an overfitted model is obtained. Such overfitting will unfortunately not be removed by the subsequent pruning. One attractive property of cross validation is that there are no parameters to determine, and the criterion seems to be relevant for most data complexities and noise levels.

### 5.2 A simulated batch fermentation process

The second problem was to identify a nonlinear state space model for a batch fermentation process where the micro-organism *Pseudomonas ovalis* transforms glucose into gluconic acid. This process has been extensively studied, and a state space model has been derived in [11]. This model uses five state variables described in Table 2 and is given by:

$$
\begin{aligned}
\dot{\chi} &= \mu_m \tfrac{sc}{k_s c + k_0 s + sc}\chi \\
\dot{p} &= k_p l \\
\dot{l} &= v_l \tfrac{s}{k_l + s}\chi - 0.91 k_p l \\
\dot{s} &= -\tfrac{1}{Y_s}\mu_m \tfrac{sc}{k_s c + k_0 s + sc}\chi - 1.011 v_l \tfrac{s}{k_l + s}\chi \\
\dot{c} &= k_l a(c^* - c) - \tfrac{1}{Y_s}\mu_m \tfrac{sc}{k_s c + k_0 s + sc}\chi - 0.09 v_l \tfrac{s}{k_l + s}\chi
\end{aligned}
\tag{19}
$$

Johansen and Foss have studied the possibility of modelling this process as a smooth interpolation of linear local models, where each local models represent different phases of the batch process (see chapter by Johansen and Foss in this book). The regimes of the local models were either manually set up according to the available process knowledge [16, 15], or an automatic identification scheme was used to determine the regimes [17].

We have used ASMOD to identify nonlinear state space models for this problem based on the same training and test data as used by Johansen and Foss. Batches were simulated for 10 hours and sampled every half hour based on (19). A training set consisting of 100 batches (2000 samples) and a test set of 30 batches (600 samples) were generated with random initial states. Random noise was added to the measurements. For details on the parameters and simulations see one of [15, 17].

As seen from the differential equations the state derivatives have a variable degree of coupling and nonlinearity in their dependency on the state variables. This was therefore a suitable and realistic problem for investigating how well the structure could be identified and modelled by the ASMOD algorithm.

Separate ASMOD models were trained for each state variable to predict the incremental change from the current to the next (half an hour ahead in time) value of the state variables based on the current state vector. The ASMOD models were initialised as empty models. New submodels were initialised with quadratic B-splines and two internal knots. SRM ($C_1 = 0.85$ and $C_2 = 4.8$) was used as the refinement criterion. No attempt was made to tune any of these parameters to improve performance.

Figure 3 shows ballistic predictions of the five state variables based upon the noisy measurement of the initial state vector. Root mean square errors for one step ahead predictions and ballistic predictions 10 hours ahead in time are given in Table 2. The errors are averages for the 30 test batches and the 20 samples in each batch. Results obtained by Johansen and Foss [17] are included for comparison.

As can be seen, the ASMOD model performs well on this problem. We also see that a very simple model was identified for modelling $\dot{p}$, which in fact is linearly dependent on one variable only. The other variables needed more complex models as was expected. The model structures reflected the dependencies existing in the model, and could thus be used to gain process knowledge.

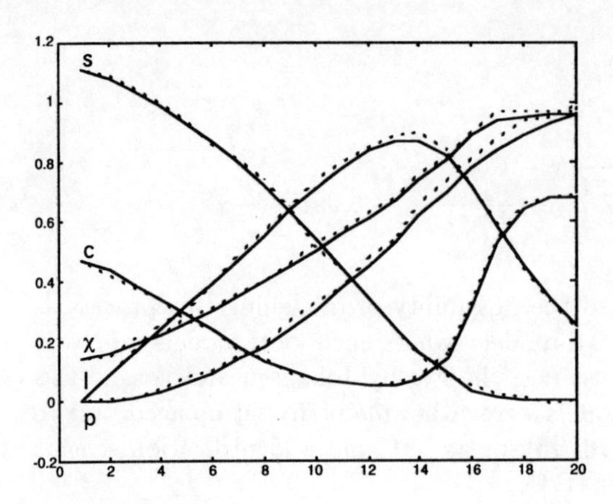


**Fig. 3.** Ballistic predictions for the first batch in test data set (solid), and the true noisy trajectories.

**Table 2.** The root mean square error for one step ahead and ballistic predictions of the 30 test batches. The given total is for all variables and all batches.

| | | ASMOD | | | Local linear | |
|---|---|---|---|---|---|---|
| Variable | Concentration of | # param. | One step | Ballistic | One step | Ballistic |
| $\chi$ | Cells | 15 | 0.0136 | 0.028 | 0.0134 | 0.034 |
| $p$ | Gluconic acid | 3 | 0.0123 | 0.020 | 0.0123 | 0.021 |
| $l$ | Gluconolactone | 16 | 0.0112 | 0.024 | 0.0157 | 0.036 |
| $s$ | Glucose | 7 | 0.0131 | 0.018 | 0.0140 | 0.020 |
| $c$ | Oxygen | 24 | 0.0095 | 0.028 | 0.0137 | 0.032 |
| Total | | | 0.0120 | 0.024 | 0.0139 | 0.029 |

## 6 Conclusions

The ASMOD algorithm was originally proposed as a method for automatic construction of empirical models using a B-spline representation. In this chapter we have further developed the theoretical foundations for ASMOD, and we have also proposed some improvements to the original algorithm.

One key feature of ASMOD is the procedure for automatic adaptation of the internal model structure to the dependencies observed in the training data. The aim of this procedure is to include only the flexibility necessary to model the observed dependencies, and to avoid all other flexibility. The resulting model is more parsimonious compared to general structures, and the model becomes more transparent to the user.

The structure identification is done through an incremental model refinement procedure in which the model in each step can either be grown to include new flexibility or pruned to remove superfluous flexibility. The sequence of model growing or pruning is determined by evaluating a set of candidate refinements for each step, and selecting the refinement which minimises an optimisation criterion. Results obtained for a simple test problem with different criteria are reported. All criteria make a tradeoff between model accuracy on the training data and model complexity, thus preferring models with high accuracy and low complexity. Several criteria work very well on our test problem, and the model structure is identified from very limited data sets with high accuracy and reliability. ASMOD shows good performance on this and other problems compared to other methods applied to the same data sets. One example application with a simulated micro-biological batch reactor showed that ASMOD is able to identify nonlinear state space models with good accuracy.

Results on the theoretical properties of ASMOD are given. The results are based on the theory of structural risk minimisation and the Vapnik-Chervonenkis dimension of the model structure. Upper bounds with a specified probability are found under weak conditions for the model prediction error, given a finite data set. Under natural conditions convergence to optimal refinement steps and model parameters with probability one can be shown in the limit when the number of data points tends to infinity.

**Acknowledgments** The authors want to thank Tor Arne Johansen at the Norwegian Institute of Technology for many valuable discussions and for supplying us with the bio-reactor data.

## References

1. H. Akaike. Fitting autoregressive models for prediction. *Ann. Inst. Stat. Math*, 21:425–439, 1969.
2. H. Akaike. A new look at the statistical model identification. *IEEE Trans. Automatic Control*, 19:716–723, 1974.
3. T. Bohlin. Derivation of a designer's guide for interactive 'gray-box' identification of nonlinear stochastic objects. *Int. Journal of Control*, 59(6):1505–1524, 1994.
4. N.A. Bridgett, M. Brown, C.J. Harris, and D.J.Mills. High-dimensional approximation using an associative memory network. In *Proc. of IEE Control'94*, volume 2, pages 1458 – 1465, 1994.
5. M. Brown, D.J.Mills, and C.J. Harris. The representation of fuzzy algorithms used in adaptive modelling and control schemes. *Submitted to "Intelligent Systems Engineering"*, 1994.
6. M. Brown and C. Harris. *Neurofuzzy adaptive modelling and control.* Prentice-Hall International (UK) Limited, 1994.
7. M. Carlin, T. Kavli, and B. Lillekjendlie. A comparison of four methods for nonlinear data modelling. *Chemometrics and Intelligent Laboratory Systems*, 23:163–177, 1994.
8. R. Franke. Recent advances in the approximation of surfaces from scattered data. In C. K. Chui and L. L. Schumaker, editors, *Topics in Multivariate Approximation*, pages 79 – 98. Academic Press, Inc., New York, 1987.
9. J. Friedman. Multivariate adaptive regression splines. *The Annals of Statistics*, 19(1):1 – 141, 1991.
10. S. Geman, E. Bienenstock, and R. Doursat. Neural networks and the bias/variance dilemma. *Neural Computation*, 4:1–58, 1992.

11. T.K. Ghose and P. Ghosh. Kinetic analysis of gluconic acid production by pseudomonas ovalis. *J. Applied Chemical Biotechnology*, 26:768 – 777, 1976.
12. T.J. Hastie and R.J. Tibshirani. *Generalized Additive Models.* Chapman and Hall, London, 1990.
13. P.J. Huber. *Robust Statistics.* John Wiley and Sons, Inc., 1981.
14. D.R Hush and B.G. Horne. Progress in supervised neural networks. what is new since Lippman. *IEEE Signal Processing Magazine*, pages 8 – 39, Jan 1993.
15. T.A. Johansen. Identification of non-linear systems using empirical data and prior knowledge - An optimization approach. *Submitted to Automatica*, Nov 1994.
16. T.A. Johansen and B.A. Foss. State-space modelling using operator regime decomposition and local models. Technical report 93-40-w, The Norwegian Institute of Technology, Trondheim, Norway, Jul 1993.
17. T.A. Johansen and B.A. Foss. Semi-empirical modelling of non-linear dynamic systems through identification of operating regimes and local models. In G. Irwin, K. Hunt and K. Warwick, editors, *Advances in Neural Networks for Control Systems*, Advances in Industrial Control. Springer-Verlag, 1995. This volume.
18. T. Kavli. ASMOD - an algorithm for adaptive spline modelling of observation data. *International Journal of Control*, 58(4):947–967, 1993.
19. T. Kavli. *Learning principles in dynamic control.* PhD thesis, University of Oslo, ISBN no. 82-411-0394-8, SINTEF, Oslo, 1993.
20. P. Lancaster and K. Šalkauskas. *Curve and Surface Fitting. An introduction.* Academic Press, Inc., London, 1986.
21. I.J. Leontaritis and S.A. Billings. Model selection and validation methods for non-linear systems. *Int. Journal of Control*, 45(1):311–341, 1987.
22. J. Moody and C.J. Darken. Fast learning in networks of locally-tuned processing units. *Neural Computation*, 1:281 – 294, 1989.
23. J. Park and I.W. Sandberg. Universal approximation using radial-basis-function networks. *Neural Computation*, 3(2):246–257, 1991.
24. T. Poggio and F. Girosi. Networks for approximation and learning. *Proceedings of the IEEE*, 78(9):1481–1497, Sep 1990.
25. M.J.D. Powell. Radial basis functions for mulivariable interpolation: A review. In J. C. Mason and M. G. Cox, editors, *Algorithms for Approximation.* Clarendon Press, London, 1987.
26. W.H. Press, B.P. Flannery, S.A. Teukolsky, and W.T. Vetterling. *Numerical Recipes in C.* Cambridge University Press, Cambridge, 1988.
27. J. Rissanen. Modelling by shortest data description. *Automatica*, 14:465–471, 1978.
28. T.D. Sanger. A tree-structured algorithm for reducing computation in networks with separable basis functions. *Neural Computation*, 3(1):67–78, 1991.
29. L.L. Schumaker. *Spline Functions: Basic Theory.* Wiley, New York, 1981.
30. V. Vapnik. *Estimation of Dependencies Based on Empirical Data.* Springer-Verlag, 1982.
31. G. Wahba. *Spline Models for Observational Data.* Society for Industrial and Applied Mathematics, Philadelphia, Pennsylvania, 1990.
32. E. Weyer. *System Identification in the Behavioral Framework.* PhD thesis, The Norwegian Institute of Technology, Trondheim, Norway, 1992.
33. E. Weyer and T. Kavli. Theoretical and practical aspects of the asmod algorithm. Internal reports (in preparation), 1994.
34. E. Weyer, R.C. Williamson, and I.M.Y. Mareels. System identification in the behavioral framework via risk minimisation. In preparation, 1994.

# Semi-Empirical Modeling of Non-Linear Dynamic Systems through Identification of Operating Regimes and Local Models

Tor A. Johansen and Bjarne A. Foss

Department of Engineering Cybernetics
Norwegian Institute of Technology
7034 Trondheim, Norway

**Abstract.** An off-line algorithm for semi-empirical modeling of non-linear dynamic systems is presented. The model representation is based on the interpolation of a number of simple local models, where the validity of each local model is restricted to an operating regime, but where the local models yield a complete global model when interpolated. The input to the algorithm is a sequence of empirical data and a set of candidate local model structures. The algorithm searches for an optimal decomposition into operating regimes, and local model structures. The method is illustrated using simulated and real data. The transparency of the resulting model and the flexibility with respect to incorporation of prior knowledge is discussed.

## 1 Introduction

The problem of identifying a mathematical model of an unknown system from a sequence of empirical data is a fundamental one which arises in many branches of science and engineering. The complexity of solving such a problem depends on many factors, such as a priori knowledge, quality and completeness of the data sequence, and required model form and accuracy.

A rich non-linear model representation based on patching together a number of simple local models into a complex global model is suggested in [30, 27, 13, 8, 9]. With this representation, the modeling problem is basically to decompose the operating range of the system into a set of operating regimes, the identification of simple local models within each regime, and the interpolation of the local models to get a global model. This is an example of the classical divide-and-conquer strategy, where a complex problem is decomposed into simple subproblems that can be solved independently, and whose solutions add up to solve the complex problem. In [9, 10] we have focused on the use of system knowledge for regime decomposition. The aim of the present paper is to report on a algorithm that automatically finds a decomposition and local models based on empirical data.

The paper is organized as follows: The problem is formulated in section 2, before an empirical modeling algorithm is developed in section 3. The algorithm is applied to simulated and real data in section 4, and in section 5 the role of prior knowledge and the transparency of the model are discussed. A comparison to related work follows, along with some concluding remarks.

## 2 Problem Formulation

We address the problem of identifying a model of an unknown non-linear system on the basis of a sequence of $l$ input/output-pairs

$$\mathcal{D}_l = ((u(1), y(1)), (u(2), y(2)), ..., (u(l), y(l)))$$

where $u(t) \in R^r$ and $y(t) \in R^m$ are the input and output vectors of the system, respectively. We denote by $\mathcal{D}_t$ the subsequence of $\mathcal{D}_l$ containing data up to and including time $t \leq l$. First, consider static models

$$y(t) = f(u(t)) + e(t) \tag{1}$$

where $e(t) \in R^m$ is zero-mean noise, and $f$ is an unknown function to be estimated. An approximation $\hat{f}$ to $f$ suggests the predictor $\hat{y}(t|u(t)) = \hat{f}(u(t))$ which gives the prediction error

$$\varepsilon(t) = y(t) - \hat{y}(t|u(t)) \;=\; \left(f(u(t)) - \hat{f}(u(t))\right) + e(t)$$

Next, consider a stable dynamic system represented by the NARMAX (non-linear ARMAX) model

$$\begin{aligned} y(t) = f(&y(t-1), ..., y(t-n_y), u(t-1), ..., u(t-n_u), \\ &e(t-1), ..., e(t-n_e)) + e(t) \end{aligned} \tag{2}$$

where $e(t) \in R^m$ is zero-mean noise, and $n_y, n_u$, and $n_e$ are non-negative integers. Given an approximation $\hat{f}$ to the function $f$, a one-step-ahead predictor $\hat{y}(t|\mathcal{D}_{t-1})$ can be formulated. The predictor and prediction error are defined by

$$\begin{aligned} \hat{y}(t|\mathcal{D}_{t-1}) &= \hat{f}(y(t-1), ..., y(t-n_y), u(t-1), ..., u(t-n_u), \varepsilon(t-1), ..., \varepsilon(t-n_e)) \\ \varepsilon(t) &= y(t) - \hat{y}(t|\mathcal{D}_{t-1}) \end{aligned}$$

The motivation behind this predictor is that while the noise sequence $e$ is unknown, $\varepsilon(t) \rightarrow e(t)$ as $t \rightarrow \infty$, if $\hat{f} = f$ and the model is invertible.

Finally, we consider state-space models

$$x(t+1) = g(x(t), u(t)) + v(t) \tag{3}$$

$$y(t) = h(x(t)) + w(t) \tag{4}$$

where $x(t)$ is a state-vector, and $v(t)$ and $w(t)$ are zero-mean disturbance and noise vectors of appropriate dimensions. In this case, the model is defined by the functions $g$ and $h$. Again, using approximations $\hat{g}$ and $\hat{h}$, it is possible to construct a one-step-ahead predictor $\hat{y}(t|\mathcal{D}_{t-1})$ using the extended Kalman-filter approach, e.g. [19]

$$\begin{aligned} \hat{x}(t|\mathcal{D}_{t-1}) &= \hat{g}(\hat{x}(t-1|\mathcal{D}_{t-2}), u(t-1)) + K(t-1)\varepsilon(t-1) \\ \hat{y}(t|\mathcal{D}_{t-1}) &= \hat{h}(\hat{x}(t|\mathcal{D}_{t-1})) \\ \varepsilon(t) &= y(t) - \hat{y}(t|\mathcal{D}_{t-1}) \end{aligned}$$

where $K(t)$ is the Kalman-filter gain matrix. This matrix will depend explicitly on the time, the functions $\hat{g}$ and $\hat{h}$, and the covariance matrices of the disturbance and noise sequences.

### 2.1 A Generalized Framework

In all these cases, we can write the model on the form

$$\eta(t) = f(\xi(t)) + e(t) \tag{5}$$

where $\eta(t) \in R^{\overline{m}}$ is a generalized output-vector, $\xi(t) \in R^{\overline{r}}$ is a generalized input-vector, and $e(t) \in R^{\overline{m}}$ is zero-mean noise. We denote the space $R^{\overline{r}}$ the input space. In the static model case (1), the input and output vectors equal the generalized input and output vectors. In the NARMAX case (2), the generalized input vector contains delayed input and output vectors in addition to delayed noise vectors, while the generalized output equals the system output. If noise terms $e(t-1), ..., e(t-n_e)$ are present, the generalized input vector is partially unknown and cannot be found exactly from the data $\mathcal{D}_t$. For state-space models (3)-(4), neither the generalized input nor the generalized output vectors can be found exactly, because they contain the unknown state vector. The purpose of the formulation (5) with the generalized input and output vectors is to write the model in a generic form with one unknown function $f$. The problem we address is to estimate this function, and since the function immediately gives the model equations, this also solves the system identification problem. Notice that the fact that the generalized inputs and outputs may not be exactly known does not complicate this problem too much, since the model parameters can still be estimated from the input/output data using a prediction error approach with the predictors described above [19].

### 2.2 Model Representation

In [30, 27, 13, 8, 9, 10] a non-linear model representation with good interpolation and extrapolation properties is described. It is based on the decomposition of the system's operating range into a number of smaller operating regimes, and the use of simple local models to describe the system within each regime. A global model is formed by interpolating the local models using smooth interpolation functions, that depend on the operating point.

We define the system's operating point at time $t$ as $z(t) = (z_1(t), ..., z_d(t))^T \in Z = R^d$, where typically $d \leq \overline{r}$ and the operating space $Z$ is a subspace or sub-manifold of the input space. It is assumed that $\xi$ and $z$ are related by a known bounded mapping $H$ so that $z = H(\xi)$. Typically, $Z$ and $H$ are designed such that the operating point $z(t)$ characterizes different modes of behavior of the system under different operating conditions. The design of $Z$ and $H$ is discussed in more detail in section 4, and in [11]. Suppose $Z$ is decomposed into $N$ disjoint sets $\{Z_i\}_{i \in I_N}$ (regimes) so that

$$Z = \bigcup_{i \in I_N} Z_i$$

for some index set $I_N = \{i_1, ..., i_N\}$ with $N$ elements. Assume that for each regime $Z_i$ we have a local model structure defined by the function $\hat{f}_i(\xi; \theta_i)$ (parameterized by the vector $\theta_i$) and a local model validity function $\rho_i(z) \geq 0$ which indicates the relative validity of the local model as a function of $z$. In addition to being smooth, $\rho_i$ is designed to have the property that $\rho_i(z)$ is close to zero if $z \notin Z_i$. Furthermore, it is assumed that for all $z \in Z$ there exists an $i \in I_N$ so that $\rho_i(z) > 0$, to ensure completeness of the model. A global model can be formed as

$$\hat{f}(\xi) = \sum_{i \in I_N} \hat{f}_i(\xi; \theta_i) w_i(z) \tag{6}$$

$$w_i(z) = \rho_i(z) \Big/ \sum_{j \in I_N} \rho_j(z) \tag{7}$$

where the functions $\{w_i\}_{i \in I_N}$ are called interpolation functions. This representation is discussed in detail in [9], and it is shown that if the local model validity functions and operating space are adequately chosen, then any continuous function $f$ can be uniformly approximated to an arbitrary accuracy on any compact subset of the input space using this representation. A model structure based on a decomposition into $N$ regimes is written

$$\mathcal{M}_N = \left\{ \left( Z_i, \rho_i, \hat{f}_i \right) \right\}_{i \in I_N} \tag{8}$$

This is somewhat redundant, since there is a close (but not necessarily one-to-one) relationship between $Z_i$ and $\rho_i$. With this representation, the modeling problem consists of the following subproblems:

1. Choose the variables with which to characterize the operating regimes, i.e. the operating space $Z$ and mapping $H$.
2. Decompose $Z$ into regimes, and choose local model structures.
3. Identify the local model parameters for all regimes.

In [9, 10, 11] it is demonstrated that in some cases, some coarse qualitative system understanding is sufficient to carry out this procedure. In the following sections we propose an algorithm that requires significantly less prior knowledge in order for us to decompose $Z$, choose local model structures, and construct interpolation functions.

### 2.3 Model Structure Identification Criteria

Let a model structure $\mathcal{M}$ of the form (8) be given. Notice that in a model structure, the model parameters $\theta^T = (\theta_{i_1}^T, \ldots, \theta_{i_N}^T)$ are considered unknown. The model structure $\mathcal{M}$ together with the admissible parameter set $\Theta_{\mathcal{M}}$ generate a model set $\{\mathcal{M}(\theta);\ \theta \in \Theta_{\mathcal{M}}\}$. In this section, we will discuss how different model structures can be compared using a sequence of empirical data to estimate their expected prediction performance. Let an unknown *future* data sequence be denoted $\mathcal{D}_t^\star$, and assume $\mathcal{D}_t^\star$ and $\mathcal{D}_l$ are uncorrelated. We introduce the notation

$$y(t) = y^\star(\mathcal{D}_{t-1}^\star) + e(t)$$
$$\varepsilon(t|\mathcal{M}, \theta) = y(t) - \hat{y}(t\,|\mathcal{D}_{t-1}^\star, \mathcal{M}, \theta)$$

where $y^\star(\mathcal{D}^\star_{t-1})$ is the deterministic (predictable) component of the system output, $e(t)$ is the stochastic (unpredictable) component, and $\varepsilon(t|S,\theta)$ is the residual. Let $\hat{\theta}_{\mathcal{M}}$ be the parameter estimate that minimizes the prediction error criterion

$$J_{\mathcal{M}}(\theta) = \frac{1}{l}\sum_{t=1}^{l} \text{trace}\left(\varepsilon(t\,|\mathcal{M},\theta)\varepsilon^T(t\,|\mathcal{M},\theta)\right) \tag{9}$$

and let $E_{\mathcal{D}}$ and $E_{\mathcal{D}^\star}$ denote expectations with respect to $\mathcal{D}_l$ and $\mathcal{D}^\star_t$, respectively. The future prediction error is given by

$$\varepsilon^\star(t|\mathcal{M},\hat{\theta}_{\mathcal{M}}(\mathcal{D}_l)) = y^\star(\mathcal{D}^\star_{t-1}) - \hat{y}(t|\mathcal{D}^\star_{t-1},\mathcal{M},\hat{\theta}_{\mathcal{M}}(\mathcal{D}_l)) + e(t)$$

where the dependence of $\mathcal{D}_l$ on $\hat{\theta}_{\mathcal{M}}$ has been written explicitly. The expected squared prediction error is defined by

$$\Sigma(\mathcal{M}) = E_{\mathcal{D}^\star}E_{\mathcal{D}}\left(\varepsilon^\star(t|\mathcal{M},\hat{\theta}_{\mathcal{M}}(\mathcal{D}_l))\right)\left(\varepsilon^\star(t|\mathcal{M},\hat{\theta}_{\mathcal{M}}(\mathcal{D}_l))\right)^T$$

Assuming $e^\star(t)$ is white noise that is uncorrelated with $\mathcal{D}^\star_{t-1}$ and $\mathcal{D}_l$, we get the following bias/variance decomposition of this expected squared prediction error

$$\begin{aligned}
\Sigma(\mathcal{M}) = E_{\mathcal{D}^\star}&\left(y^\star(\mathcal{D}^\star_{t-1}) - E_{\mathcal{D}}\hat{y}(t|\mathcal{D}^\star_{t-1},\mathcal{M},\hat{\theta}_{\mathcal{M}}(\mathcal{D}_l))\right) \\
&\cdot\left(y^\star(\mathcal{D}^\star_{t-1}) - E_{\mathcal{D}}\hat{y}(t|\mathcal{D}^\star_{t-1},\mathcal{M},\hat{\theta}_{\mathcal{M}}(\mathcal{D}_l))\right)^T \\
&+E_{\mathcal{D}^\star}E_{\mathcal{D}}\left(\hat{y}(t|\mathcal{D}^\star_{t-1},\mathcal{M},\hat{\theta}_{\mathcal{M}}(\mathcal{D}_l)) - E_{\mathcal{D}}\hat{y}(t\left|\mathcal{D}^\star_{t-1},\mathcal{M},\hat{\theta}_{\mathcal{M}}(\mathcal{D}_l)\right)\right) \\
&\cdot\left(\hat{y}(t|\mathcal{D}^\star_{t-1},\mathcal{M},\hat{\theta}_{\mathcal{M}}(\mathcal{D}_l)) - E_{\mathcal{D}}\hat{y}(t|\mathcal{D}^\star_{t-1},\mathcal{M},\hat{\theta}_{\mathcal{M}}(\mathcal{D}_l))\right)^T \\
&+E_{\mathcal{D}^\star}\left(e(t)e^T(t)\right)
\end{aligned} \tag{10}$$

The first term is the squared systematic error (squared bias) caused by a too simple model structure. The second term is the squared random error (variance) that is present because the best model in the model set $\{\mathcal{M}(\theta);\ \theta \in \Theta_{\mathcal{M}}\}$ cannot in general be identified on the basis of the finite data sequence $\mathcal{D}_l$. Finally, the third term is the unpredictable component of the system output. Notice that the first term does not depend on the data $\mathcal{D}_l$, while the third term depends neither on the data $\mathcal{D}_l$, nor on the model structure $\mathcal{M}$. It is evident that a small bias requires a complex model structure, in general. On the other hand, a small variance requires a model structure that is simple, with few parameters compared to the number of observations $l$. The perfect model is characterized both by small bias and variance, and this appears to be impossible to achieve for a small $l$. This is known as the bias/variance dilemma, cf. Fig 1.

The model set will be based on a set of functions that can approximate any smooth function uniformly on a compact subset of the input space. This is obviously a desirable property of the model set, but also a cause for some problems. The richness implies that there will exist models in the model set that

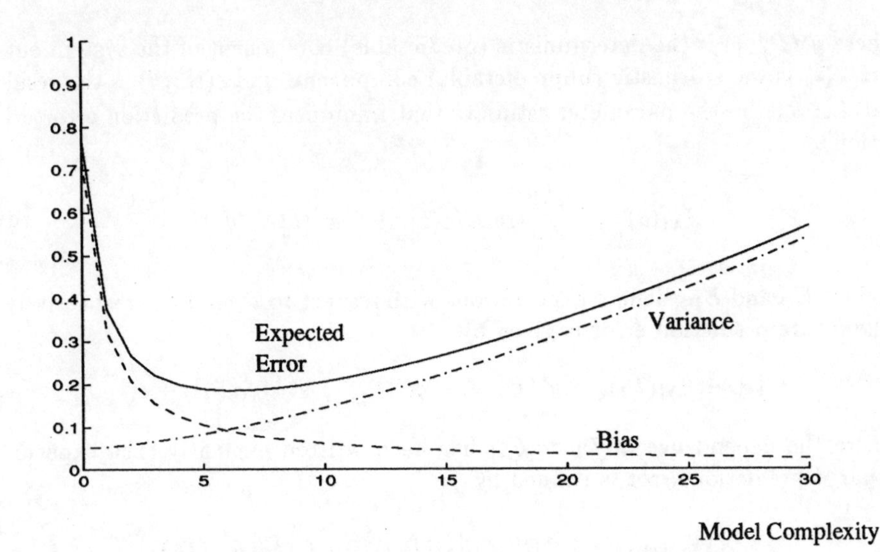


**Fig. 1.** Typical relationship between bias, variance and model structure complexity, when it is assumed that model structure complexity can be measured by a real number.

may make the bias arbitrarily small. However, the finite amount of data will give large variance for such models, and such a model will be fitted to represent not only the system, but also the particular realization of the noise. In other words, the model may give very good prediction of $\mathcal{D}_l$, but poor prediction capability when applied to $\mathcal{D}_t^*$. This is known as over-fitting, and is caused by too many degrees of freedom in the model structure. It is therefore important that a model structure with an appropriate number of degrees of freedom is found, in the sense that it balances bias and variance. We will base the model structure identification algorithm on statistical criteria that have this property.

The mean square error (MSE) criterion is defined by

$$J_{MSE}(\mathcal{M}) = \text{trace}(\Sigma(\mathcal{M}))$$

Minimizing $J_{MSE}$ will lead to a parsimonious model structure, but with a finite sequence of data $\mathcal{D}_l$, the problem is ill-posed. The reason is simply that $J_{MSE}$ cannot be computed since the probability distribution for the prediction error is unknown. An alternative would be to minimize the average squared residuals (ASR) criterion with respect to the model structure

$$J_{ASR}(\mathcal{M}) = \text{trace}\left(\frac{1}{l}\sum_{t=1}^{l}\varepsilon(t|\mathcal{M},\hat{\theta}_{\mathcal{M}})\varepsilon(t|\mathcal{M},\hat{\theta}_{\mathcal{M}})^T\right)$$

For finite $l$, $J_{ASR}$ may be a strongly biased estimate of $J_{MSE}$, since the prediction performance is measured using the same data as those to which the parameters are fitted, and the law of large numbers is not valid because this introduces a strong dependence. Hence, the use of $J_{ASR}$ for structure identification will not, in general, lead to a parsimonious model. We will in the following present several criteria that are far better estimates of $J_{MSE}$ than $J_{ASR}$.

If a separate data sequence $\mathcal{D}_l^\star$ (independent of $\mathcal{D}_l$) is known, an unbiased estimate of $J_{MSE}$ can be found by computing the empirical average squared prediction error that results when the model fitted to the data $\mathcal{D}_l$ is used to predict the data $\mathcal{D}_l^\star$. This is the simplest and perhaps most reliable procedure, but suffers from the drawback that a significantly larger amount of data is required. Experiments and collection of data are major costs for many modeling problems. We therefore proceed with some alternatives that allow the data $\mathcal{D}_l$ to be reused in order to find good estimates of $J_{MSE}$. First we consider the final prediction error criterion (FPE) [1], given by

$$J_{FPE}(\mathcal{M}) = \frac{1+p(\mathcal{M})/l}{1-p(\mathcal{M})/l}\, J_{ASR}(\mathcal{M})$$

where $p(\mathcal{M})$ is the effective number of parameters (degrees of freedom) in the model structure. $J_{FPE}$ is an estimate of $J_{MSE}$, and penalizes model complexity relative to the length of the available data sequence through the term $p(\mathcal{M})/l$. A major restriction is that the predictor is assumed to be linearly parameterized. A non-linear generalization is given in [17]. An alternative criterion can be formulated using cross-validation [28, 26]. The idea is to fit the parameters to different subsets of the data set, and test the prediction performance of the model structure on the remaining (presumed independent) data. Cross-validation may give a reasonable approximation to the use of independent data for selecting the model structure, at the cost of extra computations. The computational complexity can be considerably reduced if the predictor is a linear function of its parameters, or in general by using one of the approximate cross-validation criteria in [26] It is shown that the approximate criteria are asymptotically equivalent to FPE, as $l \rightarrow \infty$. Another approximation to cross-validation is the Generalized Cross Validation (GCV) criterion [4]

$$J_{GCV}(\mathcal{M}) = \frac{1}{(1-p(\mathcal{M})/l)^2}\, J_{ASR}(\mathcal{M})$$

which is easily seen to be asymptotically equivalent to FPE, and also assumes linear parameterization of the predictor. Any one of these criteria can be applied with the structure identification algorithm we will present in the next section.

## 3 System Identification

Let a set of candidate local model structures $\mathcal{L} = \{L_1, L_2, ..., L_{N_L}\}$ be given. $L_i$ is a parameterized function that defines a local model structure, cf. (8).

### 3.1 The Set of Model Structures Candidates

Assume the input- and output-samples in $\mathcal{D}_l$ are bounded. Then the system's operating range $Z$ can be approximated by the $d$-dimensional box

$$Z_1 = \left[z_{1,1}^{\min}, z_{1,1}^{\max}\right] \times \cdots \times \left[z_{1,d}^{\min}, z_{1,d}^{\max}\right]$$

where $z(t) \in Z_1$ for all $t \in \{1, ..., l\}$, since $H$ is a bounded mapping. Notice that the resulting model will extrapolate and can be applied for operating points outside $Z_1$. Next, we consider the problem of decomposing $Z_1$ into regimes.

Consider the possible decompositions of the set $Z_1$ into two disjoint subsets $Z_{11}$ and $Z_{12}$ with the property $Z_1 = Z_{11} \cup Z_{12}$. We restrict the possibilities by the constraint that the splitting boundary is a hyper-plane orthogonal to one of the natural basis-vectors of $R^d$, i.e.

$$Z_{11} = \{z \in Z_1 \mid z_{d_1} < \zeta_1\}$$
$$Z_{12} = \{z \in Z_1 \mid z_{d_1} \geq \zeta_1\}$$

for some dimension index $d_1 \in \{1, ..., d\}$ and splitting point $\zeta_1 \in \left[z_{1,d_1}^{\min}, z_{1,d_1}^{\max}\right]$. Local model validity functions for the two regimes are defined by the recursion

$$\rho_{11}(z) = \rho_1(z) b(z_{d_1} - \overline{z}_{11,d_1}; \lambda_{11})$$
$$\rho_{12}(z) = \rho_1(z) b(z_{d_1} - \overline{z}_{12,d_1}; \lambda_{12})$$

where $\overline{z}_{i,d_1} = 0.5\left(z_{i,d_1}^{\min} + z_{i,d_1}^{\max}\right)$ for $i \in \{11, 12\}$ is the center point of $Z_i$ in the $d_1$-direction. The function $b(r; \lambda)$ is a scalar basis-function with scaling parameter $\lambda$, and the local model validity function associated with the regime $Z_1$ is $\rho_1(z) = 1$. The scaling parameters are chosen by considering the overlap between the local model validity functions. For $i \in \{11, 12\}$, we choose $\lambda_i = 0.5\gamma\left(z_{i,d_1}^{\max} - z_{i,d_1}^{\min}\right)$ where $\gamma$ is a design parameter that typically takes on a value between 0.25 and 2.0. There will be almost no overlap when $\gamma = 0.25$, and large overlap when $\gamma = 2.0$. For each dimension index $d_1 \in \{1, ..., d\}$ we represent the interval $\left[z_{1,d_1}^{\min}, z_{1,d_1}^{\max}\right]$ by a finite number of $N_1$ points uniformly covering the interval. Now $d_1, \zeta_1, L_{1,v}$, and $L_{2,w}$ define a new model structure, where the regime $Z_1$ is decomposed according to the dimension index $d_1$ at the point $\zeta_1$, and the two local model structures are $L_{1,v}$ and $L_{2,w}$. Formally, the set of candidate model structures $\mathcal{S}_n$ with $n$ regimes is given by

$$\mathcal{S}_1 = \{\{(Z_1, \rho_1, L_j)\};\ j \in \{1, 2, ..., N_L\}\}$$
$$\mathcal{S}_2 = \left\{\left\{\left(Z_{11}^i, \rho_{11}^i, L_j\right), \left(Z_{12}^i, \rho_{12}^i, L_k\right)\right\};\ i \in \{1, 2, ..., dN_1\},\ j, k \in \{1, 2, ..., N_L\}\right\}$$
$$\begin{aligned}\mathcal{S}_3 = &\left\{\left\{\left(Z_{11}^i, \rho_{11}^i, L_j\right), \left(Z_{121}^m, \rho_{121}^m, L_k\right), \left(Z_{122}^m, \rho_{122}^m, L_n\right)\right\};\right.\\ &\qquad \left. i, m \in \{1, 2, ..., dN_1\},\ j, k, n \in \{1, 2, ..., N_L\}\right\}\\ &\cup \left\{\left\{\left(Z_{111}^m, \rho_{111}^m, L_k\right), \left(Z_{112}^m, \rho_{112}^m, L_n\right), \left(Z_{12}^i, \rho_{12}^i, L_j\right)\right\};\right.\\ &\qquad \left. i, m \in \{1, 2, ..., dN_1\},\ j, k, n \in \{1, 2, ..., N_L\}\right\}\end{aligned}$$

$$\mathcal{S}_4 = \cdots$$

The model structure set is now $\mathcal{S} = \mathcal{S}_1 \cup \mathcal{S}_2 \cup \mathcal{S}_3 \cup \cdots$. The model structure set is illustrated as a search tree in Fig. 2. Strictly speaking, the model structure set is not a tree, since different sequences of decompositions sometimes lead to the same model structure. However, we choose to represent it as a tree, for the sake of simplicity. Now the structure identification problem can be looked upon as a multi-step decomposition process, where at each step one regime from the previous step is decomposed into two sub-regimes. Such an approach will lead to a sequence of model structures $\mathcal{M}_1, \mathcal{M}_2, ..., \mathcal{M}_n$ where the model structure $\mathcal{M}_{i+1}$ has more degrees of freedom than $\mathcal{M}_i$. Due to the normalization of the local model validity functions, the model set is usually not strictly hierarchical, in the sense that $\mathcal{M}_i$ cannot be exactly represented using $\mathcal{M}_{i+1}$. However, the increasing degrees of freedom define a hierarchical structure.

### 3.2 Basic Search Algorithm

The problem is now to search through the set $\mathcal{S}$ for the best possible model structure. The estimate of the parameters in the model structure $\mathcal{M}$ is defined by a prediction error criterion

$$\hat{\theta} = \arg \min_{\theta} J_{\mathcal{M}}(\theta) \tag{11}$$

where it has been assumed that the minimum exists. This can be ensured by restricting the parameters to a compact set. Now, the chosen structure identification criterion is written $J'(\mathcal{M})$. We define for a given $n \geq 1$

$$\mathcal{M}_n = \arg \min_{\mathcal{M} \in \mathcal{S}_n} J'(\mathcal{M}) \tag{12}$$

Consider the following extended horizon search algorithm, where the integer $n^* \geq 1$ is called the search horizon.

**Search Algorithm.**

1. Start with the regime $Z_1$. Let $n = 1$.
2. At each step $n \geq 1$, find a sequence of decompositions $\mathcal{M}_n, \mathcal{M}_{n+1}, ..., \mathcal{M}_{n+n^*}$ that solves the optimization problem
$$\min_{\mathcal{M} \in \mathcal{S}_{n+n^*}} J'(\mathcal{M})$$
3. Restrict the search tree by keeping the decomposition that leads to $\mathcal{M}_{n+1}$ fixed for the future.
4. If
$$J'(\mathcal{M}_n) > \min_{k \in \{1,2,...,n^*\}} J'(\mathcal{M}_{n+k})$$
then increment $n$ and go to 2. Otherwise, the model structure $\mathcal{M}_n$ is chosen.

Referring to Fig. 2, this algorithm will search the tree starting at the top (corresponding to one local model covering the whole operating space), and selecting a decomposition at each level through a sequence of "locally exhaustive" searches of depth $n^*$. If $n^* = 1$, this is a local search algorithm.

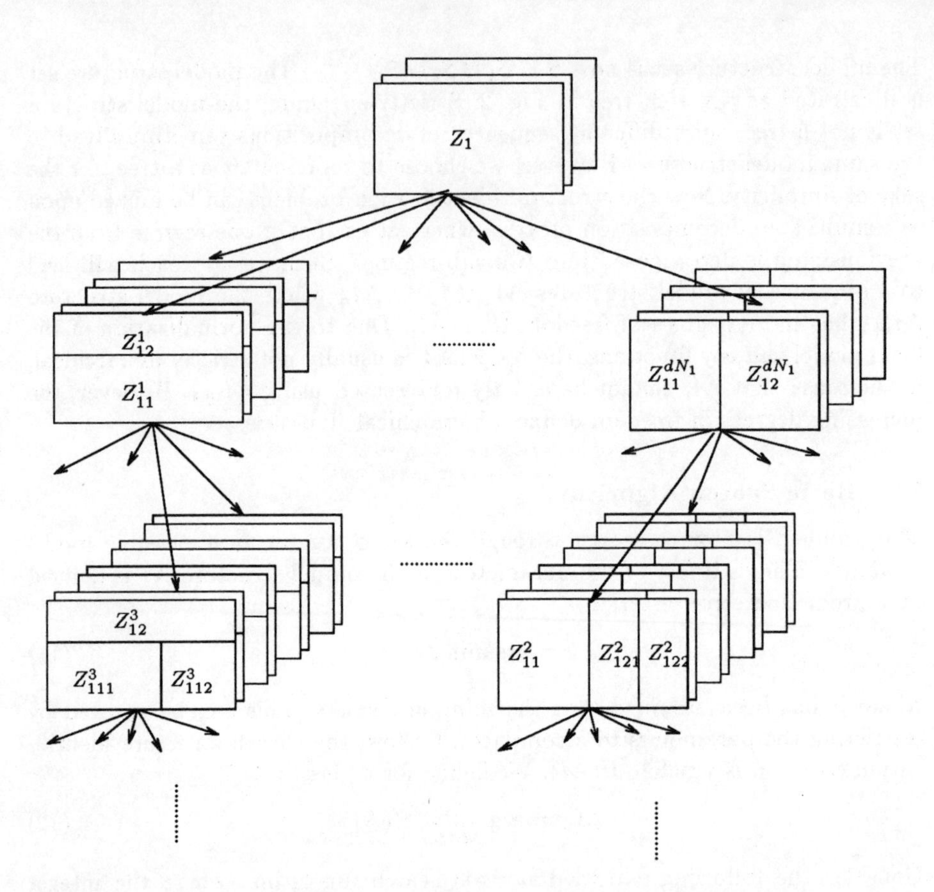


**Fig. 2.** Model structure search tree illustrating possible decompositions into regimes and choice of local model structures. Each level in the tree corresponds to the possible decompositions into one more regime than at the previous level, i.e. the model structure sub-sets $\mathcal{S}_1, \mathcal{S}_2, \mathcal{S}_3, \ldots$ The sub-set of model structures at each "super-node" in the tree corresponds to a fixed decomposition into regimes, but to different combinations of local model structures. The suggested algorithm will search this tree starting at the top (corresponding to one local model covering the whole operating space), and selecting a decomposition at each level through a sequence of "locally exhaustive" searches of depth $n^\star$.

### 3.3 Heuristic Search Algorithm

Clearly, the performance of the algorithm is expected to improve as $n^\star$ increases, but the computational complexity makes $n^\star > 3$ not feasible for any practical problem with 1994 desktop computer technology, even if the local model structure set $\mathcal{L}$ contains as few as one or two possibilities.

**Example.** Consider the problem of identifying a state-space model of the form $x(t+1) = f(x(t))$, where $\dim(x) = 5$, and we apply local models of the form

$$\begin{pmatrix} x_1(t+1) \\ x_2(t+1) \\ \vdots \\ x_5(t+1) \end{pmatrix} = \begin{pmatrix} a_1 \\ a_2 \\ \vdots \\ a_5 \end{pmatrix} + \begin{pmatrix} a_{11} \; a_{12} \cdots a_{15} \\ a_{21} \; a_{22} \quad\; a_{25} \\ \vdots \quad\;\; \ddots \;\; \vdots \\ a_{51} \; a_{52} \cdots a_{55} \end{pmatrix} \begin{pmatrix} x_1(t) \\ x_2(t) \\ \vdots \\ x_5(t) \end{pmatrix}$$

Each of the 30 parameters can be replaced by a structural zero, which gives a set of local linear model structure with $2^{30} \approx 10^9$ elements. On the other hand, even if there is only one possible local model structure that one can choose, the number of possible decompositions into no more than five regimes is

$$\#\mathcal{S}_1 + ... + \#\mathcal{S}_5 = 1 + dN_1 + 2(dN_1)^2 + 3!(dN_1)^3 + 4!(dN_1)^4$$

For $d = 2$ and $N_1 = 10$, this is approximately $4 \cdot 10^6$ candidate decompositions. With $n^\star < 4$, the model structure set is considerably reduced. In particular, $n^\star = 3$ gives about $10^5$ decompositions among which to search, $n^\star = 2$ gives about 2500, while $n^\star = 1$ gives 80 candidate decompositions.

□

Because of the combinatorial nature of the model structure set, it is clearly of interest to implement some heuristics that cut down on the computational complexity without sacrifying too much of the optimality of the algorithm. As we have seen, the number of candidate decompositions at each step in the search may be large. To reduce the number of candidates, we suggest applying the following heuristics in the "locally exhaustive" search at the second step in the search algorithm:

**Heuristic 1**. *At each level in the search tree, proceed with only the most promising candidates.*

The best candidates are of course not known a priori, so there is always a possibility that this may lead to a sub-optimal model. We suggest proceeding with the best decomposition for each of the possible splitting dimensions. Instead of trying to find the best candidates, one can often more easily single out the "least promising candidate decompositions":

**Heuristic 2**. *Discard the candidate decompositions that give an increase in the criterion from one level in the search tree to the next.*

The number of remaining candidates will typically be larger than when using Heuristic 1, but the chance of discarding the optimal decomposition may be smaller. Some candidate decompositions may give rise to regimes where no substantial amount of data is available, and may therefore be classified a priori as not feasible:

**Heuristic 3**. *Discard candidate decompositions that lead to regimes with few data points relevant to this regime compared to the number of degrees of freedom in the corresponding local model structure and local model validity function.*

Counting the number of relevant data-points associated with each regime is controversial, since the interpolation functions overlap. We use the heuristic count $l_i = \sum_{t=1}^{l} w_i(z(t))$, which has the attractive property $\sum_{i \in I_N} l_i = l$.

**Heuristic 4.** *Use a (backward or forward) stepwise regression procedure to handle local model structure sets $\mathcal{L}$ of combinatorial nature [29].*

Related to the example above, one should start with no structural zeroes, and then add one structural zero at a time, choosing the one that gives the largest improvement in the prediction performance. This should give less than $30 + 29 + 28 + ... + 1 < 30^2$ candidate model structures, which is quite different from $2^{30}$.

### 3.4 User Choices

The basis-function $b(r; \lambda)$ with scaling parameter $\lambda$ has the purpose of providing a smooth interpolation between the local models. The basis-function is assumed to have the property $b(r; \lambda) \geq 0$ for all $r \in R$ and $b(r, \lambda) \to 0$ as $|r| \to \infty$. Typical choices are kernel-functions, like the unnormalized Gaussian $\exp(-r^2/2\lambda^2)$. It may appear that the choice of this function has significant impact on the model. However, it is our experience that the algorithm and model's prediction performance are quite insensitive with respect to this choice, and the specification of this function does not require any prior knowledge about the system. What is more important is the choice of $\lambda$, which is controlled by the user-specified parameter $\gamma$.

In order to compute the criterion $J_{GCV}$ or $J_{FPE}$, the effective number of parameters $p$ in the model structure must be known. If the choice of model structure is not based on the data, then the effective number of parameters is

$$p = \sum_{i \in I_N} \dim(\theta_i) \tag{13}$$

in the case of linear regression. However, the proposed algorithm for model structure identification makes use of the data $\mathcal{D}_l$ during the search. Hence, the $p$-value given by (13) will be too small. Counting the effective number of parameters in this case is controversial. We apply the heuristic

$$p = \kappa(N - 1) + \sum_{i \in I_N} \dim(\theta_i)$$

where $\kappa \geq 0$ is a heuristic constant, which can be interpreted as a smoothing parameter, since a large $\kappa$ will put a large penalty on model complexity, and will therefore give a smooth model. A typical choice of $\kappa$ is between 0 and 4, cf. [6].

### 3.5 Statistical Properties

Consider the bias/variance decomposition (10). It has been shown in [12] that both the bias and variance will asymptotically (as $l \to \infty$) tend towards their smallest possible values, with probability one, provided

1. The parameter estimator is consistent, see [18] for conditions under which this holds.

2. The estimate $J'$ of the expected squared prediction error used for model structure identification is consistent.
3. Global minima of the parameter and structure optimization problems are found with probability one.
4. The model set $\{\mathcal{M}(\theta); \mathcal{M} \in \mathcal{S}, \theta \in \Theta_{\mathcal{M}}\}$ can be covered by a finite $\epsilon$-net.

It is known that the use of a separate validation data sequence for model structure identification gives a $J'$ that satisfies the second requirement [12]. It is also known that FPE and AIC may be slightly biased [23].

Neither the parameter optimization nor the structure optimization algorithms need result in global minima. An attractive feature of the model structure set is that it appears to have not only multiple global minima, but also many close-to-optimal local minima. It is easy to see that the restriction of the search to any sub-tree of the model structure tree does not exclude any possible decompositions into regimes. The worst thing that can happen is that the number of decompositions may be somewhat larger than necessary, or alternatively that the partition may not be as fine as desired. Obviously, this leads to suboptimality for finite amount of data, but not necessarily so asymptotically.

The fourth condition is somewhat technical, but it does in general impose a restriction on the complexity of the model set. In practise, this is not a serious restriction, as discussed in [12].

# 4 Examples

## 4.1 A Simulated Fermentation Reactor

Consider the fermentation of glucose to gluconic acid by the micro-organism *Pseudomonas ovalis* in a well stirred batch reactor. The main overall reaction mechanism is described by

$$\begin{aligned}
\text{Cells} + \text{Glucose} + O_2 &\rightarrow \text{More Cells} \\
\text{Glucose} + O_2 &\overset{\text{Cells}}{\rightarrow} \text{Gluconolactone} \\
\text{Gluconolactone} + H_2O &\rightarrow \text{Gluconic Acid}
\end{aligned}$$

The production of gluconolactone is enzyme-catalysed by the cells. We use the following state-space model to simulate the "true system" [7]:

$$\begin{aligned}
\dot{\chi} &= \mu_m \frac{sc}{k_s c + k_o s + sc} \chi \\
\dot{p} &= k_p l \\
\dot{l} &= v_l \frac{s}{k_l + s} \chi - 0.91 k_p l \\
\dot{s} &= -\frac{1}{Y_s} \mu_m \frac{sc}{k_s c + k_o s + sc} \chi - 1.011 v_l \frac{s}{k_l + s} \chi \\
\dot{c} &= k_l a (c^\star - c) - \frac{1}{Y_o} \mu_m \frac{sc}{k_s c + k_o s + sc} \chi - 0.09 v_l \frac{s}{k_l + s} \chi
\end{aligned}$$

**Table 1.** Symbols and constants in the state-space simulation model for the fermenter.

| Symbol | Description |
|---|---|
| $\chi$ | Cell concentration $[UOD/ml]$ |
| $p$ | Gluconic acid concentration $[g/l]$ |
| $l$ | Gluconolactone concentration $[g/l]$ |
| $s$ | Glucose concentration $[g/l]$ |
| $c$ | Dissolved oxygen concentration $[g/l]$ |
| $\mu_m$ | $0.39\ h^{-1}$ |
| $k_s$ | $2.50\ g/l$ |
| $k_o$ | $0.55 \cdot 10^{-3}\ g/l$ |
| $k_p$ | $0.645\ h^{-1}$ |
| $v_l$ | $8.30 mg\ UOD^{-1}\ h^{-1}$ |
| $k_l$ | $12.80\ g/l$ |
| $k_l a$ | $150.0 - 200.0\ h^{-1}$ |
| $Y_s$ | $0.375\ UOD/mg$ |
| $Y_o$ | $0.890\ UOD/mg$ |
| $c^\star$ | $6.85 \cdot 10^{-3}\ g/l$ |

with initial conditions $\chi(0) = \chi_0$, $p(0) = 0$, $l(0) = 0$, $s(0) = s_0$ and $c(0) = c^\star$. The symbols and constants are defined in Table 1.

We simulated 10 hour batches using these equations, "measuring" all states with 0.5 hour intervals, and adding sequentially uncorrelated random noise with a signal-to-noise ratio of approximately 30 dB to the states. We collected two sets of data, by randomly varying the initial conditions $\chi_0, s_0$ and the agitation speed (affecting $k_l a$). The first set contains data from 100 batches, and is used for system identification, while the second independent set is used for model testing. We define the following dimension-less normalized variables: $\chi_n = \chi/(3\ \mathrm{UOD}/ml)$, $p_n = p/(50\ g/l)$, $l_n = l/(15\ g/l)$, $s_n = s/(50\ g/l)$, $c_n = c/(0.01\ g/l)$, and the normalized state-vector $x = (\chi_n, p_n, l_n, s_n, c_n)^T$. We have specified only one possible local linear discrete-time state-space model structure $x(t+1) = a_i + A_i x(t)$ with 12 structural zeros in the $A_i$ matrices. These zeros follow directly from the reaction mechanism, see also [10]. We observe that glucose and oxygen are rate-limiting and, consequently, expected to be the main contributors to the system's non-linearities. It follows that the operating point $z = (s_n,\ c_n)^T$ captures these non-linearities and characterizes the operating conditions of the process with respect to local linear models, see also [10].

Running the identification algorithm with $n^\star = 1$, using the FPE criterion with $\kappa = 1$, and Gaussian basis-function with $\gamma = 1$, results in a model with five local models, and root average squared one-step-ahead prediction error (PE) on the test data PE=0.0139, cf. Table 2. Restricting the number of local models to three, gives PE=0.0147, while one global linear model gives PE=0.0303. This clearly indicates that there exist significant non-linearities which have been captured by the two more complex models, and not by the linear model. The five regimes are illustrated in Fig. 3. Perhaps the most interesting and attractive

**Table 2.** Root average squared prediction performance for the model based on five operating regimes.

| State | One-step-ahead prediction | Ballistic prediction |
|---|---|---|
| $\chi$ | 0.0134 | 0.0341 |
| $p$ | 0.0123 | 0.0211 |
| $l$ | 0.0157 | 0.0357 |
| $s$ | 0.0140 | 0.0204 |
| $c$ | 0.0137 | 0.0315 |
| Total | 0.0139 | 0.0293 |

feature of the method is that the identified model can be interpreted in a natural way. The five regimes correspond to the following phases in the batch

1. Initial phase.
2. Growth phase, where only the amount of micro-organisms is limiting the rate of the reactions.
3. Oxygen supply is rate-limiting.
4. Glucose is rate-limiting.
5. No glucose left, termination.

This gives a high-level qualitative description of the system. More low-level quantitative details on e.g. reaction kinetics can be added by examining the parameters of the local models corresponding to each regime. A simulation (ballistic prediction) of a typical batch from the test set is shown in Fig. 4, using the model with five local models, and the identified global linear model for comparison.

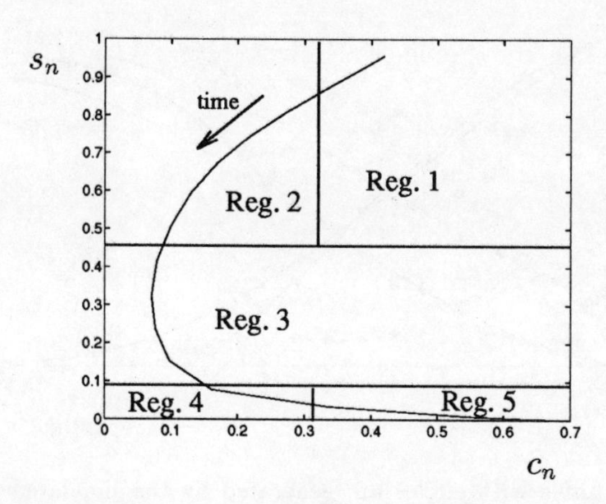


**Fig. 3.** The decomposition into five regimes using the simulated fermenter data, with a typical simulated system trajectory projected onto the $(c_n, s_n)$-plane.

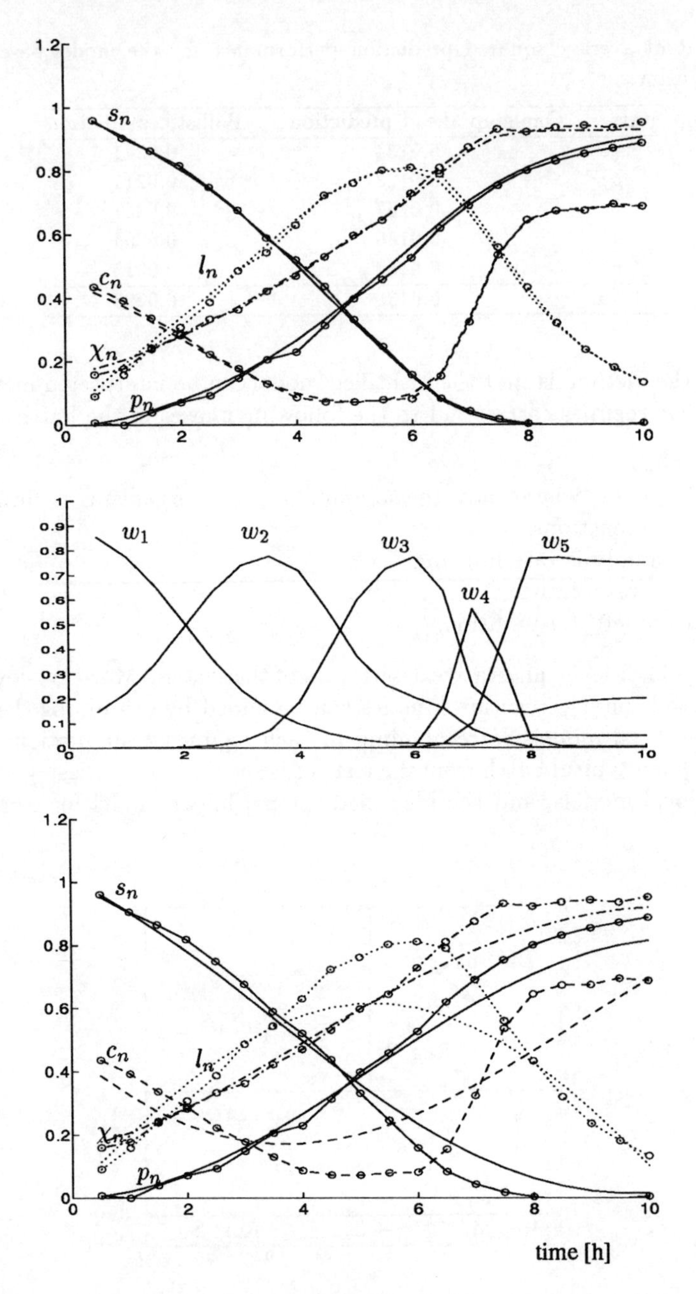


**Fig. 4.** Top: Trajectories with circles are generated by the simulated "true system", while the other trajectories are simulations of the model based on five local models. Middle: The relative weight of the various local models in the interpolation. Bottom: Simulation of the identified global linear model.

Clearly, the results favor the non-linear model. Comparison with similar models with three or four regimes, designed by hand on the basis of system knowledge [10] indicates that their performance are comparable. Although semi-empirical, the high transparency suggests that the identified model should not be viewed as a black-box.

The applied a priori knowledge is essentially the overall reaction mechanism, used for structuring the $A_i$ matrices and for selection of the two variables used for characterizing the operating space. It must be noted that the algorithm has also been applied without this knowledge, i.e. with full $A_i$ matrices and operating point $z = x$, resulting in only a slight decrease in the prediction accuracy of the identified model. In this case, it is interesting to observe that among the five components of the operating point, the algorithm chooses only $\chi$ and $c$ for decomposition into regimes. As noted in [10], due to the batch nature of the process operation, the information in the state $\chi$ is highly redundant, given the information in $s$. This is also evident from Fig. 4, which clearly shows the collinearity between these variables. Hence, the algorithm is forced to make a somewhat arbitrary choice about which variable to use for decomposition, a fact that makes the interpretation of the model more difficult. This problem is also observed with the MARS algorithm [5]. This only emphasizes the important fact that the success of empirical modeling is heavily dependent on the information in the empirical data, and that data deficiencies should to the highest possible extent be compensated for using prior knowledge.

### 4.2 Modeling of a Hydraulic Manipulator

A data sequence $\mathcal{D}_{8000}$ logged from a hydraulic TR4000 robot [15] from ABB Trallfa Robotics A/S was used to find a model describing the inverse dynamics

$$\tau(t) = f(q(t), \dot{q}(t), \ddot{q}(t)) + e(t)$$

of a joint of this robot, where $\tau(t)$ is the control signal to the servo valve, $q(t)$ is joint position, and $e(t)$ is equation error. The joint position was logged at a sampling rate of 100 Hz while the robot was moving along a randomly generated trajectory. The joint velocity and acceleration was estimated by low-pass filtering and numerical differentiations. The prediction of an estimated linear model was subtracted from the data in order to emphasize the non-linearities. According to [15], the non-linearities are mainly due to variations in the momentum arm of the hydraulic cylinder, non-linear damping, and non-linear pressure gain characteristics due to varying flow-rates in the servo valve. In addition, 1000 independent samples were used for testing the model. A number of models were identified, based on the structure identification algorithm, a least squares parameter estimation algorithm, local linear model structure, and a Gaussian basis-function with $\gamma = 1$. The results are summarized and compared to the results in [16] (marked $^{\ddagger}$), [3] (marked $^{+}$), and the MARS algorithm [6] in Table 3. The table shows that the structure identification algorithm is able to find an adequate model with a small number of parameters while maintaining the high

**Table 3.** Results of applying various identification algorithms on the hydraulic manipulator joint data. The NRMSE criterion is defined as the square root of the ratio of the average squared one-step-ahead prediction error to the variance of the output, using the independent test data.

| Model | Comments | Num. Param. | NRMSE |
|---|---|---|---|
| ASMOD‡ | Quadratic Spline Basis | 561 | 15 % |
| MARS | | 155 | 16 % |
| Local linear | $\gamma = 1$, $n^{\star} = 2$, Heuristics 1, 2, and 3 | 80 | 17 % |
| Local linear | $\gamma = 1$, $n^{\star} = 1$, Heuristic 3 | 80 | 17 % |
| MARS | | 53 | 17 % |
| Local linear | $\gamma = 1, n^{\star} = 2$, Heuristics 1, 2 and 3 | 40 | 18 % |
| Local linear | $\gamma = 1, n^{\star} = 1$, Heuristic 3 | 40 | 19 % |
| RBF+ | Gaussian radial basis-functions | 112 | 19 % |
| MARS | | 16 | 20 % |
| ASMOD‡ | Quadratic Spline Basis | 48 | 20 % |
| Local linear | $\gamma = 1, n^{\star} = 2$, Heuristics 1, 2 and 3 | 20 | 23 % |
| NN+ | Sigmoidal Neural Network (3-20-1) | 101 | 23 % |
| Local linear | $\gamma = 1, n^{\star} = 1$, Heuristic 3 | 20 | 26 % |
| NN+ | Sigmoidal Neural Network (3-5-1) | 26 | 26 % |

accuracy of the models found by MARS and ASMOD. In all cases, only the parameters corresponding to local model parameters or basis-function coefficients are counted in the table. Notice that according to the FPE criterion with $\kappa = 1$, the data sequence allows more degrees of freedom to be added to the identified model structures. This was not pursued due to the computational complexity involved. The operating point was chosen as $z = (q, \dot{q}, \ddot{q})$ although complementary identification experiments showed that $z = (\dot{q}, \ddot{q})$ was sufficient to capture most of the non-linearities, in particular when using no more than 40 parameters. Unfortunately, the large number of regimes makes the interpretation of the empirical model more difficult than in the previous example. The model should be viewed as a black box. This example mainly serves as a benchmark that shows that the accuracy achieved with the local modeling approach is comparable to some of the most popular empirical modeling algorithms from the literature.

## 5 Discussion

The amount of prior knowledge required with the proposed approach is quite reasonable. First of all, an operating point space $Z$ is required. In many cases, it is possible to choose $Z$ equal to a subspace or sub-manifold of the input space [9]. The design of $Z$ need not be based solely on a priori knowledge, but can in addition consider the distribution of the data $\mathcal{D}_l$. Quite often, there are collinearities or correlations in the data, so that $\mathcal{D}_l$ can be embedded in a subspace or sub-manifold of considerably lower dimension than the input space. In that case, $z$ need not be of higher dimension than this embedding. Some prior knowledge will often make it possible to reduce the dimension of $z$ considerably. This is important, since it may both reduce the complexity of the model and improve its transparency, and also reduce the computational complexity for the empirical modeling algorithm considerably.

A set of local model structure candidates must be specified. If no a priori knowledge exists to support one choice over the other, one will typically choose local linear model structures of various orders and possibly with structural zeros as default, since linear models are well understood and possible to interpret. Moreover, a linear model will always be a sufficiently good approximation locally, provided the system is smooth, and the regimes are small enough. On the other hand, if there is substantial a priori knowledge available in terms of mechanistic local model structures, these can be included as illustrated in [11]. Such local model structures may for example be simplified mass- and energy-balances.

The purposes of a model can be diverse, e.g. system analysis, design, optimization, prediction, control, or diagnosis. In many applications, it is important that the model can be easily interpreted and understood in terms of the system mechanisms. With empirical models, which are often based on black-box model representations, this is often a hard or impossible task. However, the approach presented in this paper gives highly transparent empirical models because

- local models are simple enough to be interpreted,
- the operating regimes constitue a qualitative high-level description of the system that is close to engineering thinking.

Notice that the interpretation of local linear models as linearizations of the system at various operating points is valid only if the model validity functions do not overlap too much. This point is stressed in [20], and the use of local identification algorithms for each local model is one possibility that will improve the interpretability.

### 5.1 Related work

Local linear models are applied in [13] and [27] together with a clustering algorithm to determine the location of the local models. A parameterized regime description and a hierarchical estimator used to estimate the regime parameters simultaneously with the local model parameters is described in [8] and [14]. An algorithm based on local linear models and decomposition of regimes in which the system appears to be more complex than the model, is suggested in [21]. A model representation based on local polynomial models and smooth interpolation is proposed in [22]. The structure identification algorithm is based on an orthogonal regression algorithm that sequentially discards local model terms that are found to have small significance.

When the interpolation functions are chosen as the characteristic functions of the regime-sets $Z_i$, a piecewise linear model results. The resulting model will not be smooth, and may not even be continuous, which may be a requirement in some applications. Also, we have experienced that smooth interpolation between local linear models usually gives better model fit compared to a piecewise linear model with the same number of parameters. The local linear modeling approach combined with a fuzzy set representation of the regimes also leads to a model representation with interpolation between the local linear models [30]. In that case, it is the fuzzy inference mechanism that implicitly gives an interpolation. Structure identification algorithms based on clustering [2, 32, 31] and local search [29] have been proposed in this context. In this case, the $\rho_i$-functions are interpreted as membership functions for fuzzy sets.

The algorithm of Sugeno and Kang [29] is perhaps the closest relative to the present algorithm. The main difference is the extra flexibility, and effort is begin applied to find a closer to optimal model with the present algorithm. A statistical pattern recognition approach with multiple models leads to a similar representation based on a piecewise linear model and discriminant functions to represent the regime boundaries [24]. Finally, in [25] it is suggested a model representation with neural nets as local models and a structure identification algorithm based on pattern recognition. The pattern recognition algorithm will detect parts of the input space in which the model fit is inadequate, and refine the model locally.

With this large body of literature in mind, one may ask: What are the contributions and improvements represented by the present approach? We have attempted to take the most attractive features from the algorithms in the literature and combined these into one algorithm. We have emphasized interpretability of the resulting model, flexibility with respect to incorporation of prior knowledge, and a transparent modeling and identification process that is close to engineering thinking. The price we have to pay is a computer intensive algorithm. Some may also argue that the algorithm is too flexible and not completely automatic, and as a result it may be difficult to apply for inexperienced users. However, it is our view that real world applications require perhaps even more flexibility and a less automated approach.

## 6 Concluding Remarks

The proposed empirical modeling and identification algorithm is based on a rich non-linear model representation which utilizes local models and interpolation to represent a global model. With this representation, the semi-empirical modeling problem is solved using a structure identification algorithm based on a heuristic search for decompositions of the system's operating range into operating regimes. This algorithm is the main contribution of this work. We want to emphasize an important property of the modeling method and identification algorithm, namely the transparency of the resulting model. The transparency is linked with the possibility to interpret each of the simple local models independently, but more importantly with the fact that the identified regimes can often be interpreted in terms of the system behavior or mechanisms.

In general, the fundamental assumptions behind empirical modeling are that 1) the empirical data is not too contaminated by noise and other unmodeled phenomena, and 2) that the data set is complete in the sense that it contains a sufficient amount of information from all interesting operating conditions and system variables. Unfortunately, these assumptions are often not met in practical applications. The proposed algorithm should therefore be applied with care, and as a part of a computer aided modeling environment that allows flexible incorporation of prior knowledge, not as an automatic modeling algorithm. Moreover, one should undertake a study of the robustness of the algorithm with respect to contaminated, sparse, and incomplete data, in particular for high dimensional and otherwise complex modeling problems.

## Acknowledgments

This work was partly supported by The Research Council of Norway under doctoral scholarship grant no. ST. 10.12.221718 given to the first author. We want to thank Dr. Tom Kavli and several of his colleagues at SINTEF-SI for valuable discussions and making the hydraulic manipulator data available.

## References

1. H. Akaike. Fitting autoregressive models for prediction. *Ann. Inst. Stat. Math.*, 21:243–247, 1969.
2. J. C. Bezdek, C. Coray, R. Gunderson, and J. Watson. Detection and characterization of cluster substructure. II. Fuzzy c-varieties and complex combinations thereof. *SIAM J. Applied Mathematics*, 40:352–372, 1981.
3. M. Carlin, T. Kavli, and B. Lillekjendlie. A comparison of four methods for nonlinear data modeling. *Chemometrics and Int. Lab. Sys.*, 23:163–178, 1994.
4. P. Craven and G. Wahba. Smoothing noisy data with spline functions. Estimating the correct degree of smoothing by the method of generalized cross-validation. *Numerical Math.*, 31:317–403, 1979.
5. R. D. De Veaux, D. C. Psichogios, and L. H. Ungar. A tale of two nonparametric estimation schemes: MARS and neural networks. In *Proc. 4th Int. Conf. Artificial Intelligence and Statistics*, 1993.
6. J. H. Friedman. Multivariable adaptive regression splines (with discussion). *The Annals of Statistics*, 19:1–141, 1991.
7. T. K. Ghose and P. Ghosh. Kinetic analysis of gluconic acid production by pseudomonas ovalis. *J. Applied Chemical Biotechnology*, 26:768–777, 1976.
8. R. A. Jacobs, M. I. Jordan, S. J. Nowlan, and G. E. Hinton. Adaptive mixtures of local experts. *Neural Computation*, 3:79–87, 1991.
9. T. A. Johansen and B. A. Foss. Constructing NARMAX models using ARMAX models. *Int. J. Control*, 58:1125–1153, 1993.
10. T. A. Johansen and B. A. Foss. State-space modeling using operating regime decomposition and local models. In *Preprints 12th IFAC World Congress, Sydney, Australia*, volume 1, pages 431–434, 1993.
11. T. A. Johansen and B. A. Foss. A dynamic modeling framework based on local models and interpolation – combining empirical and mechanistic knowledge and data. Submitted to Computers and Chemical Engineering, 1994.

12. T. A. Johansen and E. Weyer. Model structure identification using separate validation data - asymptotic properties. Submitted to the European Control Conference, Rome, 1995.
13. R. D. Jones and co-workers. Nonlinear adaptive networks: A little theory, a few applications. Technical Report 91-273, Los Alamos National Lab., NM, 1991.
14. M. I. Jordan and R. A. Jacobs. Hierarchical mixtures of experts and the EM algorithm. Technical Report 9301, MIT Computational Cognitive Science, 1993.
15. T. Kavli. Nonuniformly partitioned piecewise linear representation of continuous learned mappings. In *Proceedings of IEEE Int. Workshop on Intelligent Motion Control, Istanbul*, pages 115–122, 1990.
16. T. Kavli. ASMOD – An algorithm for adaptive spline modelling of observation data. *Int. J. Control*, 58:947–967, 1993.
17. J. Larsen. A generalization error estimate for nonlinear systems. In *Proc. IEEE Workshop on Neural Networks for Signal Processing, Piscataway, NJ*, pages 29–38, 1992.
18. L. Ljung. Convergence analysis of parametric identification methods. *IEEE Trans. Automatic Control*, 23:770–783, 1978.
19. L. Ljung. *System Identification: Theory for the User*. Prentice-Hall, Inc., Englewood Cliffs, NJ., 1987.
20. R. Murray-Smith. Local model networks and local learning. In *Fuzzy Duisburg*, pages 404–409, 1994.
21. R. Murray-Smith and H. Gollee. A constructive learning algorithm for local model networks. In *Proceedings of the IEEE Workshop on Computer-Intensive Methods in Control and Signal Processing, Prague, Czech Republic*, pages 21–29, 1994.
22. M. Pottmann, H. Unbehauen, and D. E. Seborg. Application of a general multimodel approach for identification of highly non-linear processes – A case study. *Int. J. Control*, 57:97–120, 1993.
23. R. Shibata. Selection of the order of an autoregressive model by Akaike's information criterion. *Biometrica*, 63:117–126, 1976.
24. A. Skeppstedt, L. Ljung, and M. Millnert. Construction of composite models from observed data. *Int. J. Control*, 55:141–152, 1992.
25. E. Sørheim. A combined network architecture using ART2 and back propagation for adaptive estimation of dynamical processes. *Modeling, Identification and Control*, 11:191–199, 1990.
26. P. Stoica, P. Eykhoff, P. Janssen, and T. Söderström. Model-structure selection by cross-validation. *Int. J. Control*, 43:1841–1878, 1986.
27. K. Stokbro, J. A. Hertz, and D. K. Umberger. Exploiting neurons with localized receptive fields to learn chaos. *J. Complex Systems*, 4:603, 1990.
28. M. Stone. Cross-validatory choice and assessment of statistical predictions. *J. Royal Statistical Soc. B*, 36:111–133, 1974.
29. M. Sugeno and G. T. Kang. Structure identification of fuzzy model. *Fuzzy Sets and Systems*, 26:15–33, 1988.
30. T. Takagi and M. Sugeno. Fuzzy identification of systems and its application to modeling and control. *IEEE Trans. Systems, Man, and Cybernetics*, 15:116–132, 1985.
31. R. R. Yager and D. P. Filev. Unified structure and parameter identification of fuzzy models. *IEEE Trans. Systems, Man, and Cybernetics*, 23:1198–1205, 1993.
32. Y. Yoshinari, W. Pedrycz, and K. Hirota. Construction of fuzzy models through clustering techniques. *Fuzzy Sets and Systems*, 54:157–165, 1993.

# On Interpolating Memories for Learning Control

H. Tolle[1] and S. Gehlen[2] and M. Schmitt[3]

[1] Institute of Control Engineering, Technical University Darmstadt
[2] Zentrum für Neuroinformatik GmbH, Bochum
[3] R. Bosch GmbH, Dept. Motor Vehicle Safety Systems, Schwieberdingen

**Abstract** To imitate human flexibility in controlling different complex non-linear processes on the basis of process observation and/or trial and error, learning control has been developed. The main elements of such control loops are interpolating memories. The chapter deals after an introduction to learning control loops with such devices by putting forward different alternatives, discussing their behaviour in general and going into details of recent research work on mathematically inspired interpolating memories. The respective improvements are motivated and results of applications in the areas of biotechnology and automotive control are presented. In a conclusion some further application areas and realisation aspects are discussed and a critical assessment of status and usefulness of learning control is given.

## 1 Learning Control

A still unmatched challenge in control engineering is human flexibility to control easily very different, complex non-linear processes, as does an operator, who controls at first some industrial process, e.g. a chemical production, and then on going home, his car on some strongly curved road. Actually, part of the respective ability are the unrivalled sensors of human beings, like the eye, and the high number of actuator degrees of freedom, e.g. in the human hand-arm-system. But on top of this exists some information processing scheme, which is till now not understood in detail. However, modern computer technology has given us tools, to perform very fast complex computations allowing step by step better imitations of human performances. From an engineers point of view it is in this context only of secondary importance, whether a considered approach is similar to the information processing in the human brain or not, as long as the results are satisfactory. Therefore, a certain number of ideas to create schemes, which can cope flexibly with different processes and/or process situations have evolved over the time. In general one can combine them under the heading "computer assisted on-line process identification and adequate controller modification". Fig. 1 shows the respective structure.

In all cases, some minimal knowledge about the process is necessary:

– The process must be controllable by its inputs (qualitative controllability).
– The process behaviour must express itself in the measured values (qualitative observability).

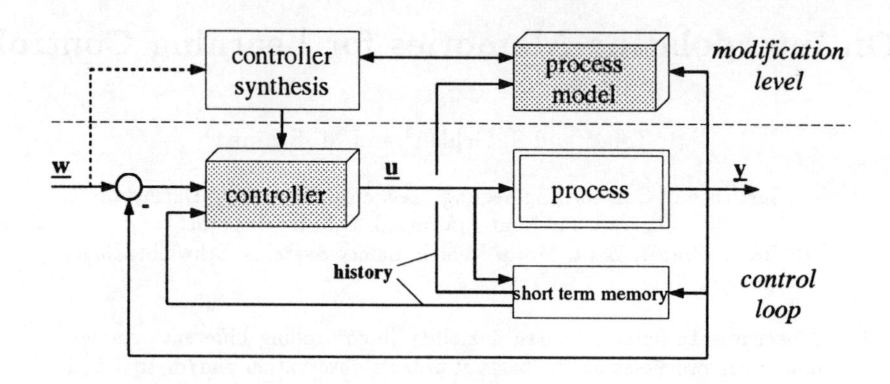


**Figure1.** General scheme of adaptive/self-organizing learning control

- The velocity of dynamic changes must be known to select an adequate sampling time $T_0$ for the digital information processing, to find out, whether dead times exist or not, and how much changes, change rates and so on are of importance. In general that means how much process-output and process-input history $\underline{y}[(k-1)T_0]\cdots\underline{y}[(k-\alpha)T_0]$, $\underline{u}[(k-1)T_0]\cdots\underline{u}[(k-\beta)T_0]$ has to be considered at each time instant $t = kT_0$ for adequate process handling (qualitative dynamics).

The dynamic elements to be selected appropriately – the controller and the process model marked as shaded blocks in fig. 1 – can generally be characterized as manifolds over the sum of the input and output history as a n-dimensional input space for each element of the output vector, that means by linear or non-linear manifolds in $n+1$ dimensional spaces. The different schemes to get a flexible control loop are given by the different ways to build up the manifolds in $\mathbb{R}^{n+1}(\underline{x})$.

The first approach, coming into the focus of research from 1970 on, is "parameter adaptive control" in which mainly linear manifolds in $\mathbb{R}^{n+1}$ are considered – e.g. models of the form

$$\frac{y(z)}{u(z)} = \frac{b_{\nu-m}z^{-(\nu-m)} + \cdots + b_\nu z^{-\nu}}{1 + a_1 z^{-1} + \cdots + a_\nu z^{-\nu}}\,;(m+1+\nu = n)\,. \tag{1}$$

The parameters are updated in all cases, when a setpoint change takes place which pushes the plant behaviour out of the region, in which the identified linear plant model is acceptable and therefore the respectively optimized linear controller works satisfactorily. For further details see e.g. [10]. A problem with this approach can be the transition between the old and the new linear descriptions. Furtheron there exists a major difference vis-à-vis human operators, since new adaptations take place even for set points, which where handled successfully earlier (the actual tangent planes to the in reality non-linear manifolds are used and their parameters are not stored with some situation description).

A second, probabilistic approach, called "self organizing control" and summarized in [23] didn't get as much attention. However, herein now the required nonlinear manifolds are generated by using weighted sums of globally defined given support functions $\varphi_i(\underline{x})$, the manifolds being shaped by adaptation of the constant weights $w_i$ ($x_{n+1} = \sum_{i=1}^{k} w_i \varphi_i(x_1, x_2 \cdots x_n)$). This can be considered as a Ritz-approach from variational calculus – see e.g. [7] – and the main problems connected with it are to choose the number and kind of supporting functions. In addition in many cases a high sensitivity in the weights is observed.

These problems have been solved by the "learning control loop LERNAS" – see [27] – in 1983 where the cerebellum model CMAC put forward in [1], [2] was used to represent the nonlinear manifolds. In this case fixed but just locally defined simple supporting functions are used, approximating the nonlinear manifold by interpolation. The number of used functions is directly linked with the envisaged numerical accuracy and the approximation scheme is similar to the finite element method if one disregards the overlapping of the supporting functions, which reduces the number of necessary supporting functions heavily. A first successful application of LERNAS to a real plant – control of a two input - two output chemical test bed – took place in 1984 [6].

Surprisingly, the CMAC application in control loops did not attract many researchers in spite of its clear advantages and a number of interesting employments in robotics – see e.g. [18]. Instead feedforward neural nets with back propagation adaptation are today the most popular devices for the representation of the $\mathbb{R}^{n+1}$ manifolds in process control. Even in the case, that therein not the globally defined and by that very change-sensitive sigmoids are used as supporting functions but the locally defined radial basis functions, the problem of selecting the number of supporting functions and/or hidden units remains as a disadvantage vis-à-vis the CMAC approach.

Let us turn to applications of the manifold storage. Although this section is entitled "learning control" we cannot deal with the whole spectrum of respective structures. We will just discuss three examples of the usefulness of non-linear $\mathbb{R}^{n+1}$ manifolds for learning.

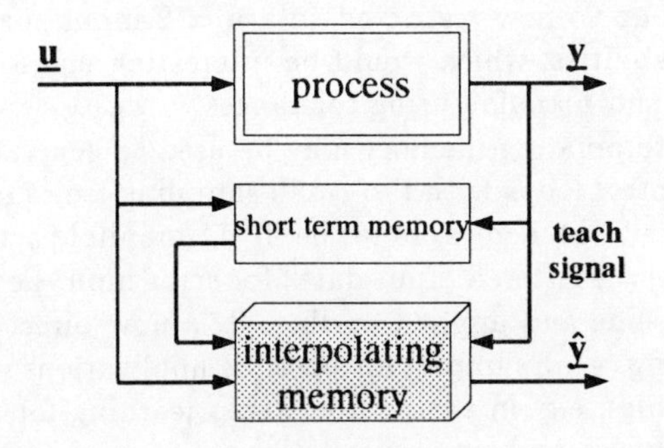


**Figure2.** Scheme for learning of a process model

The first and most simple application is the learning of a process model for off-line investigations – see fig. 2 –. Here a predictive process model $\{\underline{y}(k-1)\cdots\underline{y}(k-\alpha);\underline{u}(k-1)\cdots\underline{u}(k-\beta)\}\rightarrow\hat{\underline{y}}(k)$ – $\hat{\underline{y}}$ being an estimate of $\underline{y}$ – is generated by running a nonlinear manifold storage device (interpolating memory) in parallel to the process and/or training it off-line with collected process data. The learned process model is an alternative to a complicated mathematical model and may be used e.g. for control optimization. Especially for biotechnological processes, where fermentations may need days or even weeks and mathematical modelling is practically not possible, this allows an enormous reduction of otherwise necessary trial and error effort.

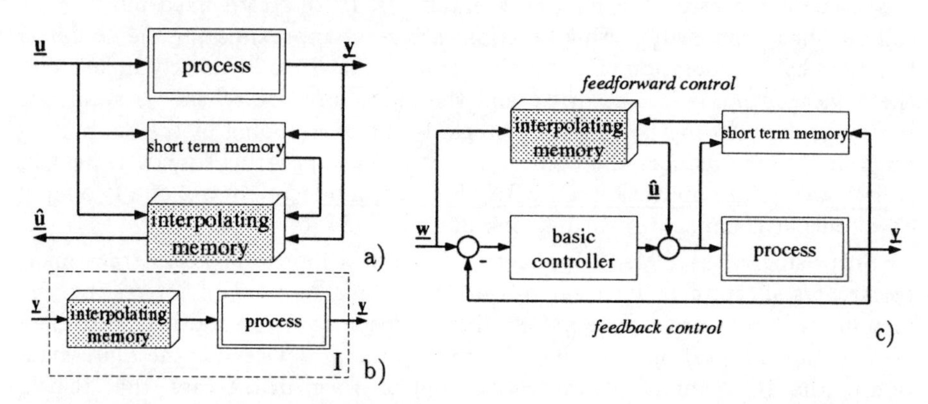


**Figure3.** Learning -a), theoretical application -b), and practical application -c) of an inverse process model $(P)^{-1}$

A second application is learning of an inverse process model and using it for linearization and decoupling of complex multi-input/multi-output processes. In practical use the inverse is not brought directly into the feedback loop as part of the controller, but used as a feedforward control element to simplify the task of a basic linear controller – see fig. 3 –. Feedforward control using non-linear characteristic manifolds is a common tool in different areas of automotive control, however, up to now restricted to a $n = 2$ input space and without on-line learning capabilities, which would be interesting e.g. for the compensation of wear and tear and manufacturing tolerances.

Finally the interpolating memory may be used to generate very quickly control actions in control loops with too small sampling times for on-line computation of complex control schemes. Here the $\mathbb{R}^{n+1}$ manifold acts as a look-up table (based on learning = scattered input data) for error input-actuator response output generated off-line and applied on-line. It is now directly inserted into the control loop (see fig. 4). Examples for possible applications may be found in the area of mechatronics, e.g. in the area of grasp learning for multifingered robot hands where otherwise problems exist ([12]).

Naturally, there are a lot of further possible learning control structures and

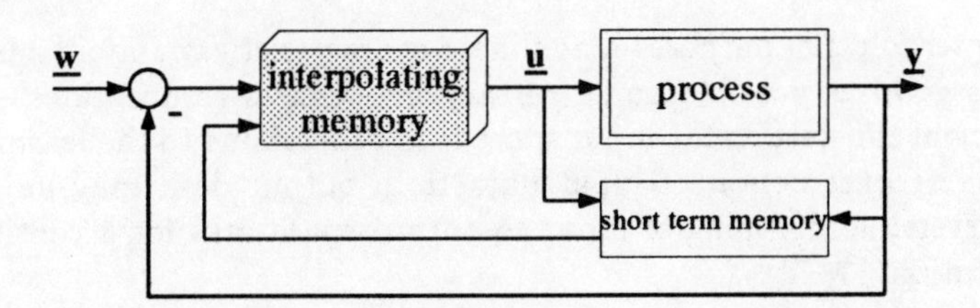


**Figure4.** Off-line trained, on-line applied learned controller

application areas (e.g. the automatic take over of operator strategies). A respective discussion can be found among others in [26].

# 2 Interpolating Memories

## 2.1 Neural Networks

As already mentioned neural networks[1] are today the mainly considered tool for constructing an interpolating memory.

Actually, two different tasks can be performed by neural networks

- interpolation and
- classification.

The reason, why both tasks can be performed by this tool is, that in both cases the real world is represented by a reduced set of data. However the intentions of interpolation and classification are quite different.

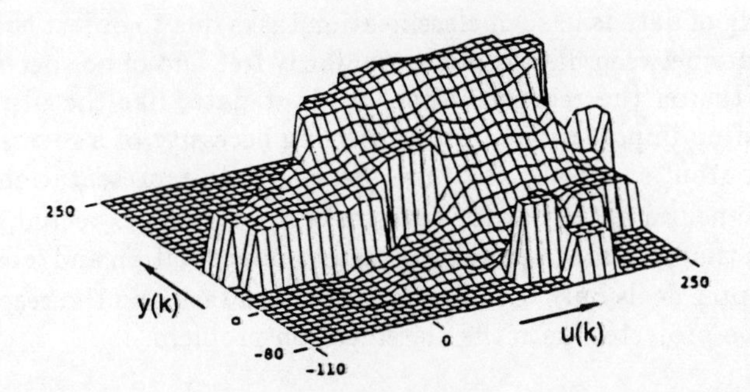


**Figure5.** Learned interpolation of a non-linear surface in $\mathbb{R}^3$.

[1] There exist different kinds of neural networks. One of the most frequently used models will be discussed in section 2.2 and is depicted in fig. 8.

For interpolation purposes a nonlinear characteristic manifold has to be presented as good as possible under certain smoothness assumptions – see fig. 5. Clustering of the data in the input space is in general due to the learning scheme. Furtheron reconstruction of input data from output data may be of interest. Such a reversible mapping of input to output space asks for a continuous data representation.

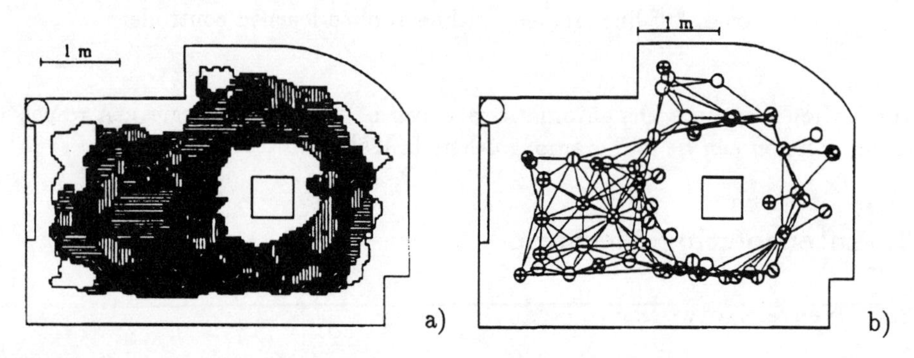


**Figure6.** Situation classification by ultrasonic sensor data – a) – to build up a free space graph – b) –for a mobile robot.

In classification one wants to reduce the data to a handable amount of discrete marks for decision and planning purposes. Fig. 6 gives a respective example of building up a graph representation of free space from ultrasonic sensor data generated by a mobile robot with a curiosity component. The sensor data are classified into general situations by a neural net – here a Kohonen map – so that e.g. all corner areas are marked in the same sense, namely being white, and can by that be condensed to similar situations in the graph representation used for trajectory planning (for details see [15]).

Clustering of data is in such classification tasks due to object characterisation and the border between data is in general fairly free and of no special significance – fig. 7. Furtheron the reconstruction of input data, like the ultrasonic sensor pattern, is of no importance and that missing necessity of a reversibly mapping of input to output space allows to use discrete data representations.

Although neural nets work for both tasks it has to be expected, that different models from this area may be best adapted to interpolation and/or classification.

This chapter deals only with the interpolation task and the respective results should not be projected onto the classification problem.

## 2.2 Other Approaches and Comparison

For automatic interpolation of multidimensional manifolds based on scattered inputs four main approaches can be distinguished, two neuronally inspired –

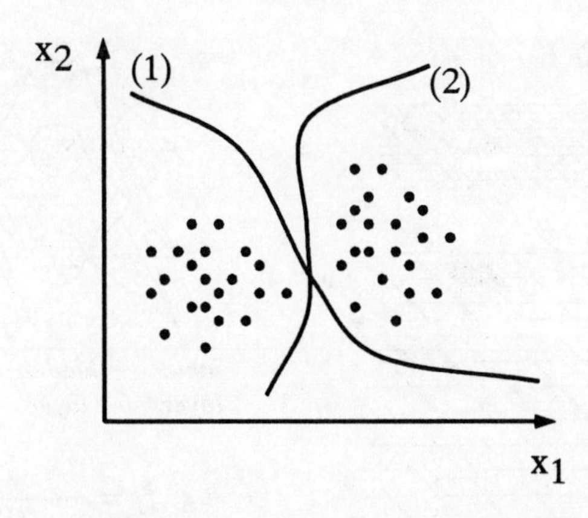


**Figure7.** For classification exact borders are often unnecessary (example).

CMAC/AMS with locally defined supporting functions and Feedforward Networks with globally defined sigmoid non-linearities and backpropagation error correction – and two of them mathematically inspired – MIAS, employing locally defined planes for approximation of the manifold and globally defined basic polynomials, appropriately weighted.

CMAC/AMS – CMAC being the name given in [2] to the method, AMS being the name for a very effective implementation of the method in [5] – starts from the observation, that in a digital implementation one has always some minimal resolution $\epsilon_i$ for an input $s_i$, so that one gets in the input area given by $[s_{i\,min}, s_{i\,max}]$ basic hypercubes with the lateral sizes $\epsilon_i$, $i = 1, 2...n$. Now instead of using these hypercubes as input space for each output value $p_j$, $\rho$ larger hypercubes with the lateral sizes $\rho\epsilon_i$ are used, shifted against each other by $\epsilon_i$ and being connected with output values $p_j \,/\, \rho$. By adding up the respective output values over the input space, one gets graded pyramids of height $p_j$ as supporting functions – see fig. 8 a with two inputs $s_1 = y(k)$, $s_2 = u(k)$, $p = \hat{y}(k+1)$ and $\rho = 3$ – which are interpolating if their results are added up for nearby values, but do not give any influence far away. Fig. 5 is the result of such an interpolation with $\rho = 16$. For further details see e.g. [27].

Fig. 8b depicts a feedforward neural net with one hidden layer, in which as well as in the output layer the weighted inputs $\underline{w}_j^T \underline{x}$ are sent over the globally nonzero sigmoid non-linearity $f_j$. Training this net with given input-output pairs means to change the weights $w_{ji}$ by a backward gradient error distribution. Details can be found e.g. in [16].

MIAS (McLain-type Interpolating Associative Memory) was developed in 1986 by J. Militzer, using ideas from surface representation in computer graphics, to cope with the problem, that in the usual interpolating memory approaches the interpolating scheme is independent of the local density of input data. Fig. 9 presents the main ideas of this method: for a certain point $\underline{s}$ in the input space,

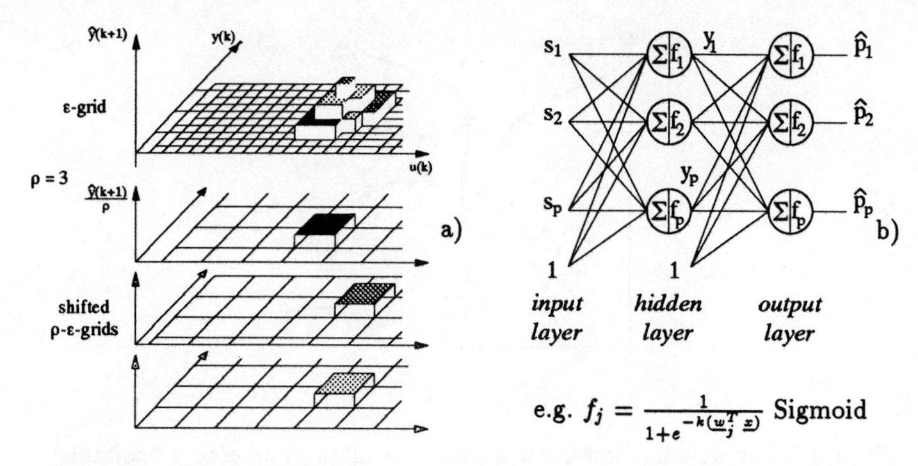


**Figure8.** CMAC/AMS-scheme – a) – and feedforward net structure –b).

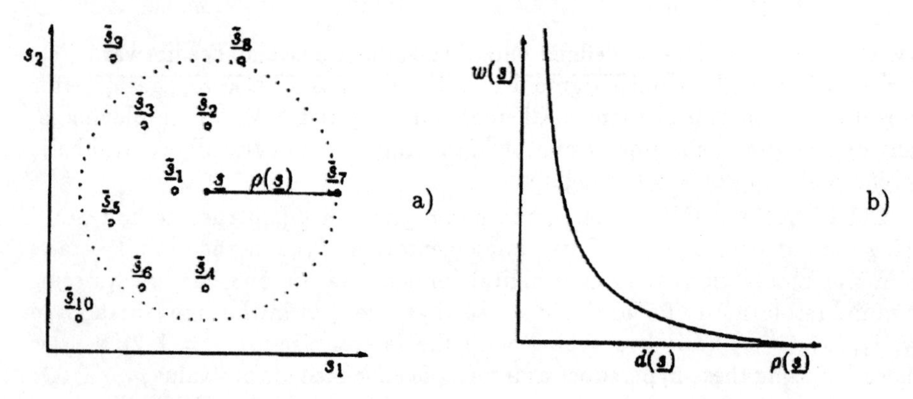


**Figure9.** MIAS-scheme – a) – and point weighting therein – b).

for which the output answer $p$ shall be generated, the $q+1$ nearest training points $\tilde{s}_l$[2] are searched for. Through the outputs $\tilde{p}_l$ of these points a balancing hyperplane is put by $p(\underline{s}) = a_0 + \sum_{k=1}^{n} a_k s_k$, the balancing being achieved by calculation of the coefficient vector $\underline{a}$ via the performance criterion

$$J = \sum_{l=1}^{q} w_l \left[p(\tilde{\underline{s}}_l - \underline{s}) - \tilde{p}_l\right]^2 \Rightarrow \min_{\underline{a}} \tag{2}$$

with weighting factors $w_l$ giving higher priority to the values of points nearer to $p_l$ than to points further of: $w_i = [\frac{1}{d_l(s)} - \frac{1}{\rho(s)}]^2$ – $d_l(s)$ = distance between $\underline{s}$ and $\tilde{\underline{s}}_l$, $\rho(s)$ = maximum considered distance (distance between $\underline{s}$ and $\tilde{\underline{s}}_{q+1}$). Results achieved with this scheme are presented again e.g. in [27].

[2] The ˜ are indicating a renumeration of the inputs in relation to the considered point $\underline{s}$.

Naturally, manifolds can be approximated by a weighted superposition of polynomials of different degree, the problem being herein to select a manageable number of elements from

$$p = \beta_0 + \sum_{i=1}^{n} \beta_i s_i + \sum_i \sum_j \beta_{ij} s_i s_j + \cdots + \sum_i \cdots \sum_q \beta_{i \ldots q} s_i \cdots s_q \tag{3}$$

Respective automatic approaches have been given firstly in [11] and recently in [14].

| | neuronally inspired | | mathematically inspired | |
|---|---|---|---|---|
| | CMAC/AMS | Sigmoid net with backprop. | MIAS | Polynom approximation |
| training indication | yes | no | yes | no |
| response time | ≈ const | O(r) | O(ν)* | $O\left(\frac{(n+q)!}{n!\,q!}\right)$ |
| calculation effort | low | medium | medium* | high |
| convergence velocity | high | low | high | low |
| variable generalization | no | no | \|\|yes\|\| | no |
| noise filtering | possible | possible | difficult* | possible |
| tolerance ⇒ part. destruct. | high | high | high | low |

**Table1.** Qualitative comparison of basic interpolation approaches, (r = number of neurons used in the neural net, ν = number of trained points, n = number of inputs, q = highest polynomial degree; * properties to be discussed (modified in III).

A discussion of relative merits of these basic four schemes can be found in [19] from which the qualitatve comparison of table 1 is taken[3]. In this table training indication means the possibility to know, whether nearby to a considered point in input space already enough information has been gained, that the produced output value can be considered as reliable or whether this is not the case. Variable generalization is the above mentioned property of MIAS, to interpolate over a small area, when a lot of information is available near the considered input point and to interpolate over a broader area (eventually limited by a maximum), if only by this measure enough supporting points can be found. E.g. in control loop applications much information is (training points are) available for a certain set point, but relatively few information for transitions between set points ([17]). The other categories in table 1 do not need further explanations and are not discussed here.

[3] Some further discussion including B-splines as supporting functions for interpolation can be found in [9]. Therein the close relationship between CMAC/AMS, B-spline interpolation and certain fuzzy membership function approaches is pointed out, also.

Only some general remarks may be of interest: Feedforward networks with sigmoid non-linearities and the polynominal approximation are fairly sensitive to changes and need therefore a high learning effort, due to the fact, that the sigmoids and polynomials are non-zero everywhere, which means, that a slight change due to new training information influences not only the nearby system answers but all system answers for the whole input space. This is not the case for CMAC/AMS and MIAS working with just locally defined supporting functions, which leads to clearly much less training effort for the same quality of system answers[4].

For neural networks the problem of sensitivity and learning effort can be reduced by using radial basis functions instead of sigmoids as non-linear threshholds [5]. However, the right number of hidden units has to be determined as a major design effort still – e.g. by trial and error – since the free parameters of CMAC/AMS and MIAS have a direct phyical meaning, so that they can be chosen on the basis of knowledge about the considered application.

### 2.3 Developments Considered

Since in the last decade mathematically inspired approaches to automatic interpolation for scattered data in input spaces extending further than $n = 2$ have got less attention than neuronally inspired approches, the article concentrates on the mathematically inspired methods and – due to the sensitivity problem therein – on locally supported manifold representations.

Actually two recent developments will be put forward together with respective application areas:

- MIAS modifications to reduce response times, improve the variable generalization through avoiding extrapolations and to allow noise filtering (see [8]).
- MIAS simplifications to get a new method for very fast access and small memory requirements for applications with as well very high frequency sampling as limited computational power and memory ([25]).

---

[4] In [3] it was recently pointed out, that Feedforward Neural Nets with Back Propagation have certain inherent advantages in reachable accuracy for dimensions $n > 2$ vis-a-vis approches like CMAC/AMS with their fixed supporting functions due to the fact, that the supporting functions in the Feedforward Neural Nets are adapted, too. However, this is a theoretical advantage not directly linked to the practical training effort. Actually, required manifold surfaces are i.g. much faster modelled by CMAC/AMS than e.g. by Feedforward Neural Nets with sigmoids – see [19] and/or [27].

[5] Details about this approach can be found e.g. in [4], as well as further discussions on interpolation with locally defined supports.

# 3 Mathematical Interpolation in a Biotechnological Application

## 3.1 Control of Biotechnological Processes

Comparing biotechnological fermentations with technical devices one can state, that in the biotechnological case it is much more difficult to find a satisfactory mathematical model. Although the general elements and basic course in a biochemical process is mostly well known as also the elementary behaviour, like e.g. Monod-kinetics, there exist strong couplings and many not exactly known parameters, which make modelling uneffective and uncertain. Therefore, the number of reliable models in this area is very limited.

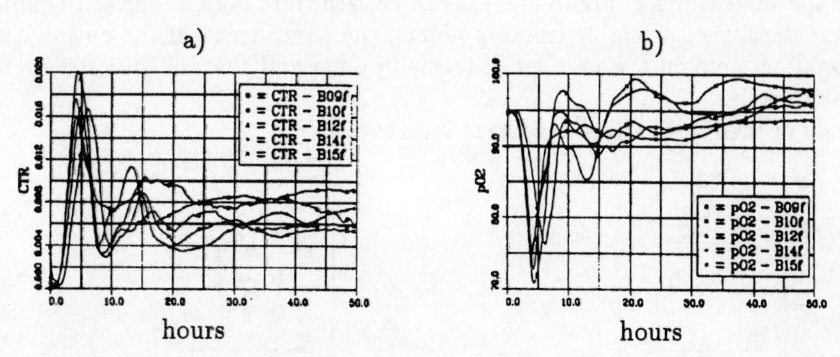


**Figure10.** CTR-and $pO_2$ courses for seemingly similar initial conditions and cultures.

Furtheron, one has to fight with difficult measurement problems. This is as well due to missing appropriate sensors for on-line measurements as due to the sensitivity of living bacteria and/or cells to interventions from the outside: The respective cultures can easily be poisoned. This leads to the fact, that one has to live with limited and/or rare information about the actual process status and especially about the detailed culture composition: The behaviour of apparently identical cultures is fairly different over time. Fig. 10 shows this for the here considered process example, the production of $\alpha$-amylase by bacillus subtilis. $\alpha$-amylase finds its industrial application in starch hydrolysis. The measurable characteristic describing the cell growth are CTR, the $CO_2$-transfer rate and $pO_2$, the oxygen partial pressure. Respective measurements were taken each minute. Although the initial conditions and cultures used were seemingly identical for all five fermentations B09f, ..., B15f, one gets different CTR-courses (fig. 10a) and/or $pO_2$-courses (fig. 10b) over the fermentation time of roughly 50 hours.

On the basis of this background, two problems exist

- efficient on-line process prediction to allow adequate control actions to be taken

– automatic process model generation for off-line set point profile optimization.

Both problems can be successfully attacked by storing different experienced process behaviours in an interpolating memory as a predictive process model. In the following results on process behaviour prediction with $\alpha$-amylase production in a 20 l fermenter and MIAS as interpolation unit will be shown[6]. For results on set point profile optimization, details of the experimental set up and on reached performances and/or a discussion of further aspects to improve fermentation control see [8].

### 3.2 MIAS Modifications

The special features of MIAS are – as can be seen from table 1 – its self-adaption to the density of available training points, the dependence of the output value generation time on the number of trained points and the difficulty, to swallow noise directly.

All of these properties have been improved in [8].

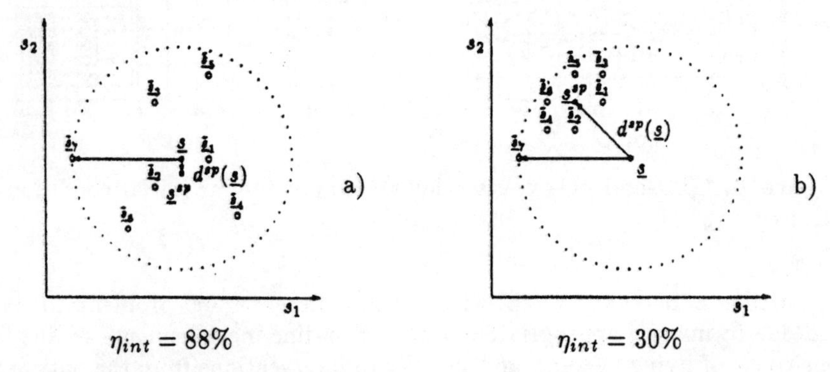


**Figure11.** MIAS-example, two inputs, $q+1 = 7$ with different supporting point distributions giving different $\eta_{int}$ values.

Actually, the interpolation scheme of fig. 9 may fail, if the points used to compute the interpolating plane are distributed in a wrong manner. Fig. 11b shows such a case: All respective points are crowded in a small area of the hypersphere through $\tilde{\underline{s}}_{q+1}$ around the considered point $\underline{s}$ – compare 2.2 for the respective meaning –, so that one extrapolates by using these points instead to achieve the interpolation aimed for. This problem was solved by introduction of a second "training indicator" The original training indicator from table 1 is a mean to ensure locality. Interpolation by a balanced plane is inadequate in case

[6] The fermentations have been performed by the Institute of Biochemistry of the Technical University Darmstadt

of strong curvature and fairly distant supporting points. Therefore some $\rho_{max}$ has to be specified for the radius $\rho(\underline{s})$ of the hypersphere around $\underline{s}$. Militzer did this by requesting

$$\eta_{loc} = \frac{200}{200 + \rho(\underline{s})} \geq \kappa_{loc}\% \text{ (e.g. } \kappa_{loc} = 60) \tag{4}$$

which restricts $\rho(\underline{s})$ to $\rho(\underline{s}) \leq 200(1 - \kappa_{loc})/\kappa_{loc}$ with $\kappa_{loc}$ being a parameter for adaptation to the expected curvature of the considered non-linear manifold. The newly introduced second training indicator, to be fulfilled in addition for acceptable interpolation reads:

$$\eta_{int} = 1 - \frac{d^{c.g.}(\underline{s})}{\rho(\underline{s})} \geq \kappa_{int}\% \text{ (e.g. } \kappa_{int} = 70) \; . \tag{5}$$

In this formula $d^{c.g.}(\underline{s})$ represents the distance of the center of gravity of the $q$ points used to calculate the parameters of the interpolating plane, so that $\eta_{int}$ guarantees, that this center of gravity is not too far off from $\underline{s}$. As one can see from fig. 11, one gets as to be expected a high value for $\eta_{int}$ for a fair distribution of $\underline{\tilde{s}}_1, \underline{\tilde{s}}_2, \cdots \underline{\tilde{s}}_q$ and a low value for an unacceptable distribution (large extrapolation).

To avoid a fast growing of points to be searched through to find the $q + 1$ nearest neighbours $\underline{\tilde{s}}_i$ to the considered point $\underline{s}$ in the input space, furtheron some $\rho_{min}$ was introduced, some small hypersphere, in which all input values are considered to be the same. This means, if $|\underline{s}_k - \underline{s}_i| < \rho_{min}$, the output value of $\underline{s}_k$ is attributed to $\underline{s}_i$ as a new value and the list of trained input-output values is not extended by the introduction of a new point.

Adding a counter to the information stored with each point and using the output value updating rule:

$$p_i^{new} = \frac{p_k + \nu_i p_i^{old}}{\nu_i + 1} \tag{6}$$

$\nu_i$ being the value in the counter for $\underline{s}_i$, one gets as well a limitation of the trained input-output relationships to significantly different input situations as a noise filtering procedure: White noise in the output values $p_i$ will be filtered out by the above procedure over time, if similar situations are encountered frequently.

### 3.3 Results with MIAS and Sigmoid Feedforward Networks

Naturally, the addition of $\eta_{int}$, $\rho_{min}$ and a counter for each $\underline{s}_i$ increases the computational effort and the memory places required for each individual MIAS response calculation, although the general computational effort and memory requirement is reduced by the reduction of inputs accepted as being different. But such questions do not play any role in the considered application, since sampling times of some minutes, being representative for chemical and/or biochemical process control, are excessively higher, than the answering time of automatically interpolating devices[7]. So sampling times are here no limiting factor to apply to

[7] The response calculation for a selected configuration of $q = 100$ points is less than 200 msec for MIAS (SUN/SLC Workstation).

MIAS in any form.

However, results of research should comprise in general as well new approaches and evidence of their successful application as some comparison with other, frequently used methods. Therefore in the work on process control for $\alpha$-amylase production ordinary sigmoid feedforward networks with back propagation were used in addition to MIAS and CMAC/AMS.

As a means of comparison two quality measures[8] were used

- $E_{max}$,the maximal error in % and

- $E_{bm}$,the mean absolute error in % between predicted and real output values.

Actually, trajectory prediction from two different predictive process models will be used here to illustrate the result, which is generally applicable.

As predictive process models the following two give both acceptable results ($k$ = sampling instant: $t = k \cdot T_0$ ; $T_0$ = sampling time = 30 minutes):

$$1) \left\{ \begin{array}{l} pO_2(k) \\ pO_2(k-1) \\ CTR(k) \end{array} \right\} \rightarrow \left\{ \begin{array}{l} pO_2(k+1) \\ CTR(k+1) \end{array} \right\}$$

$$2) \left\{ \begin{array}{l} pO_2(k) \\ pO_2(k-1) \\ CTR(k) \\ CTR(k-1) \end{array} \right\} \rightarrow \left\{ \begin{array}{l} pO_2(k+1) \\ CTR(k+1) \end{array} \right\}$$

a)

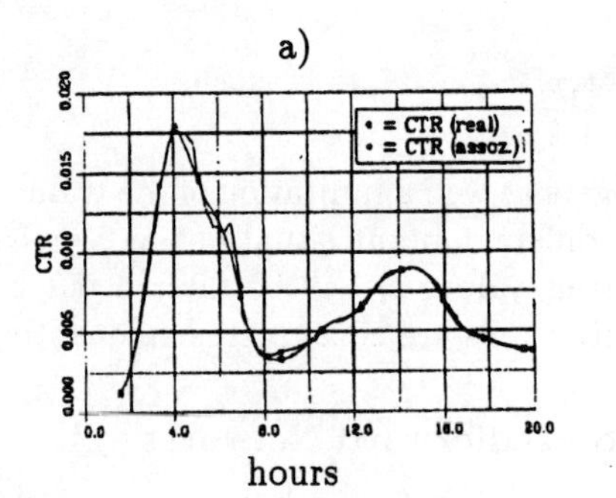


b)

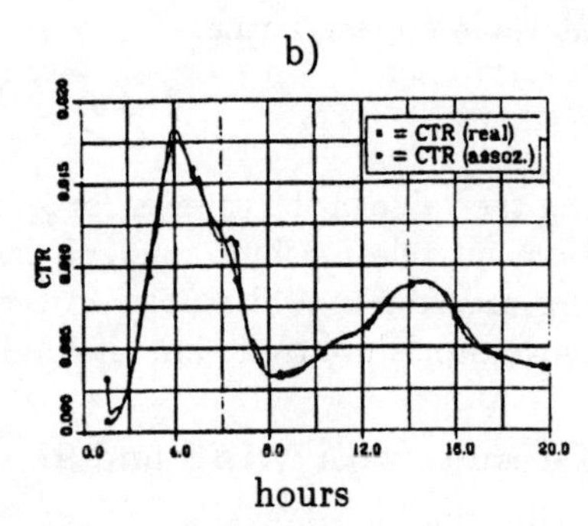


**Figure12.** MIAS prediction of CTR for a) model 1 and b) model 2, first 20 hours (most dynamical phase).

The procedure used was: Four of the courses from fig. 10, B09, B10, B12, B15 were trained into the interpolating memories, the fifth one, B14, was predicted stepwise, that means the expected values at $k+1$ were predicted on the basis of the measured values at $k$, $k-1$. Fig. 12 is showing graphically the results

[8] For a precise mathematical definition see either [8] or [27].

achieved with MIAS for CTR and the most dynamic phase (first 20 hours). One cannot see much difference. Actually, the more complicated model 2 gives slightly less good predictions, as can be seen from table 2.

| | MIAS | | | | Backprop. Net | | | |
|---|---|---|---|---|---|---|---|---|
| | $pO_2$ | | CTR | | $pO_2$ | | CTR | |
| | $E_{max}$ | $E_{bm}$ | $E_{max}$ | $E_{bm}$ | $E_{max}$ | $E_{bm}$ | $E_{max}$ | $E_{bm}$ |
| model 1 | 4.49 | 0.38 | 13.6 | 2.70 | 11.0 | 1.01 | 21.4 | 6.18 |
| model 2 | 6.24 | 0.41 | 15.0 | 2.82 | 10.9 | 1.48 | 21.6 | 8.55 |

**Table2.** Performance of MIAS and BackpropNet for model 1 and model 2.

In this table results are given taking into account the whole courses (50 hours), the prediction of $pO_2$ and CTR and the two different approaches MIAS and Backpropagation Feedforward Sigmoid Net. The respective parameters have been for MIAS $q = 80$, $\eta_{loc} = 40\%$, $\eta_{int} = 10\%$, $\rho_{min} = 12 \cdot 10^{-4}$, $\eta_{int}$, $\eta_{loc}$ not being very important due to the fact, that the respective manifolds have a limited curvature and that the training points do not tend to give extrapolation situations in the considered case (experience). For the backpropagation net the structure was determined by a trial and error procedure to have two hidden layers with 7 neurons in the first one, 5 neurons in the second one. Not considering this design effort, to get the achieved results, the mere training effort was for the Feedforward Sigmoid Net at least 10 times higher than for MIAS. Nevertheless are the interpolation results in the average by 100% worth than the results achieved with MIAS.

In conclusion one can state, that the mathematically motivated interpolating memory MIAS is for chemical and/or biochemical applications clearly to be preferred to the frequently used Feedforward Nets with sigmoid threshholds and backpropagation error adaptation. The modifications made in [8] did remove open points in the MIAS design, so that MIAS seems to be the best choice, if the possible number of training situations can be limited and moderate calculation times and memory sizes are at hand.

# 4 Characteristic Manifolds for Automotive control

## 4.1 Problems in Automotive Control

Modern cars are integrating more and more electronic support to improve performances. However, at least up to now two basic opinions restrict the hereby given possibilities:

- firstly, new additional sensors are mostly considered to be eventually unreliable and too expensive for standard cars with the consequence, that only limited online measurements are available in general;

- secondly, high performance computers are considered as an unnecessary, too costly device for standard cars, so that on-line computation of somehow sophisticated control algorithms is not possible.

But pre-compensation of certain nonlinear behaviour is already state of the art and is performed by feedforward control using two input - one output preset characteristic fields with a regular grid support - similar to a look up table for logarithms - which allows very quick and easy interpolation. Application areas are e.g.

- for the motor: fuel injection control, ignition control,
- for the car behaviour: antilock-breaking (ABS), traction control (ASR).

Although no learning capabilities are included, certain adaptation possibilities by shifting of the characteristic field as a whole through parameter changes are possible, partly, mainly to take into account unconsidered influences, like air pressure change during hill climbing.

However, higher dimensional characteristic fields are of interest – e.g. to include up to now neglected dynamics – and also learning (in view of differences in individual cars and component ageing). Actually, this has not been taken up seriously due to the following reasons:

- the number of supporting points grow exponentially with the dimension in case of support on regular grids,
- there exist no detailed investigations concerning the raise in computing time and/or necessary memory places and the practical improvements bought by this additional effort;
- the current simple algorithms for characteristic field adaptation have to be reconsidered: they may become very complicated.

To address this problem a research project has been started at the Institute of Control Engineering, Department of Control Systems Theory and Robotics of the Technical University Darmstadt in 1990 – suggested and supported by the Robert Bosch GmbH, Stuttgart – from which reflections and results of the next section stem.

## 4.2 Characteristic Manifolds for Limited Calculation Effort

A basic requirement for considerations on improved non-linear feedforward characteristic field control is, not to deviate too much from currently used hardware. That means, that the necessary response computation time limitation of $1msec$ – 6000 rotations/minute of a motor means 1 rotation in $10msec$ – has to be achieved without fancy computers and that not more than 100 to 150 data points should be necessary for the characteristic field representation in view of RAM-space limitations[9].

[9] RAM = Random-Access Memory

These restrictions made it necessary to deviate from the CMAC/AMS and/or MIAS-approaches, where computational effort and memory space were not design drivers, and to look for minimal effort solutions, being derivable most efficiently by mathematical reflections, so that the result may be considered as a highly simplified MIAS-approach.

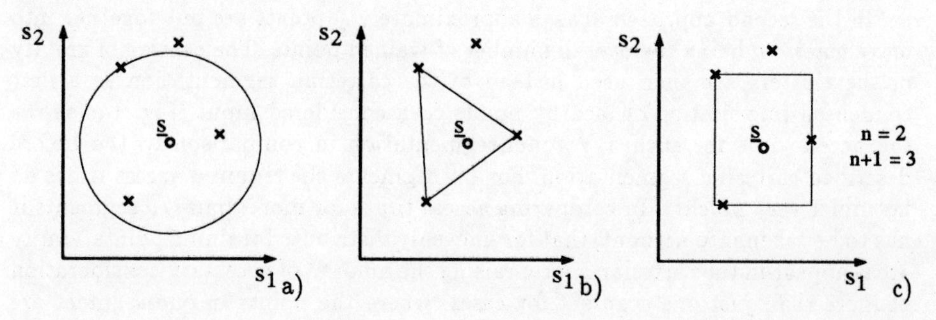


**Figure13.** Comparison of different approaches to find the supporting points for an interpolating plane to estimate p($\underline{s}$); a) sphere, b) simplex, c) cartesian hull

A first step was to go down from the $q+1$ outputs to compute a balancing hyperplane to the minimal required points to represent a hyperplane. These are in a $n$ input - one output $\mathbb{R}^{n+1}$-space just $n+1$ points. Furtheron to avoid extrapolation and to achieve as well fast computation times, the sphere used in cartesian space was changed to a cartesian hull, the obvious solution of using an enclosing simplex being rejected due to the necessary computational effort – see fig. 13 for a pictorial representation in the two input case.

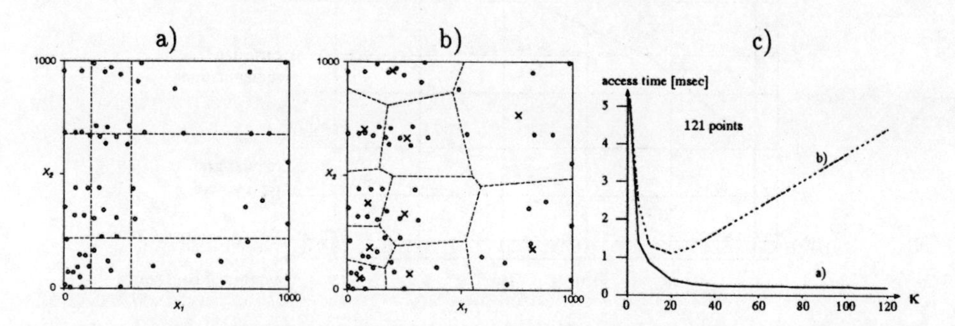


**Figure14.** Segmentation schemes and access times belonging to them: a) cartesian segmentation, b) Voronoi segmentation, c) access time over the number of segments $\kappa$.

To reduce the time to search through all training points to find out the $n+1$ nearest points for creation of the cartesian hull a two step hierarchisation scheme is advantageous. Actually two different approaches have been considered – fig. 14 –.

The first approach is using a cartesian segmentation, trying to enclose the training points into subsegments of the input space, which contain all roughly the same number of training values, by hyperplanes orthogonal to the n coordinates spanning the input space.

In the second approach always approximately $\frac{\nu}{\kappa}$ points are put together into one cluster – $\nu$ being the overall number of trained points. The centers of gravity of the clusters are then used instead of the cartesian segmentation as a first condensed information on nearby points to a considered input. Fig. 14c shows the access time for such a Voronoi-segmentation in comparison to the before described cartesian segmentation. For $\sqrt{\nu}$ segments the required access times do not differ very much[10]. In comparing access times for more than $\sqrt{\nu}$ segments it has to be taken into account, that for unevenly distributed training points, empty areas appear in the cartesian space, raising the amount of necessary consideration of more than just one segment for cases, where the points in one segment are not enough to calculate the cartesian hull of fig. 14c for the considered input value. This reduced efficiency - growing with the number of segments due to the missing flexibility in the cartesian approach has led to the experience, that the Voronoi-segmentation with $\sqrt{\nu}$ segments seems to be the best hierarchisation approach in general.

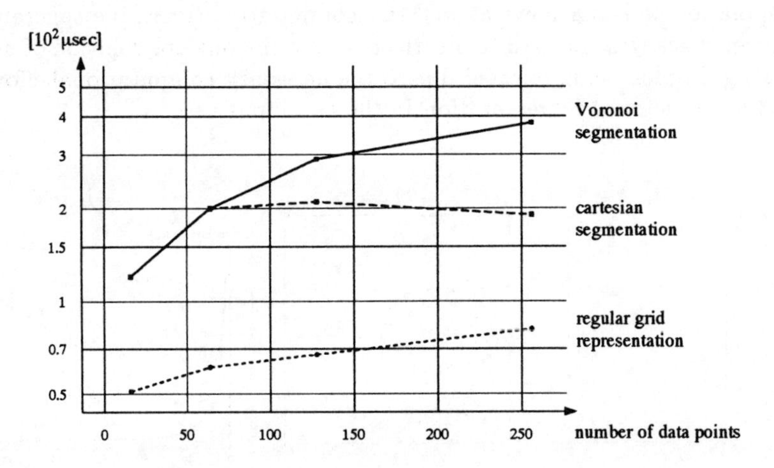


**Figure15.** Access (response generation times) for $n = 2$, $\sqrt{\nu}$-segmentation on scattered data and for regular grid representation.

[10] If a learning strategy is used that changes not only the output values of the stored datapoints, but also their locations, the segmentation is also updated, in order to maintain short access times, an effort not being included in fig. 14c, since this does not take place continously on-line.

Fig. 15 shows for real motor data used in traction control – see [24] – the necessary access and/or response calculation times to get answers in characteristic fields with two inputs for regular grid data fields as used today and scattered data fields with $\sqrt{\nu}$ segments in dependence of the number of training points. One can see, that with actually used hardware on-line access times of less than $0.5msec$ can be reached, while regular grids are better by a factor 3, so that on one hand at least for non-locally-adaptable fields and two inputs the regular grid is the simplest and by that the fastest solution, but on the other hand scattered data fields can be considered as a possible solution, too.

By looking at the potential of scattered data characterisation of characteristic fields one has, however, furtheron to include specific properties of scattered data support, which changes the situation in favour of scattered data already for $n = 2$, but especially for $n > 2$.

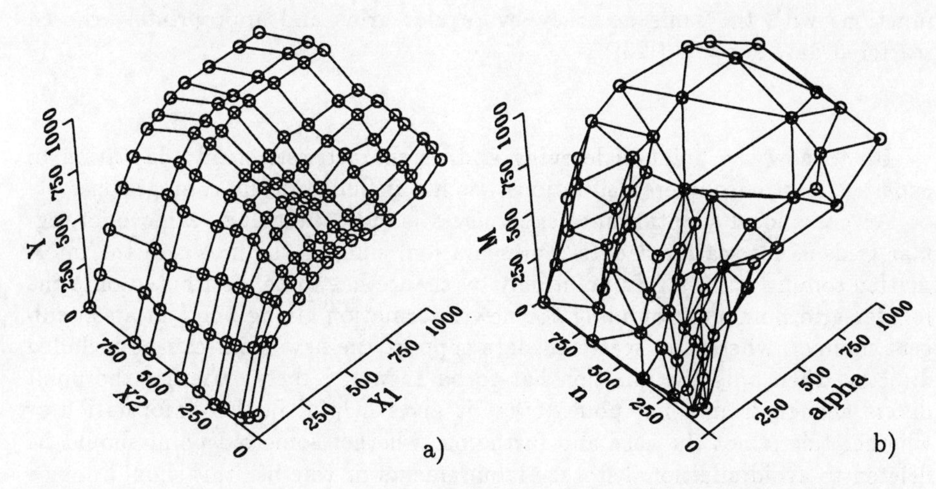


**Figure16.** Real motor torque data; a) regular grid support needing 120 points, b) scattered data support needing 64 points to achieve the same mean accuracy (same $E_{bm}$)

Fig. 16 points out, that mostly characteristic fields can be described by scattered data supports with much less supporting points as with regular grid supports achieving the same mean accuracy. Here the real motor data used for fig. 16 are represented by scattered inputs with practically half of the necessary supporting points on a regular grid.

For the assumptions and formulas, on which fig. 15 and 16 are based as well as on further approaches with even better results the reader is referred to [25].

The advantages of scattered data support raise dramatically with the dimension $n$. Table 3 shows the amount of datapoints that are needed to store a multidimensional sigmoidal function. The number of datapoints is the most interesting fact, because if the output values of the datapoints have to be updated to compensate a changed process-behaviour, at least one memory word in the expensive RAM is needed for each datapoint. For $n = 2$ the reduction factor is 0.56 while for $n = 4$ less than 10 % of the datapoints used in a lattice-like datafield are necessary.

| n | 2 | 3 | 4 |
|---|---|---|---|
| Regular grid | 36 | 216 | 1296 |
| Scattered Data | 20 | 71 | 103 |

**Table3.** Number of data points necessary to represent $n$-dimensional sigmoid functions with the same accuracy by regular grids and appropriately chosen scattered data support ([25]).

Up to now the point of learning and/or characteristic field adaptation to experiences or encountered situations was not included in the comparisons.

Actually, local adaptation to experiences is more difficult to achieve for regular grids as for scattered data representation, since in the first case the information coming up at a grid point only by chance has to be distributed onto the regular grid points surrounding the new information giving point in an intelligent manner, whereas in scattered data support the new point can be included directly. Then only the question has to be answered there, whether the point differs enough from other points, that it gives helpful further information or whether this is not the case and furtheron, whether some old point should be deleted to avoid additional storage requirements or whether this should not be done and, in case that some point has to be deleted, which point should be chosen for deletion.

Adaptation to encountered situations may be handled globally, that means e.g. by shifting the whole characteristic field. For example for a sensor calibration error, a field shifting could be the correct counter-measure. In this case no differences exist for regular grid supported and scattered data supported characteristic manifolds.

In connection with scattered data supported manifolds two strategies have been developed to deal with new information, the strategy of "Structure Adaptation of the Nearest Datapoints – SAND" and the strategy of "Minimum Information Loss Learning – MILL". In both cases

– all data used are assumed to be prefiltered (on-line noise elimination);

and the adaptation is

– restricted to stationary phases (to be detected automatically, see [25]).

SAND imitates the proceeding necessary in regular grids by not storing the new point but distributing the information onto the $n+1$ points taken into account for the cartesian hull in case of interpolation for the considered point. They are corrected at the same time with an amount according to their cartesian distance in such a way, that they give afterwards the right interpolation output, that means the output given by the point used for adaptation. This strategy gives good results in case of not too strong changes of the stored process characteristic. A drawback is that oscillations may turn up in case of frequent changes between the same two stationary phases, as can happen with air pressure influences in a hilly landscape.

In MILL a prefixed number of points $\kappa$ is considered and individual points are substituted by new points. To avoid too small corrections as well the distance of the point used for learning to the nearest of the existing supporting points $d_{min}$ is considered as the amount $\Delta p$ of change between the output value, which would be found by interpolation for this point and the output value of the point used for learning. If $\Delta\varphi = \Delta p/d_{min}$ is greater than a given threshold $\Delta\varphi_0$, than the nearest point is substituted by the new point with its output value.

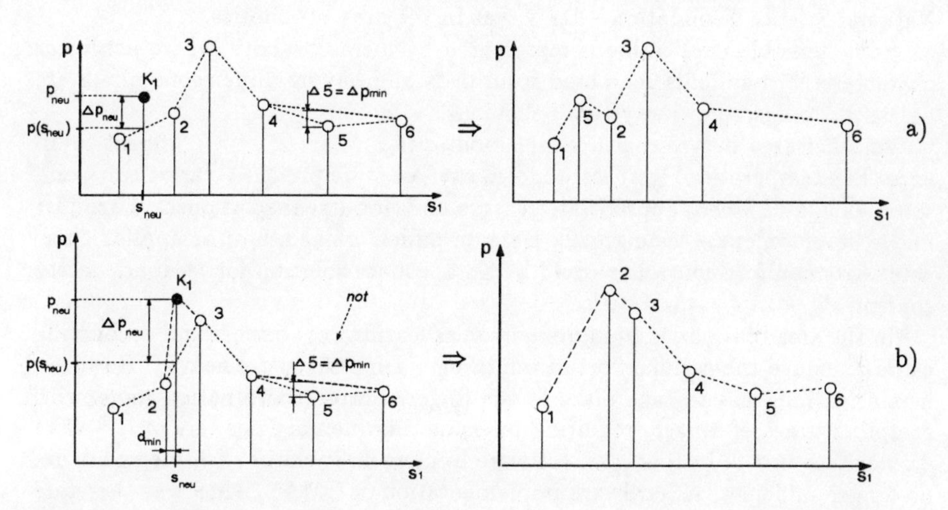


**Figure17.** Sketch of MILL-scheme for n = 1: a) some old point is deleted in view of the new information, b) only a correction of point 2 takes place.

If $\Delta\varphi \leq \Delta\varphi_0$, then for all the considered $\kappa$ points the change $\Delta p$ is calculated for each individual point, if this point is deleted and the respective output value is interpolated from the cartesian hull using the appropriate remaining points. Now $\Delta p_{min}$ is taken: If this minimal value is the value of the point used for learning, the respective information is deleted. If this minimal value is resulting from some other point, this point is deleted and the point used as learning candidate put into the list of supporting points. The procedure is depicted for an input dimension of $n = 1$ in fig. 17. It is especially suited for strong local changes, e.g. due to component ageing. However, for its application it has to be taken into account that it needs more calculation effort than SAND.

In practical applications one would use both procedures, SAND and MILL, selected on some decision mechanism. Also it would have to be considered not to work with an adaptable characteristic manifold only but with a feedforward control signal generated by adding up the output of a basic fixed characteristic manifold and some changes incorporating second adaptive characteristic manifold to as well reduce the learning effort as improve the system safety.

# 5 Outlook

## 5.1 Further Applications and Realisation Aspects

There exist surely a lot of application areas for scattered data supported characteristic manifolds. Here only some from the area of mechatronics shall be mentioned due to the fact of the engagement of the Department of System Control Theory and Robotics as well in the "Sonderforschungsbereich 241 – IMES – Integrated Mechanical and Electronical Systems", supported by the German National Science Foundation – DFG – as in the area of robotics.

Some possible application is represented by memory-alloys, where nonlinear characteristic manifolds generated from tests and having dimensions $n > 2$ are necessary to describe the system behaviour.

Another area is tyre-road contact monitoring, where a lot of different influences like temperature, wetness, road coarseness, tyre pressure, car velocity and so on exist and where appropriate sensors and non-linear relationships are just under development – see e.g. [22]. Here an offline trained, on-line applied characteristic manifold controller could be an adequate solution for feedback safety control.

In the area of robotics from mechatronics learning of stiffness and decoupling can e.g. reduce the design effort in multifinger gripper control heavily. However, herein learning has to take place in the finger control/coordination loops with sampling times of $1msec$, requiring interpolating memory answers of less than $0.1msec$ for $n \geq 9$. In this case, software implementations of such memories are no longer sufficient. A hardware implementation of CMAC/AMS was therefore started – compare [20] –, the first experimental chips being due at the end of 1994.

Finally it should be mentioned, that learning control loops and by that automatically interpolating devices may also reduce the design effort for robot arm coordination, where stiffness parameters difficult to measure but learnable play an important role.

For realisation an important point is to include knowledge and/or basic requirements into the system design directly, avoiding to learn these aspects, too. An example of what can be gained in this way is given in fig. 18, which is concerned with the control of one finger of the experimental three-fingered robot gripper developed at the department – see e.g. [21]. Actually a non-linearily deformed PI-controller is best suited to give the wanted performance – [13]. If the

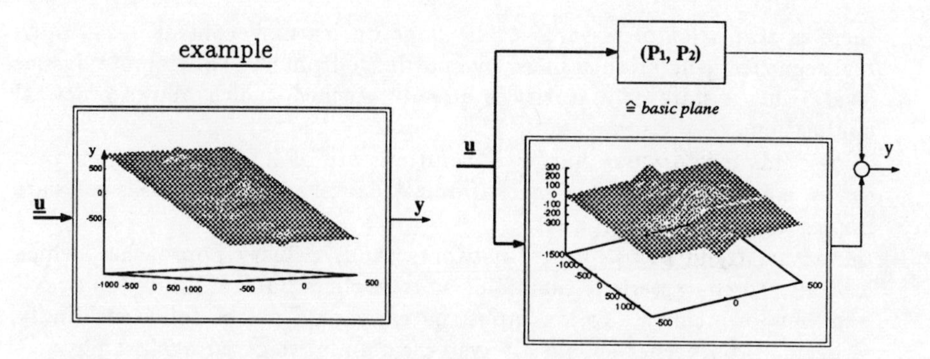


**Figure18.** Example of advantages in learning deviations from some basic characteristic only: finger control in a multifingered robot gripper: a) total learning, 50 training cycles; b) nonlinear deviation learning (basic characteristic = linear PI-controller), now 30 training cycles to achieve same accuracy (CMAC/AMS), suffice.

whole respective behaviour has to be learned, one needs 50 training cycles to reach a sufficient status (fig. 18a), if one learns only the deviation from a linear PIcontroller, 30 training cycles suffice, that means one can live with 60 % of the otherwise necessary training effort.

## 5.2 Conclusions and General Remarks

The Department of Control Theory and Robotics of the Institute of Control Engineering, Technical University Darmstadt, works now for 15 years in the area of autonomous systems including especially learning control and artificial intelligence methods to imitate the power and flexibility of human information processing. On the other hand also since more than 20 years appropriate conventional control engineering approaches to handle complex tasks – e.g. multivariable process control, robust control, optimization – have been investigated and further developed. This allows to judge on the relative importance of learning control using the herein discussed automatically interpolating memories. As a result, it can be stated:

- there exist a number of control problems in which learning seems to be advantageous, either by reducing the design effort or by supplying non-linear solutions in a computationally very efficient way.
- in low level control is the main problem of learning control automatic interpolation in high dimensional, non-linear manifolds, for high level control – like supervision and planning – is the main problem the classification (not dealt within this article).
- the selection of used interpolation techniques should be adjusted to the application area.

- there is still a lot of research to be done on learning control – e.g. optimal sequence generation assisted by knowledge from an artificial intelligence level – but a state of maturity is already reached, which allows practical application.
- some fields exist, where hardware solutions are of interest (very fast processes, e.g. from the field of mechatronics), however, in most areas software implementations are sufficient.
- the actual trend to use neural networks (and/or fuzzy approaches, which lead also to characteristic manifolds as controllers but are limited in praxis – in view of lucidity – to low input spaces) should not be followed blindly, respective approaches should always be carefully weighted against physico-mathematical solutions.
- especially problem constraints may be taken into account better by mathematically derived improvements of neuronally inspired approaches than by blind use of simple neural nets.
- in general combinations of learning approaches, conventional control methods and coded knowledge are the best choice to handle complex systems.

## Acknowledgement

The support of certain parts of the presented research work by the German Science Foundation - DFG - and by the Robert Bosch GmbH is greatfully acknowledged.

## References

1. J.S. Albus. *Theoretical and Experimental Aspects of a Cerebellar Model.* PhD thesis, University of Maryland, Maryland, 1972.
2. J.S. Albus. A new approach to manipulator control: The cerebellar model articulation controller. *Transactions ASME*, 97(3), 1975.
3. Andrew R. Barron. Universal approximation bounds for superpositions of a sigmoidal function. *IEEE Transactions on Information Theory*, 39(3):930–945, May 1993.
4. Martin Brown and Christopher J. Harris. *Neurofuzzy Adaptive Modelling and Control.* Prentice Hall, ISBN 0-13-134453-6, 1994.
5. E. Ersü and J. Militzer. Software implementation of a neuron–like associative memory system for control applications. In *2nd IASTED Conference on Mini– and Micro–Computer Applications – MIMI '82.* Davos, Switzerland, March 1982.
6. Enis Ersü and Jürgen Militzer. Real-time implementation of an associative memory-based learning control scheme for non-linear multivariable processes. In *IEE-Symposium: Application of Multivariable System Techniques*, Plymouth, UK, 31. Oktober - 2. November 1984.
7. P. Funk. *Variationsrechnung und ihre Anwendung in Physik und Technik.* Springer Verlag, 2 edition, 1970.

8. Stefan Gehlen. *Untersuchungen zur wissensbasierten und lernenden Prozeßführung in der Biotechnologie.* PhD thesis, TH Darmstadt, FG Regelsystemtheorie & Robotik, 1993. Fortschritt-Berichte VDI, Reihe 20, Rechnerunterstützte Verfahren, Nr. 87, VDI-Verlag, ISBN 3-18-148720-1.
9. C. J. Harris, C. G. Moore, and M. Brown. *Intelligent Control – Aspects of fuzzy logic and neural nets.* World scientific, 1993.
10. Rolf Isermann. *Digital Control Systems.* Springer, 1981.
11. A. G. Ivankhenko. Heuristic self-organization in problems of engin. cybernetics. *Automatica*, 6, 1970.
12. K. Kleinmann, M. Hormel, and W. Paetsch. Intelligent real–time control of a multifingered robot gripper by learning incremental actions. In *IFAC/IFIP/IMACS Int. Symp. on Artificial Intelligence in Real–Time Control.* Delft, The Netherlands, June 1992.
13. K. Kleinmann and R. Wacker. On a selftuning decoupling controller for the joint control of a tendon driven multifingered robot gripper. In *IEEE/RSJ International Conference on Intelligent Robots and Systems (IROS '94).* Munich, 1994.
14. M. Kortmann and H. Unbehauen. Ein neuer Algorithmus zur automatischen Selektion der optimalen Modellstruktur bei der Identifikation nichtlinearer Systeme. *Automatisierungstechnik (at)*, 1987.
15. A. Kurz. Building maps based on a learned classification of ultrasonic range data. In D. Charnley, editor, *1st IFAC Workshop on Intelligent Autonomous Vehicles.* Pergamon Press, Southampton, Southampton, UK, April 1993.
16. R. P. Lippmann. An introduction to computing with neural nets. *IEEE ASSP Magazine*, April 1987.
17. Jürgen Militzer and Henning Tolle. Vertiefungen zu einem Teilbereiche der menschlichen Intelligenz imitierenden Regelungsansatz. In *Jahrestagung der Deutschen Gesellschaft für Luft- und Raumfahrt*, München, 1986.
18. W. Thomas Miller III, Filson H. Glanz, and L. Gordon Kraft III. Application of a general learning algorithm to the control of robotic manipulators. *The International Journal of Robotics and Control*, 6(2):84–98, 1987.
19. W. S. Mischo, M. Hormel, and H. Tolle. Neurally inspired associative memories for learning control. A comparison. In *ICANN – 91, International Conference on Artificial Neural Networks.* Espoo, Finland, June 1991.
20. Walter Sebastian Mischo and Henning Tolle. Ein assoziativer VLSI-Prozessor zur schnellen Informations-/Stellsignalgenerierung. In *Fachtagung Integrierte mechanisch-elektronische Systeme*, number 179 in VDI Fortschrittsberichte, pages 263–278. VDI Verlag, 1993.
21. W. Paetsch and M. Kaneko. A three fingered, multijoined gripper for experimental use. In *IROS '90, Int. Workshop on Intelligent Robots and Systems.* Tsuchiusa, Ibaraki, Japan, July 1990.
22. Jürgen Roth, Bert Breuer, and Jörg Stöcker. Kraftschlußerkennung im rotierenden Reifen. In *Fachtagung Integrierte mechanisch-elektronische Systeme*, number 179 in VDI Fortschrittsberichte, pages 132–143. VDI Verlag, 1993.
23. G. N. Saridis. *Self-Organizing Control of Stochastic Systems.* M. Dekker, 1977.
24. M. Schmitt and H. Tolle. Das Assoziativkennfeld, eine lernfähige Standardkomponente für Kfz–Steuergeräte. *ATZ (Automobiltechnische Zeitschrift)*, 94(1), 1994.
25. Manfred Schmitt. *Untersuchungen zur Realisierung mehrdimensionaler, lernfähiger Kennfelder in Großserien-Steuergeräten.* PhD thesis, TH Darmstadt, 1994. – in preparation –.

26. H. Tolle, P. C. Parks, E. Ersü, M. Hormel, and J. Militzer. Learning control with interpolating memories – general ideas, design-lay-out, theoretical approaches and practical applications. *Int. J. Control*, 56, 1992.
27. Henning Tolle and Enis Ersü. *Neurocontrol.* Number 172 in Lecture Notes in Control and Information Sciences. Springer-Verlag, 1992. ISBN 3-540-55057-7.

# Construction and Design of Parsimonious Neurofuzzy Systems

K.M. Bossley, D.J. Mills, M. Brown and C.J. Harris

Image, Speech and Intelligent Systems Research Group
Department of Electronics and Computer Science,
University of Southampton, UK.

**Abstract.** Static fuzzy systems have been extensively applied in the Far East to a wide range of consumer products whereas researchers in the west have mainly been concerned with developing adaptive neural network that can learn to perform ill-defined, difficult tasks. Neurofuzzy systems attempt to combine the best aspects of each of these techniques as the transparent representation of a fuzzy system is fused with the adaptive capabilities of a neural network, while minimising the undesirable features. As such, they are applicable to a wide range of static, design problems and on-line adaptive modelling and control applications. This chapter focuses on how an appropriate structure for the rule base may be determined directly from a set of training data. It provides the designer with valuable qualitative information about the physics of the underlying process as well as improving the network's generalisation abilities and the condition of the learning problem.

## 1 Introduction

The ability to construct an appropriate model from a data set and provide the designer with knowledge about the underlying physical process is a fundamental problem in much of systems theory. Such a model may be used in the design of a controller or a fault detection system, or else it could be used to gain a better understanding of the order and form of a complex input-output mapping. Linear system identification is a mature technique [16] but even this is recognised as an "art", where human intervention is a critical part of the design process.

Fuzzy and neurofuzzy models and controllers have been widely applied in the Far East in a variety of consumer products including washing machines, autofocus cameras, automobiles and even toilets! The reason for their success is probably not because they are inherently robust and perform well, rather the transparent[1] or glass box models that are produced enable the designer to incorporate and modify the behaviour of the system in a "natural" manner. The behaviour of a fuzzy system is explained using vague rules of the form:

[1] The ability to understand the knowledge stored in a network.

IF (*error is positive small* AND *error change is almost zero*)
THEN (*output is positive small*) 0.5

where fuzzy sets are used to represent the vague linguistic statements such as *small* and the number associated with the rule denotes the *confidence* or strength with which it fires. However, many fuzzy systems are partially *opaque* to the designer as complex fuzzification, inferencing and defuzzification operations are performed on the original input data.

Neurofuzzy systems are a particular type of fuzzy network which uses specialised fuzzy set shapes and operators, and it is this simplification which means that the neurofuzzy systems are amenable to a thorough mathematical analysis and are transparent to the designer. Links can be made with standard model identification algorithms and tests, and the "novelty" of a neurofuzzy approach can be quantified in terms of well-understood concepts. The training problem is also considerably simplified as the network is *linearly* dependent on its set of adjustable parameters. However, as with a lot of modelling techniques, neurofuzzy systems suffer from the well known *curse of dimensionality*, as the size of the rule base grows exponentially with the input dimension. This, coupled with the fact that experts find it hard to correctly articulate rules that depend on a large number of inputs, means it is essential to develop construction algorithms which can determine their structure automatically, given a set of training data.

Neurofuzzy construction algorithms should embody a number of basic data modelling principles, for instance:

- **Principle of data reduction:** the smallest number of input variables should be used to explain a maximum amount of information.
- **Principle of network parsimony:** the best models are obtained using the simplest possible, acceptable structures that contain the smallest number of adjustable parameters.

By selecting the most relevant input variables and by choosing an appropriate structure for the fuzzy rule base, it is possible to significantly reduce the system's memory requirements. This is true for a wide variety of modelling algorithms, and it is interesting to note that some of the techniques which were originally proposed to overcome the curse of dimensionality (such as projection pursuit) are found in neural network architectures, see section 3.3.

It is important to realise however, that fuzzy rules are not the only, or necessarily the best, method for representing knowledge. They provide a new way for designers to make their black-box models more transparent (so called *glass box* modelling), but if it is known that there is a linear relationship between an input variable and the output, then a designer would understand this best by stating this dependency explicitly. Fuzzy rules can implement a linear mapping exactly, but on this point the authors agree with Nakamori and Royke [19] who stated that:

> *Modelling is an act that clears up our vague knowledge, not an act that tries to express vagueness imprecisely.*

The most intelligent and parsimonious modelling algorithms will be those techniques that make the best use of *all* the available knowledge, whether this is fuzzy or not.

## 2 Conventional Neurofuzzy Systems

Fuzzy and neurofuzzy networks can be represented in a manner similar to expert systems, as shown in figure 1. It consists of a (fuzzy) knowledge base and an

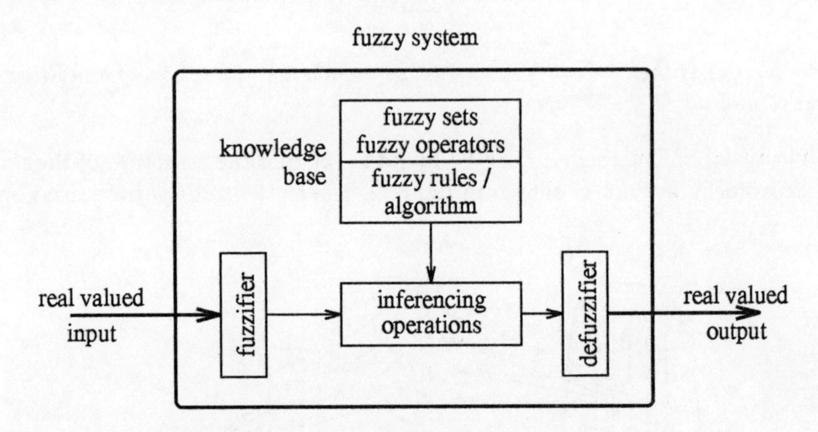


**Fig. 1.** The basic components of a fuzzy system.

inference engine for making decisions about previously unseen data, but there are also two interface modules which *fuzzify* the real-valued input data and *defuzzify* the calculated fuzzy output set. All of the inference operations: pattern matching (ANDing), inferencing (IF THEN) and aggregation (ORing) are performed on fuzzy input and output sets, so the real-valued input must first be represented as a fuzzy set and the information contained in the resulting fuzzy output set must be compressed to a single real-valued number. Despite the apparent simplicity of the fuzzy rules, this is potentially an opaque procedure, but neurofuzzy systems simplify it considerably.

### 2.1 Structure

The structure of a neurofuzzy system is determined by the functions used to represent the fuzzy sets, the operators used for implementing the logical operations and the fuzzification and defuzzification strategies. Generally, a neurofuzzy system represents a particular type of fuzzy system that uses singleton fuzzification and centre of gravity defuzzification algorithms and *algebraic* (sum and product) rather than the conventional *truncation* (min and max) operators. This implementation strategy has several advantages as the system's output tends to be smoother and this makes it possible to analyse its behaviour theoretically. In fact, the following theorem shows how the shape of the fuzzy input sets is related to the network's output in a neurofuzzy system which uses algebraic operators.

**Theorem 1** *When B-splines are used to implement the fuzzy membership functions, the real-valued inputs are represented using singleton fuzzy sets, algebraic operators are chosen to implement the fuzzy logical functions, a centre of gravity defuzzification operator is used and the rule confidences are normalised, then the output of a neurofuzzy system is given by:*

$$y(\mathbf{x}) = \sum_{i=1}^{p} \mu_{\mathbf{A}^i}(\mathbf{x})\, w_i \tag{1}$$

*where* $\mu_{\mathbf{A}^i}(\mathbf{x})$ *is the* $i^{th}$ $(i = 1, \ldots, p)$ *fuzzy membership function of a multivariate input* $\mathbf{x}$ *and* $w_i$ *is the* $i^{th}$ *weight.*

The proof of this theorem can be found in [3], and the structure of the resulting neurofuzzy system is shown in figure 2, where the multivariate fuzzy input

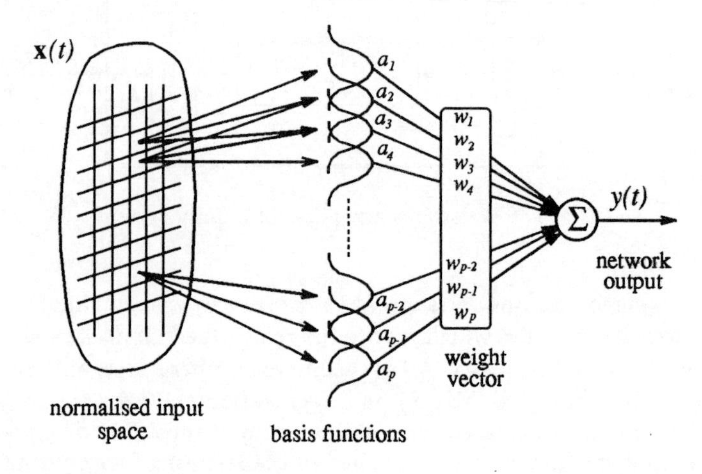


**Fig. 2.** A typical neurofuzzy system's architecture where the fuzzy input membership functions are termed basis functions.

sets have been termed *basis functions* $(\mu_{\mathbf{A}^i}(\mathbf{x}) \equiv a_i(\mathbf{x}))$, and are defined on the lattice in the input space which is generated from the projection of the univariate basis functions. The weight, $w_i$, that scales each basis function represents a local estimate of the value of the network's output given that the input lies within the set. However, before describing how these basic systems are inadequate for dealing with a large number of inputs, it is worth describing some of the terms mentioned in Theorem 1 in greater detail.

**B-spline Fuzzy Sets:** B-spline basis functions are piecewise polynomials of order $k$ which have been widely used in surface fitting applications. They can be used to represent fuzzy membership functions [3] as they have several important properties:

- A simple and stable recurrence relationship can be used to evaluate the degree of membership.
- The basis functions have a compact support[3] which means that knowledge is stored locally across only a small number of basis functions.
- The basis functions form a *partition of unity* as: $\sum_{i=1}^{p} \mu_{\mathbf{A}^i}(\mathbf{x}) \equiv 1$ hence producing accurate smooth approximations [29].

B-spline basis functions of different orders are shown in figure 3, and it can be seen that they can be used to implement binary, *crisp* fuzzy sets ($k = 1$) or the standard triangular fuzzy membership functions ($k = 2$) as well as smoother representations.

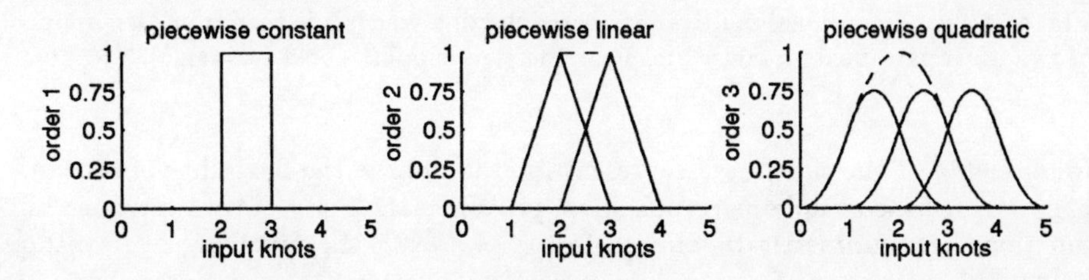


**Fig. 3.** B-spline fuzzy membership functions of orders $1 - 3$ and their additive forms (dashed lines) which correspond to trapezoidal and $\Pi$ fuzzy sets.

These univariate B-spline basis functions can be used with multivariate inputs by generating a *tensor product* representation where every basis function is multiplied by every other defined on the remaining variables. Unfortunately, the number of multivariate basis functions is directly related to the number of points in the lattice on which they are defined. For instance, when there exist $(r_i + 1)$ intervals on the $i^{th}$ input axis, the number of multivariate basis functions is:

$$p = \prod_{i=1}^{n} (r_i + k) \tag{2}$$

and this corresponds to a fuzzy rule base where the antecedents of each rule contain all the input variables and every possible combination of univariate linguistic statements is taken. This is illustrated in Fig. 4, where a two-dimensional basis function is formed by taking the product two univariate basis functions.

**Weights and Rule Confidences:** Many fuzzy and neurofuzzy systems do not consider rule confidences as an integral part of the theory. However, they are

[3] The output of the basis function is non-zero in only a small part of the input space.

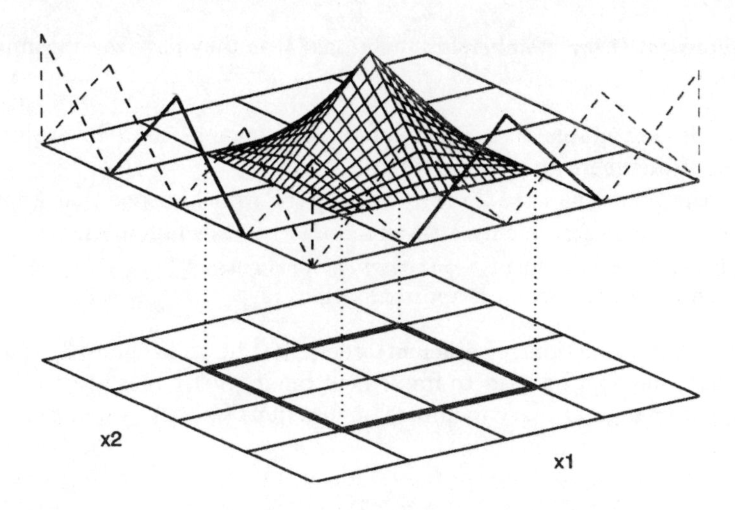


**Fig. 4.** A two-dimensional multivariate basis function which is formed from the product of two univariate bold triangles and its associated support (bold rectangle).

fundamental if the rule based representation is to have the flexibility of the simpler weight based implementation strategy. When the fuzzy output membership functions are symmetric B-splines of order $k$ ($\geq 2$), the following relationship holds [3]:

$$w_i = \sum_{j=1}^{q} c_{ij}\, y_j^c$$

where:

$$c_{ij} = N_j^k(w_i)$$

$N_j^k(.)$ is the $j^{th}$ fuzzy output membership function ($j = 1, \ldots, q$), $y_j^c$ is its corresponding centre and $c_{ij}$ is the rule confidence which relates the $i^{th}$ fuzzy input set to the $j^{th}$ fuzzy output set and when it is zero, the rule does not fire. This means that a weight can be *fuzzified* to produce a rule confidence vector $\mathbf{c}_i$, which can then be *defuzzified* and the original weight is obtained. Therefore a fuzzy, linguistic algorithm can be generated from the neurofuzzy output expression given in Equation 1. This method for generating the rule confidences also has some desirable properties:

- At most $k$ rule confidences are non-zero as the B-spline output sets have a compact support. Therefore the rule base is sparse.
- Each rule confidence vector sums to unity, *i.e.* $\sum_{j=1}^{q} c_{ij} \equiv 1$, and every rule confidence lies in the unit interval, *i.e.* $c_{ij} \in [0,1]$.

Any learning rule that adapts the rule confidences directly should satisfy a similar set of constraints, otherwise the rule base is difficult both to interpret linguistically and to analyse theoretically. To the authors' knowledge, this is the only such method.

## 2.2 Training Algorithms

The neurofuzzy systems described in equation 1 have one important advantage: the network's output is *linearly* dependent on the weight set. As has just been shown, there is a direct relationship between the weight vector and the rule confidence matrix so adapting the weights is equivalent to modifying the strength with which each rule fires. *Any* linear training algorithm can be used to adapt the weight vector and measuring the network's performance $J$ with respect to the Mean Squared output Error (MSE):

$$J = \sum_{t=1}^{L} (\widehat{y}(t) - y(t))^2 \tag{3}$$

for $L$ training pairs (where the desired output is $\widehat{y}(t)$), means that a closed form optimal solution always exists.

When the training data are known *a priori* (this is termed batch learning), the optimal weight vector can be found by inverting the so-called normal equations:

$$\mathbf{R}\mathbf{w} = \mathbf{p} \tag{4}$$

where $\mathbf{R}$ is the autocorrelation matrix whose $ij^{th}$ element is given by $r_{ij} = E\left(a_i(t)a_j(t)\right)$, and $\mathbf{p}$ is the cross-correlation vector whose $i^{th}$ element is $p_i = E\left(a_i(t)\widehat{y}(t)\right)$ (it has been implicitly assumed that the discrete form of the expectation operator is represented by the averaged sum).

The form of the autocorrelation matrix determines which training algorithm is "best" for these neurofuzzy networks. The aim of this chapter is to develop small to medium sized neurofuzzy systems which have local supports, hence $\mathbf{R}$ will be of moderate size and fairly sparse. In these situations, the best learning algorithm is generally conjugate gradient [25] which iteratively modifies the weight vector using optimal, $\mathbf{R}$-orthogonal updates. This training algorithm was used in all of the examples described in this chapter.

## 2.3 Curse of Dimensionality

For a standard B-spline network where the multivariate basis functions (fuzzy sets) are defined on a lattice, the number of parameters in the network is an *exponential function* of the input space dimension. Therefore, the cost of both implementing the rule base and obtaining an output increases exponentially fast as the input space dimension grows (see equation 2), limiting the number of neurofuzzy inputs to effectively four. This was termed the *curse of dimensionality* by Bellman [2] and is a common problem in many nonlinear modelling techniques. In fact, there has been a large research effort devoted to developing algorithms that can partially overcome this problem. There are implications for the amount of training data which need to be generated, as well as for the size and number of the basis functions and the network's ability to generalise. One way to overcome the curse of dimensionality is to try to exploit structural information which is known (either *a priori* or is discovered during the training process) about the form of the desired function. This can then be used to build parsimonious neurofuzzy systems with the following properties:

- It has a smaller number of parameters to train, and a reduced implementation cost.
- The system has improved generalisation abilities because its structure is appropriate.
- The rule base is more transparent because it is smaller.

All of these points are relevant even for small neurofuzzy networks (1 or 2-dimensional), so the construction algorithms may be useful for any sized neurofuzzy modelling problem.

# 3 Alternative Neurofuzzy Representations

Section 2 has discussed the curse of dimensionality issues associated with conventional lattice based neurofuzzy systems. The curse of dimensionality is common to many model building techniques and so it seems sensible to look at how researchers in other disciplines have attacked the same problem. In particular, there has been much progress in the field of statistics that can be applied directly to a neurofuzzy system. This section will describe some of the more promising approaches. They all involve some method of exploiting structure to reduce the required number of fuzzy rules but without affecting the quality of the approximation. It leaves the designer with a choice of which model to use. They all have their merits so it should be remembered that no one model will be appropriate for every problem.

## 3.1 Additive Decompositions

A straightforward attempt to alleviate the curse of dimensionality is to search for smaller dimensional lattice subnetworks. This is an extremely popular approach within the statistics community where it is termed an ANalysis Of VAriance (ANOVA) decomposition. For an $n$ dimensional function, the ANOVA decomposition is expressed as:

$$f(\mathbf{x}) = f_0 + \sum_{i=1}^{n} f_i(x_i) + \sum_{i=1}^{n} \sum_{j=i+1}^{n} f_{ij}(x_i, x_j) + \cdots + f_{1,2,\ldots,n}(x_1, \ldots, x_n)$$

where $f_0$ is the bias and the remaining terms represent the combinations of univariate, bivariate, *e.t.c.* subfunctions that *additively* decompose the function $f$. An ANOVA decomposition describes the relationship between the different input variables but it is only useful if the interactions involving more than say four inputs are identically zero. This constraint could limit the potential for applying an ANOVA representation but often an adequate approximation is still obtained. The question must be asked: are all the nonlinear features important or will a model composed of simple subfunctions suffice?

An immediate advantage of the ANOVA representation is that each subfunction can be a lattice based neurofuzzy system and so network transparency is retained. Also, the output is a *linear* function of the concatenated weight vectors for each subfunction which means the training algorithms derived for conventional neurofuzzy systems still apply. The potential reduction in the number of fuzzy rules can be illustrated by considering the approximation of Powell's function, a system with four inputs $x_1, \ldots, x_4$ and one output $y$

$$y = (x_1 + 10x_2)^2 + 5(x_3 - x_4)^2 + (x_2 - 2x_3)^4 + 10(x_1 - x_4)^4. \tag{5}$$

This function is an ideal candidate for an ANOVA decomposition since it contains four, two-dimensional subfunctions. If seven basis functions are used on each axis, the ANOVA system will use approximately 200 rules whereas a conventional neurofuzzy system uses over 2000, the majority of which are redundant.

An algorithm called Adaptive Spline Modelling of Observational Data (ASMOD) [13] has been recently proposed, that searches for an ANOVA decomposition using the information obtained from a set of training data. The algorithm is not yet mature but it does represent a promising approach to the construction of high dimensional neurofuzzy models.

### 3.2 Non-lattice representations

Unfortunately, the lattice representation adopted by both conventional neurofuzzy systems and the submodels of an additive decomposition may be restrictive, it fails to parsimoniously model local functional behaviour. This is because the properties of a lattice mean that including a new rule into a rule base leads to the production of redundant rules. To demonstrate, consider the two dimensional, lattice partitioned fuzzy rule base shown in figure 5a. If the new rule

IF ($x_1$ *is almost zero* AND $x_2$ *is almost zero*) THEN ($y$ *is positive medium*)

is added to the rule base, the fuzzy input set distribution shown in figure 5b is obtained. The definition of a lattice means that including this new rule results in the creation of twelve redundant ones. This is a direct consequence of the curse of dimensionality, a property that still exists in the submodels of the additive decomposition described in section 3.1.

This drawback can be overcome by attempting a more local representation, which can be achieved by abandoning the lattice representation and using a strategy that *locally partitions* the input space. Local partitioning would allow the inclusion of the single rule in the above example. To produce fuzzy interpretable basis functions, an appropriate partitioning strategy that produces rectangular partitions has to be employed. Popular non-lattice representations include k-d trees and quad trees, 2-dimensional examples of which are illustrated in figure 6. K-d trees were originally developed as an efficient data storage mechanism. They result from a succession of axis-orthogonal splits made across the entire domain of an existing partition. Each split produces new regions which themselves can potentially be split. The k-d tree representation has been employed by several researches in both the neural network and fuzzy logic communities [26, 27].

In the neural network community decision tree structures have been used to recursively partition the input space [7, 12, 24]. Typically these approaches

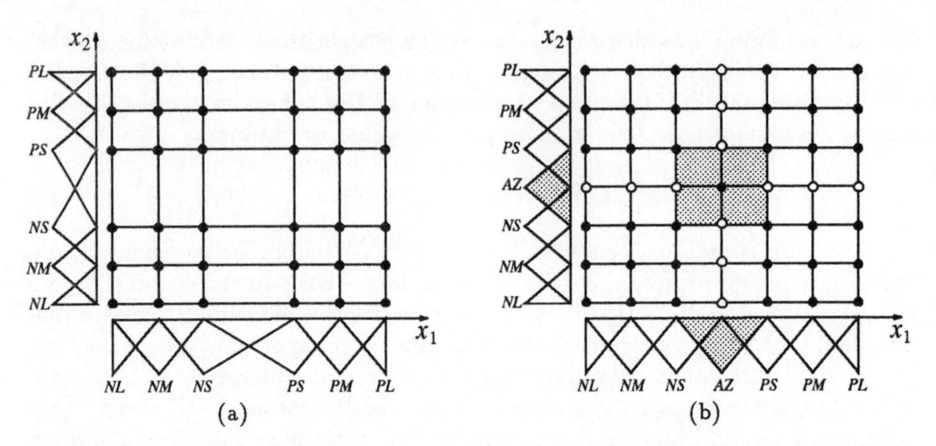


**Fig. 5.** The distribution of fuzzy input sets is illustrated before (a) and after (b) the insertion of a new fuzzy rule (the shaded region in (b)). Each o in (b) represents an additional rule created as a result.

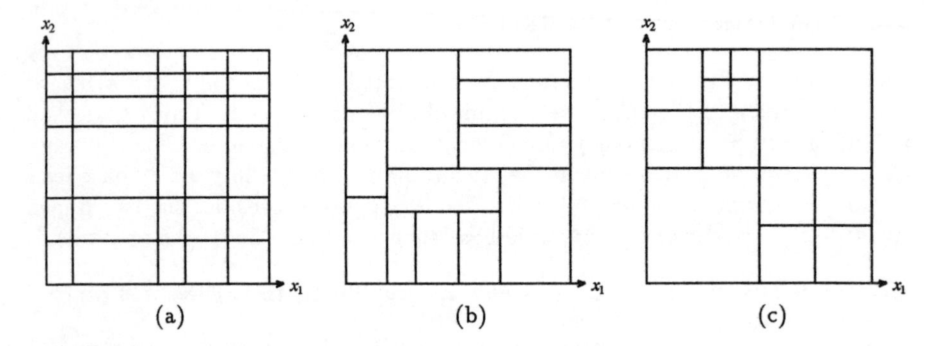


**Fig. 6.** Examples of three different of the input space partitions: (a) lattice, as used by ASMOD; (b) k-d tree; (c) quad tree.

produce partitions that resemble k-d trees. The first and arguably most popular approach is Multivariate Adaptive Regression Splines (MARS) developed by Friedman [7]. MARS produces a high dimensional approximation by forming the product of truncated spline basis functions. These multidimensional splines are recursively identified by a stepwise construction algorithm, producing a parsimonious model. Despite the success of this algorithm, it constructs models with some undesirable properties: ill-conditioned training; non-sparse representation; weight crosstalk; and no logically consistent fuzzy interpretation. For these reasons, the MARS algorithms can not be directly applied to a neurofuzzy system. Also Kavli [12], in an extension to the ASMOD model, proposed a similar algorithm for the construction of B-spline models that use both global and local partitioning. The algorithm was called Adaptive B-spline Basis function Modelling of Observational Data (ABBMOD) and it results in a k-d tree partition of the input space on which B-splines are independently defined. The use of B-splines networks in this way makes ABBMOD directly applicable to the construction of neurofuzzy systems. This approach is adopted in section 4.

An alternative but similar approach to exploiting local model flexibility is to use quad trees, as shown in figure 6c. Quad trees allow less flexible partitions of the input space, giving a more structured representation. Rigorous structures of this form tend to produce models with better transparency. The use of quad trees in fuzzy systems, termed *fuzzy boxtrees*, was proposed by Sun [27]. In this scheme, regions of the input space where the model performs inadequately are split giving greater model flexibility. Splits are repeated until an adequate model is obtained. This multi-resolution approach is very similar to that of Moody [17], who proposes a hierarchy of B-splines with increasing resolution. The motivation behind this work is to produce a model that combines the generalisation ability of a coarse resolution with the ability to model fine details with the fine resolution. A natural extension of this work would be to construct a hierarchy of local multi-resolution lattice based neurofuzzy models. This representation is particularly attractive due to the desirable properties of lattice based models: transparency and a simple, efficient implementation.

### 3.3 Input Space Projections

The idea behind input space projection techniques is to project the original input space onto a new space, where the modelling task is made easier. Typically the input space is projected onto a smaller space in such a way that the resulting model has reduced complexity. Input space projection techniques applied to modelling are generally of the form:

$$f(\mathbf{x}) : \Re^n \xrightarrow{g(\mathbf{x})} \Re^{n_p} \xrightarrow{h(\mathbf{z})} \Re^m \qquad (6)$$

where $n$, $n_p$, and $m$ are dimensions of the original input space, the newly defined intermediate space and the output space respectively, and $\mathrm{g}(\mathbf{x})$ and $\mathrm{h}(\mathbf{z})$ are arbitrary functions that need to be defined.

Input space projections offer several advantages when applied to neurofuzzy systems. One major drawback of conventional neurofuzzy systems is that they only perform axis orthogonal splits of the input space. This problem can be addressed by the use of irregular basis functions [18], but such an approach has no practical fuzzy interpretation. If the projection approach summarised by mapping 6 is employed, a fuzzy interpretation is maintained. The power of locating good projections can be demonstrated by considering the first term of Powell's function 5:

$$y = (x_1 + 10x_2)^2.$$

Let the input space be transformed by the projection $g(\mathbf{x}) = x_1 + 10x_2$, then the modelling task is simplified to the problem of learning the univariate function $h(z) = z^2$. The mapping has been illustrated in figure 7.

In recent years several different modelling techniques based on projections of the input space have been developed. They can classified as either *unsupervised* or *supervised*. Unsupervised techniques produce the projections of the input

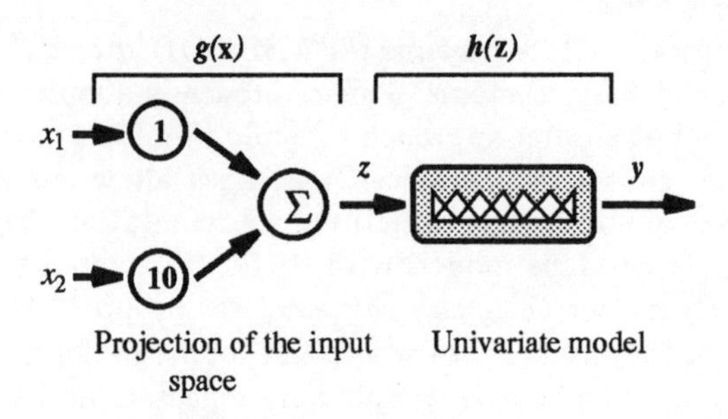


**Fig. 7.** Example of an input space projection.

space according to the distribution of the input data, the dependent output behaviour is not considered. Hence, these techniques are not directly applicable to the modelling task but are used extensively as powerful methods for data compression. Supervised methods however, produce projections that are governed by the dependent output behaviour making modelling of the output easier. Despite the inadequacies of unsupervised projections, they have been used to aid modelling and so examples of both supervised and unsupervised techniques will now be reviewed and their relevance to neurofuzzy systems described. When applying a projection technique, it should always be remembered that the computational cost in producing the projection must always be outweighed by the reduction in cost of constructing the model and/or the improvement in the quality of the approximation.

**Principle Component Regression:** This is an unsupervised projection technique based on the most common statistical algorithm used in multivariate data analysis, Principal Component Analysis (PCA). In PCA, a high dimensional input space is projected onto a lower dimensional intermediate space via an orthogonal linear transformation that maximises the information content of this lower dimensional space. The transformation identifies the main directions of data variation and allows weak directions to be discarded. Linear PCA can be generalised to the nonlinear case, where a nonlinear coordinate system is defined on which the variance of the data are maximised [9]. The difficulty is in choosing the form of the nonlinear projections; which function is appropriate? This problem was tackled by Wedd [28] who used RBF networks to produce the nonlinear projections. Neurofuzzy systems could also be applied here.

If PCA is used to reduce the size of the input space from which a model is constructed, the technique is termed Principle Component Regression (PCR). In its simplest form, a linear multiple regression is taken based on the linear principle components producing a model of the form:

$$y = \sum_{i=1}^{n_p} w_i \mathbf{x}^T \mathbf{u}_i$$

where $w_i$ is the $i^{th}$ weight of the regression and $\mathbf{u}_i$ is $i^{th}$ principle component of the input data. However, PCR is not restricted to forming a standard linear regression, the output can be a nonlinear function of the principle components. The principle components can also be either linear or nonlinear but due to their simplicity, linear projections tend to be exploited. PCR has been applied by both the neural network and fuzzy logic researchers. It can be used to preprocess the input space before presenting the input to an algorithm for automatically generating fuzzy rules, hence reducing the size of the input space and consequently the rule-base [22]. Peel et al.[20] use PCA in a similar way, to aid the training and construction of a feedforward neural network. The input space is first transformed and then fixed as the hidden layer of the network and so training is reduced to solving a simple linear optimisation problem.

The major drawback of PCR is that the output behaviour on which $\mathbf{x}$ is dependent is not considered in the synthesis of the projections. There is a danger that projections with small variance in the input space but which are highly correlated with the output may be discarded. The projections produced by PCA are not directly applicable to the task of modelling and should be treated with caution. More security would be obtained by incorporating PCA into a neurofuzzy construction algorithm. The set of $n$ principle components (none are discarded) can be used by an algorithm such as ASMOD, which can select the appropriate inputs. Any redundant projections (those not highly correlated with the output) are ignored [22].

**Partial Least Squares Regression:** This algorithm is also a linear projection technique, but unlike PCR, Partial Least Squares (PLS) regression is a *supervised* method, where the influence of $y$ is incorporated into the formulation of the orthogonal projections. The projections are chosen to maximise the covariance of the input data. PLS may also be generalised to a nonlinear modelling technique. The simplest form of this is to produce a nonlinear regression on the PLS projections. The idea was first suggested by Wold [30] who used a spline based model to produce the nonlinear regression. A similar strategy was adopted by Lines [15], where a construction algorithm for the splines is used to locate the optimal regression. However, the problem with the general approach is that the projections are formed by assuming a linear regression on $y$ and this assumption may be inappropriate for the nonlinear regression.

A natural extension of this work is to incorporate a neurofuzzy system into the nonlinear PLS regression. It may be also be advantageous to perform nonlinear projections similar to those employed by nonlinear PCA, where the covariance of the data are maximised. A series of $n_p$ neurofuzzy models can be used to produce the nonlinear PLS projections on which a parsimonious neurofuzzy model can be constructed. This is an important and interesting area for future research.

**Projection Pursuit Regression:** Friedman and Stuetzle [8] first proposed Projection Pursuit Regression (PPR) as a multivariate regression method immune from the curse of dimensionality. The form of PPR fits nicely into the framework of mapping 6. For a single output, it assumes the model:

$$y = f_0 + \sum_{i=1}^{n_p} w_i f_i(\alpha_i^T \mathbf{x})$$

where $\alpha_i^T \mathbf{x}$ denotes a one dimensional projection of the input $\mathbf{x}$. On each projection, an arbitrary smooth univariate function $f_i$ is defined and the sum over all these functions gives the output. As with all input space projection techniques, the usefulness of this approach is that for many functions the required number of projections is relatively small. Note that if the allowable projections and univariate functions are restricted then PCR and PLS are both special cases of PPR.

The PPR model is constructed iteratively. The model is initialised with a single projection of the input space and $\alpha_1$, $f_1(.)$ and $w_1$ are optimised. A new projection is then included and the parameters of each projection are optimised cyclically to minimise a loss function. This approach to training is similar to the backfitting algorithm [10]. If the contribution of the new projection is insignificant then construction is halted and the previous PPR is used as a model. A drawback is that each construction requires the solution to a nonlinear optimisation problem.

It should also be noted that the Multi-Layer Perceptron (MLP) is just a special case of PPR where the univariate smooth functions are sigmoid functions. A comparison was made by Hwang et al.[11] who showed that projection pursuit learning outperforms back propagation by producing more parsimonious models. The results are to be expected as the PPR activation functions ($f_i$) are more flexible than those in an MLP. The similarity between the two approaches does suggest that PPR will suffer from the same drawbacks as an MLP, *i.e.* the model has poor interpretability (opaque) and requires the solution of a global, ill-conditioned, nonlinear optimisation problem, although this approach can overcome the curse of dimensionality [1]. Lane et al.[14] propose combining the generalisation properties of MLPs with the computational efficiency and learning speed of B-spline models. In this approach, univariate B-spline networks are used as the activation functions in the nodes of the MLPs. Neurofuzzy models could be used in this type of approach producing a hierarchical type model. The usefulness of these types of models still require investigation.

### 3.4 Product Nodes

Rather than use a full product submodel to represent the interaction between input variables, it may be wise to experiment with more parsimonious representations. For example, product decomposition can be tested where the output of several lower dimensional submodels are multiplied to produce the higher dimensional mapping. The idea can be illustrated by considering the approximation of $f(x_1, x_2) = \sin(x_1) * \cos(x_2)$ which would normally be modelled by *one* two-dimensional submodel. Employing product decomposition means that *two* one-dimensional submodels are used whose outputs are multiplied, as illustrated

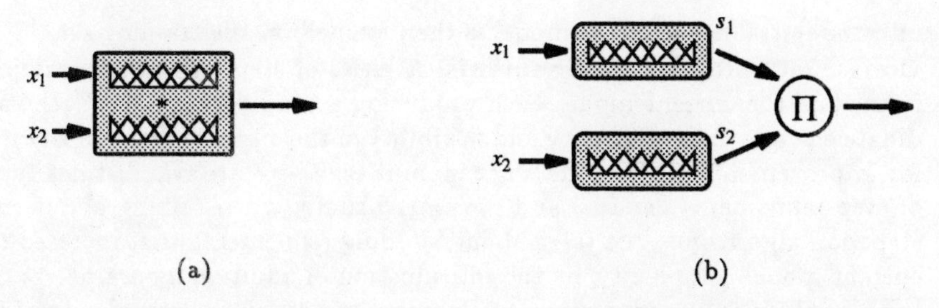


**Fig. 8.** A tensor product (a) and a product node (b) network.

in figure 8. The product node doesn't replace the two univariate subnetworks which would still contribute to the output, rather it adds a very simple interaction term whose flexibility is influenced by their form. If there existed 7 fuzzy sets on each input variable, the full tensor product network would generate 49 basis functions, whereas the two univariate subnetworks and the product node would only produce 28.

A potential disadvantage with considering product nodes is that the optimisation problem is no longer linear and a closed-form optimal solution does not exist. However, initial simulations have indicated that the parameters converge to their optimal values, especially when the input variables are uncorrelated [4].

## 4 A Generic Neurofuzzy Construction Algorithm

Section 3 has described a number of different neurofuzzy representations that can be exploited to produce a parsimonious model. Irrespective of which representation is used, the specific fuzzy rules of the system must be identified. Traditionally model identification is performed by a human expert, where limited subjective knowledge about the system is used, although an *ad hoc* approach of this kind is very restrictive, relying on the availability of linguistic domain knowledge. Also, no attempt is made to avoid overfitting or underfitting of the data. This deficiency is prominent in other modelling techniques such as regression analysis, neural networks and fuzzy logic, and it has motivated the development of construction algorithms that can automatically generate their model structure using information contained in a training set. One of the earliest iterative construction algorithms is the Group Method of Data Handling (GMDH) [5], developed by Ivakhnenko in 1966. GMDH automatically builds a parsimonious regression model using a training set. Other algorithms based on the same, simple ideas have been proposed and so as an overview, a generic neurofuzzy iterative construction algorithm is presented. This COnStruction and MOdel Selection algorithm, named COSMOS, can be applied to any of the alternative neurofuzzy representations.

1. **Initialise the algorithm with a simple model:** As a neurofuzzy model can be described by a rule base, available *a priori* knowledge can be encoded

into the initial model. This model is then trained on the training set.

2. **Construct candidate refinements:** A series of step refinements are performed on the current model, each producing a candidate model. Each candidate changes the complexity and flexibility of the current model in a search for improvements to its modelling capabilities. There are two distinct types of step refinement, *building* and *pruning*. Usually when talking about construction algorithms, one talks about building refinements that increase the current model complexity, by the introduction of additional parameters. As the algorithm should identify a parsimonious structure, the complexity of the model must be kept to a minimum. For this reason, it is intuitively appealing to allow for pruning refinements which reduce the model complexity, giving the ability to *delete* superfluous rules. These destructive steps are equally as important as the building steps, and to allow for the construction of truly parsimonious models, every building refinement must have an opposite pruning refinement. Pruning refinements are particularly important in one-step ahead construction algorithms, to allow erroneous steps to be corrected.
3. **Model selection:** From the set of candidate models one must be chosen to update the current model. The selection must achieve a compromise between the accuracy and the complexity of the approximation. This compromise can be viewed as the *bias/variance* dilemma. Reducing the bias can be achieved by increasing model complexity, but that tends to produce high variance such that the expected approximation error of the model across all possible training sets increases. There are a number of techniques that try to address this problem by assuming the availability of an independent test set but, in cases where data are scarce, the luxury of a test set can not be afforded. Therefore, methods that strike a compromise between bias and variance using only the information from a training set must be used. They include:
   i Regularisation techniques [21] that embed *a priori* information about the smoothness of the function to be approximated by incorporating a constraint. The constraint appears as a term added to the MSE 3:

   $$E = \text{MSE} + \lambda E^r$$

   where $\lambda$ controls the degree to which the approximation is smoothed. $E^r$ can take many forms; for example, it can penalise large weight values or large curvatures. When second derivatives are available, curvature driven smoothing is more popular.

   ii Statistical significance measures derived from information theory such as: Akaike's Information Criterion (AIC), Structural Risk Minimisation (SRM), or the Minimum Description Length (MDL). These techniques tend to combine the MSE with some measure of network complexity. Models with the same MSE but fewer free parameters will have a smaller statistical significance will be selected first.

   Either or both methods can be employed to identify the best candidate model which, if the approximation quality has improved, is then used as a basis for further refinements (by returning to step 2).

When trying to identify a representative model, it may be advantageous to combine the strategies presented in section 3. But to avoid loss of model transparency, more opaque models must be weighted so that they are only selected if a sufficient improvement in approximation quality is obtained. This type of weighting could be incorporated as a penalty in the model selection criteria, where lattice neurofuzzy models incur no penalty.

To demonstrate the advantage of a combined approach, consider the two-dimensional Gaussian function given by

$$f(\mathbf{x}) = \exp\left(-5\left[(x_1 - 0.5)^2 + (x_2 - 0.5)^2\right]\right) + e(t)$$

where $e(t)$ are noise samples drawn from a normal distribution with a mean of zero and a variance of 0.02. This function was used to generate a uniformly distributed training set of 440 points, as shown in figure 9.

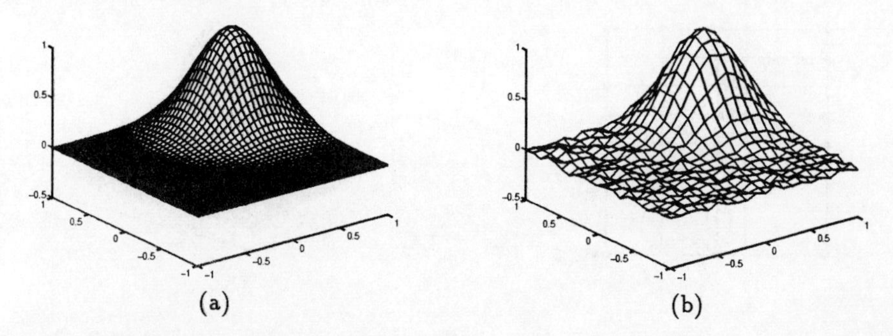


**Fig. 9.** The 2-dimensional Gaussian function, (a) shows the true surface and (b) shows the same surface corrupted by noise.

It can be seen that the function possesses local behaviour, and hence it is expected that employing some form of local partitioning would be advantageous. The ABBMOD algorithm was therefore tried and it produced the input space partition shown in figure 10a and the corresponding output surface of figure 10b. Unfortunately, the input space partitioning is not particularly local because the model consists of 3 separate tree networks. This results in overlapping basis functions that degrade the transparency of the final model.

If ASMOD is applied to the same training set, a better approximation is obtained. The resulting model and output surface are shown in figures 11a and 11b respectively. Despite the improved performance of the lattice based model, it still possesses several redundant rules suggesting that local partitioning would be more appropriate.

To illustrate the idea of combining different strategies, the above ASMOD model was assumed to have located the correct input dependencies. It was then used to initialise a *restricted* ABBMOD algorithm which first prunes any redundant basis functions, and then only allows the insertion and deletion of new basis functions. This hybrid approach produced a more parsimonious model, with a

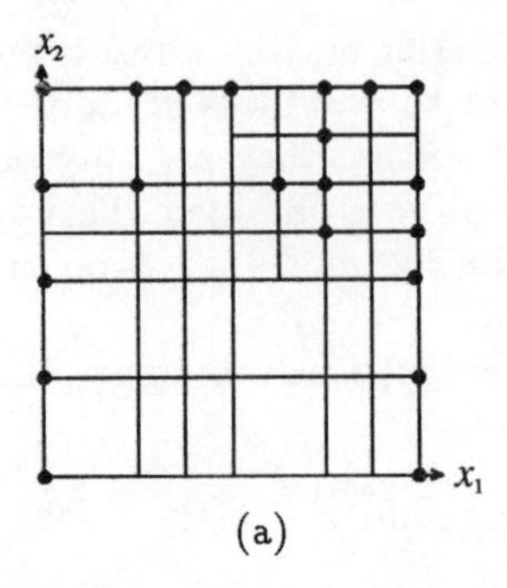


(a)

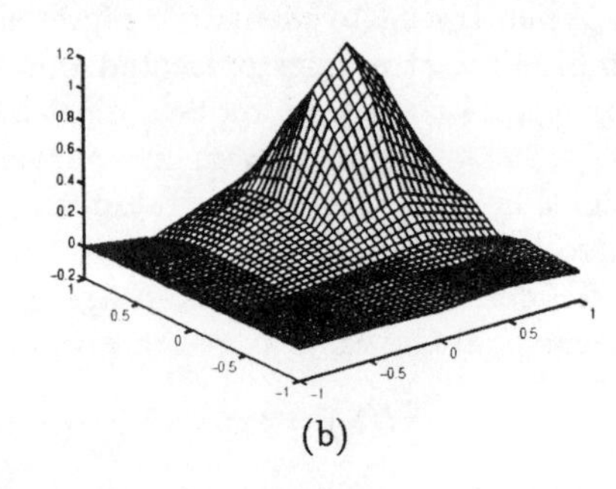


(b)

**Fig. 10.** The model produced by the ABBMOD algorithm, (a) shows the partitioning of the input space produced by three overlapping tree networks and (b) shows the output surface.

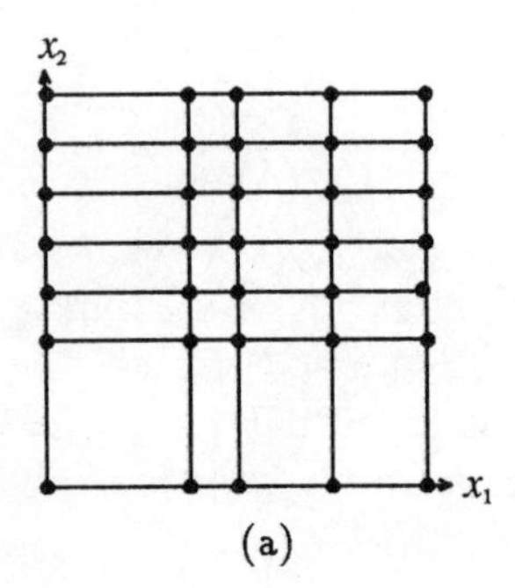


(a)

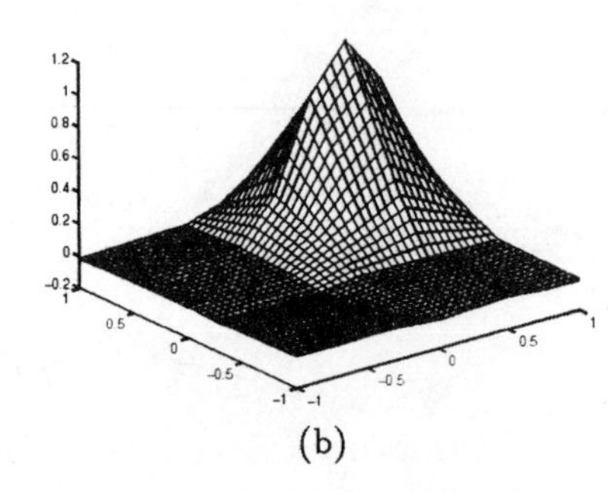


(b)

**Fig. 11.** The model produced by the ASMOD algorithm, (a) shows the partitioning of the input space and (b) shows the output surface.

better output response as shown in figure 12. During the construction of this model, one of the selected refinements produced a small, *ill-conditioned* basis function because the training data in its support were not sufficiently exciting to correctly identify the associated weights. It meant that the resulting model generalised poorly. To prevent the inclusion of such basis functions, a check on the number and distribution of training data in the support of new basis functions should be made. For the purpose of this example, construction was halted when a basis function with these properties was encountered. This approach produced a model with 23 parameters and a MSE on an uncorrupted test set of 0.0107853. It performed better than the ASMOD model which consisted of 35 parameters and had a MSE of 0.0121455. Note that further refinements could still improve the hybrid model. The question as to whether the model is transparent still remains and it is suggested that due to the large number of overlapping basis functions, maybe some form of hierarchical, multi-resolution, lattice based neurofuzzy model should be used to produce local partitioning.

This approach can be easily generalised to incorporate the different neurofuzzy representations defined in section 3. Firstly, the additive decomposition

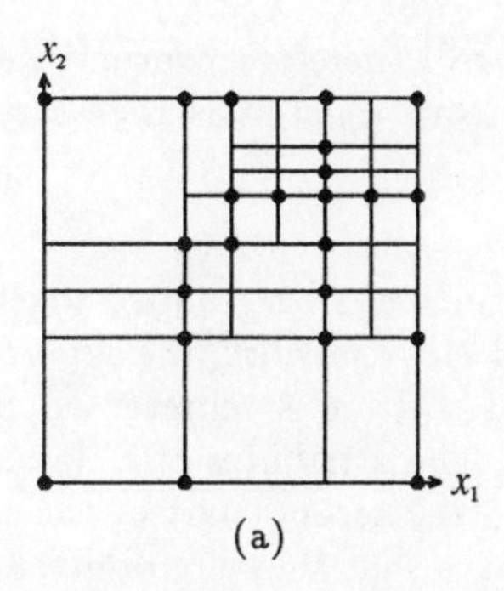


(a)

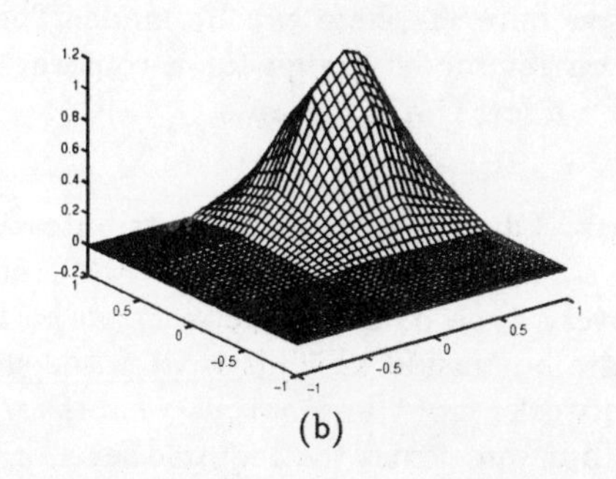


(b)

**Fig. 12.** The model produced by the hybrid algorithm, (a) shows the partitioning of the input space and (b) shows the output surface.

of the model is identified by the ASMOD algorithm and then the submodels are further refined. These further refinements can incorporate the alternative representations: local modelling; input preprocessing (input space projections); product networks, to try to locate better more representative models.

## 5 Object Oriented Design

The construction algorithms described in section 3 have been implemented using an object oriented design (OOD) methodology. An OOD methodology was chosen since it is one means for efficiently handling complexity and it was obvious from the outset that the system to be implemented was inherently complex. Object orientation reduces the gap between the problem statement and its model in a software system by the use of objects. Objects normally appear as nouns in a problem statement and they combine attributes with operations. An attribute is a data value held the object and an operation is a transformation that may be applied to the object. An OOD also promotes other factors such as extensibility, correctness, robustness and reusability [6]. The support for code reuse is particularly important since it can significantly reduce the amount of code that needs to be written and tested.

There are many popular OOD methodologies but perhaps the easiest to understand and apply is the Object Modelling Technique (OMT) [23]. OMT supports three orthogonal views of the system: the object, dynamic and functional models. The *object model* describes the static structure of the objects in the system and states their relationships. It is represented by an object diagram, examples of which will be given later. An alternative view of the system is provided by the *dynamic model* which describes aspects of the system that change with time. Finally, the *functional model* describes how data are transformed as they transfer around the system. Although all three models are required for a complete system description, the object model is the most fundamental since it is the basis on which the two other models are designed. The objects must first be defined before a description of how they change state or how information is passed between them can be made. This section will therefore concentrate on the object model design for a construction algorithm. First it is necessary to define a few OMT concepts.

**Class:** Objects with similar attributes and operations can be grouped together into a class. Figure 13 shows the OMT notation for class network, the superclass of every type of neural network to be implemented. It it is represented by a square box inside which is written the class name. The attributes of a class, *i.e.* a networks input and output vectors, are listed in the second part of the class box but sometimes the required level of detail means that they are omitted.

**Associations:** Relationships between classes are called associations and they normally appear as verbs in a problem statement. The OMT notation for an association is a line drawn between the classes being linked. This is illustrated in figure 13 where class network has been linked to class training set and class learning rule. The action performed by the association is often captured in the association name, *e.g.* trains, although the name may be omitted if a pair of classes has a single association whose meaning is obvious. If many instances of one class are related to instances of another, the OMT notation is a circle placed at the end of the association line. The multiplicity can be quantified by writing a number or interval next to the circle. For example, the same training set might be used to train many instances (1+) of class network.

**Inheritance:** The sharing of attributes and operations among classes based on a hierarchy is a fundamental concept in object orientation and is called inheritance. A superclass can be defined and then successively refined into more specific subclasses. Each subclass inherits the properties of its superclass and then adds its own unique properties. The OMT notation for inheritance is a triangle connecting the superclass to the subclass. For example, figure 13 shows that subclasses linear, fuzzy and nonlinear all inherit the attributes (and operations) of superclass network. This is because all types of network require an input and output vector. Sometimes the distinction between subclasses is less clear in which case the OMT notation dictates that the inheritance triangle must be filled in.

There are of course many more OMT concepts but the above are sufficient to understand the object diagrams that follow. The system to be described is an OMT implementation of a construction algorithm for producing neurofuzzy approximations. It begins with the object model shown in figure 13, the more important classes of which will now be described.

Network is the superclass of every type of neural network to be implemented. It has an input vector and an output vector, attributes common to all networks, and it links to class training set and learning rule. The class training set has attributes input and desired outputs which are both matrices that contain the input-output data pairs used to train the network. Operations allowed on objects

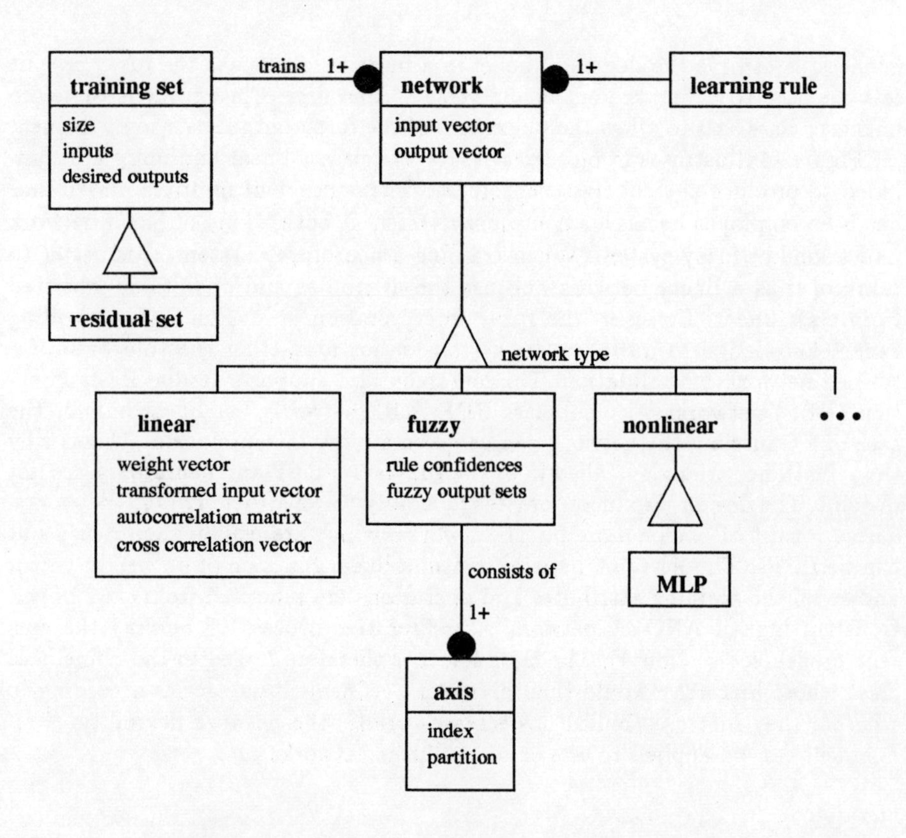


**Fig. 13.** The inheritance hierarchy for class network.

of class training set include the ability to preprocess data pairs to aid training. A subclass of training set is residual set which is sometimes used during model construction to alleviate the amount of computation required to evaluate each model enhancement. Its only attribute is a desired output matrix formed by removing the contribution of the current model from the training set.

The subclasses derived from network include linear, fuzzy and nonlinear although the possibility exists to add more network types later. Subclass linear contains all those networks whose output is a linear function of the weight vector $\mathbf{w}$. Linear networks can be trained by solving the linear system of normal equations 4. Both $\mathbf{R}$ and $\mathbf{p}$ are calculated using the vector of transformed inputs $\mathbf{a}(\mathbf{x})$. Hence $\mathbf{R}$, $\mathbf{w}$, $\mathbf{p}$ and $\mathbf{a}(\mathbf{x})$ are all attributes of linear. Linear networks are also the subject of each refinement category and so a many-to-one association is therefore made with the class refinement.

The subclass fuzzy contains types of networks based on fuzzy logic concepts. Several types of fuzzy network could be implemented, *i.e.* those based on truncation (*min/max*) or algebraic (*product/sum*) operators. The attributes of class fuzzy are a set of rule confidences and a fuzzy output set distribution. Rule confidences determine the degree of belief in a fuzzy rule whereas the fuzzy output sets are used to define the output universe of discourse. Class fuzzy is therefore linked to class axis to allow the placement of the fuzzy output sets to be defined.

Figure 14 illustrates in more detail how the classes linear and fuzzy are combined to produce the subclass neurofuzzy. The concept of multiple inheritance has been employed because a neurofuzzy system is both a kind of linear network *and* a kind of fuzzy system. When training a neurofuzzy system, it is better to think of it as a linear network and use the attributes and operations inherited from class linear. However, the fuzzy interpretation is invaluable for encoding expert knowledge to initialise the system or for extracting the rule base of a trained network for validation. The hierarchy also supports Radial Basis Function (RBF) networks with subclass RBF. RBF networks are linear but for the case of a Gaussian, the network can be given a fuzzy interpretation. This is why the inheritance triangle is filled in as the subclasses RBF and neurofuzzy are not disjoint. The design also incorporates the ability to construct ANOVA networks. These consist of one or more linear networks, which are usually neurofuzzy but can be RBFs. The ANOVA network is still a linear function of its weight vector and so all the training attributes and operations are inherited from class linear. Constructing an ANOVA network is an iterative process of *refining* the current model (see section 4). The class linear is therefore linked to the refinement class whose hierarchy would then divide the refinement categories according to whether they increase (build) or decrease (prune) the network flexibility. A refinement can be applied to one or many linear networks.

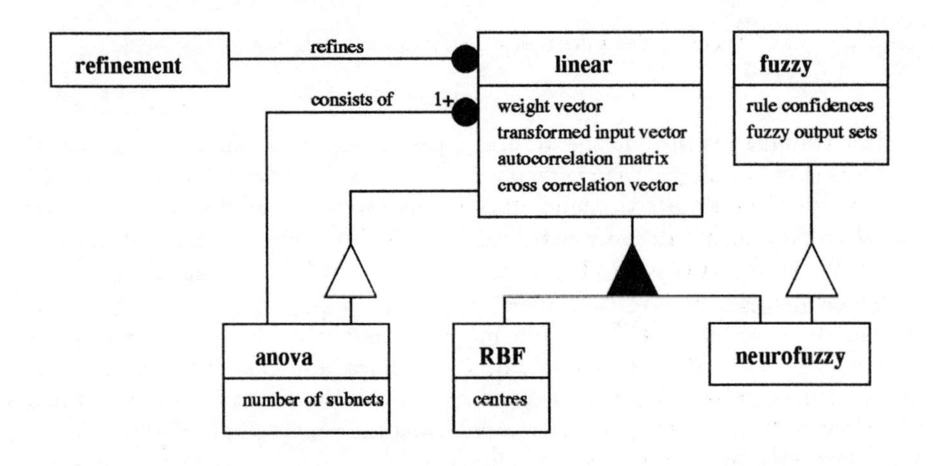


**Fig. 14.** The inheritance hierarchies for classes linear and fuzzy.

Figure 15 shows how class neurofuzzy is further refined to introduce subclasses lattice and tree which each describe a different method of implementation. In lattice, a regular structured grid is used to define the position of the Gaussian RBF or B-spline basis functions (a suitable strategy for ASMOD) whereas in class tree, the neurofuzzy network partitions the input space using decision and leaf nodes (a suitable strategy for ABBMOD). Leaf nodes will define the multidimensional basis functions.

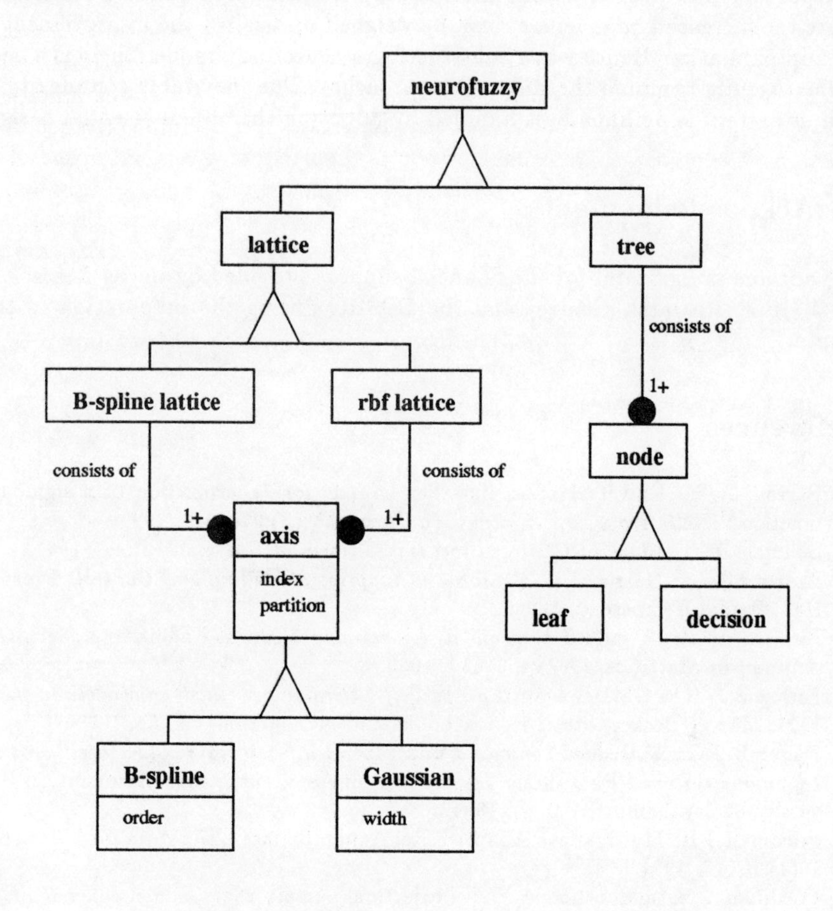


**Fig. 15.** The inheritance hierarchy for class neurofuzzy.

## 6 Conclusions

Neurofuzzy systems combine the positive attributes of a neural network and a fuzzy system by providing a fuzzy framework for representing linguistic rules with a neural network's well defined modelling and learning characteristics. Unfortunately, as with most other modelling techniques, their application is limited by the curse of dimensionality, to problems involving a small number of input variables. This chapter has described several promising approaches to alleviate this curse that all apply standard theory from other disciplines to a neurofuzzy system. All the techniques involve some method of exploiting structure to reduce the required number of fuzzy rules but without degrading the quality of the approximation. The most straightforward lattice based approaches incur no loss of transparency but a conflict sometimes arises with more sophisticated techniques where the increased opaqueness must be weighed up against the improvement in the approximation. Hence when constructing a neurofuzzy model, there is a need to *intelligently* combine the different approaches. The inevitable complexity of such as system is perhaps best handled by adopting the object oriented design.

## 7 Acknowledgements

The authors are grateful for the financial support provided by Lucas Aerospace, GEC (Hirst Research Centre) and the EPSRC during the preparation of this chapter.

## References

1. Barron, A.R. Universal approximation bounds for superposition of a sigmoidal function. *IEEE Trans. on Information Theory*, 39(3):930–945, 1993.
2. Bellman R.E. *Adaptive Control Processes.* Princeton University Press, 1961.
3. Brown M. and Harris C.J. *Neurofuzzy Adaptive Modelling and Control.* Prentice Hall, Hemel Hempstead, 1994.
4. Buja A., Hastie T. and Tibshirani R. Linear smoothers and additive models. *The Annuals of Statistics*, 17(2):453–535, 1989.
5. Farlow S.J. The GMDH algorithm. In *Self-Organising Methods in Modelling*, pages 1–24. Marcel Decker, Statistics:textbooks and monographs vol. 54, 1984.
6. Fraser R.J.C. *Embedded command and control infrastructures for intelligent autonomous systems.* PhD thesis, Department of Aeronautics and Astronautics, University of Southampton, U.K., 1994.
7. Friedman J.H. Multivariate Adaptive Regression Splines. *The Annals of Statistics*, 19(1):1–141, 1991.
8. Friedman J.H. and Stuetzle W. Projection pursuit regression. *Journal of the American Statistical Association*, 76(376):817–823, 1981.
9. Gnanadesikan R. *Methods for Statistical Data Analysis of Multivariate Observations.* John Wiley And Sons, New York, 1977.
10. Hastie T.J. and Tibshirani R.J. *Generalized Additive Models.* Chapman and Hall, 1990.
11. Hwang J., Lay S., Maechler M., Douglas R. and Schimet J. Regression model in back-propagation and projection pursuit learning. *IEEE Transactions on neural networks*, 5(3), 1994.
12. Kavli T. *Learning Principles in Dynamic Control.* PhD thesis, University of Oslo, Norway, 1992.
13. Kavli T. ASMOD: an algorithm for Adaptive Spline Modelling of Observation Data. *International Journal of Control*, 58(4):947–968, 1993.
14. Lane S., Flax M.G., Handelman D.A. and Gelfand J. Multi-layered perceptrons with b-spline receptive field functions. *NIPS*, 3:684–692, 1991.

15. Lines G.T. *Nonlinear Empirical Modelling Using Projection Methods.* PhD thesis, Department of informatics, University of Oslo, 1994.
16. Ljung L. *System Identification: Theory for the User.* Information and System Sciences Series. Prentice Hall, Englewood Cliffs, NJ, 1987.
17. Moody J. Fast learning in multi-resolution hierarchies. In *Advances in Neural Information Processing Systems I*, pages 29–39. Morgan Kaufmann, 1989.
18. Murray-Smith R. *A local model network approach to nonlinear modelling.* PhD thesis, Department of Computer Science, University of Strathclyde, 1994.
19. Nakamori Y. and Ryoke M. Identification of fuzzy prediction models through hyperellisoidal clustering. *IEEE Transactions on systems, man, and cybernetics*, 24(8):1153–1173, 1994.
20. Peel C., Willis M.J. and Tham M.T. A fast procedure for the training of neural networks. *Journal of Process Control*, 2(4):205–211, 1992.
21. Poggio T. and Girosi F. Neural networks for approximation and learning. *Proceedings of the IEEE*, 78(9):1481–1497, 1990.
22. Roberts J.M. and Mills D.J. and Charnley D. and Harris C.J. Improved kalman filter initialisation using neurofuzzy estimation. *submited to: 4th IEE International Conference on Artificial Neural Networks*, 1994.
23. Rumbaugh J., Blaha M., Premerlani W., Eddy F. and Lorensen W. *Object-oriented Modeling and Design.* Prentice Hall, Englewood Cliffs, New Jersey, 1991.
24. Sanger T.D. Neural network learning control of robot manipulators using gradually increasing task difficulty. *IEEE Trans. on Robotics and Automation*, 10(3):323–333, 1994.
25. Shewchuk J.R. An introduction to the conjugate gradients method without the agonizing pain. Technical Report CMU-CS-94-125, School of Computer Science, Carnegie Mellon University, 1994.
26. Sugeno, M. and Kang, G.T. Structureed identification of fuzzy model. *Fuzzy Sets and Systems, North-Holland*, 28:15–33, 1988.
27. Sun C. Rule-base structure identification in a adaptive network based inference system. *IEEE Transactions on Fuzzy Systems*, 2(1), 1994.
28. Wedd A.R. An approach to nonlinear princple component analysis using radial basis functions. Technical Report memo 4739, Defence Research Agency Malvern, 1993.
29. Werntges H.W. Partitions of unity improve neural function approximation. *IEE International Conference on Neural Networks*, 2:914–918, 1993.
30. Wold S. Nonlinear partial least squares modelling. II spline inner relation. *Chemometrics and Intelligent Laboratory Systems*, 14:71–94, 1992.

# Fast Gradient Based Off-Line Training of Multilayer Perceptrons

Seán McLoone and George Irwin

Control Engineering Research Group
Department of Electrical and Electronic Engineering
The Queen's University of Belfast
Belfast BT9 5AH

**Abstract.** Fast off-line training of Multilayer Perceptrons (MLPs) using gradient based algorithms is discussed. Simple Back Propagation and Batch Back Propagation, follow by viewing training as an unconstrained optimization problem. The inefficiencies of these methods are demonstrated with the aid of a number of test problems and used to justify the investigation of more powerful, second-order optimization techniques such as Conjugate Gradient (CG), Full Memory BFGS (FM) and Limited Memory BFGS (LM). Training is then at least an order of magnitude faster than with standard BBP, with the FM algorithm proving to be vastly superior to the others giving speed-ups of between 100 and 1000, depending on the size of the problem and the convergence criterion used.

Possibilities of parallelisation are investigated for both FM and LM based training. Parallel versions of these routines are proposed and shown to give significant speed-ups over the sequential versions for large problems.

# 1 Preliminaries

## 1.1 Network Topology

The Multilayer Perceptron (MLP) is a feedforward network consisting of an input layer, one or more hidden layers and an output layer of simple processing units called neurons (Fig.1(a)). The input to each neuron in a given layer is the weighted sum of the outputs of all the neurons in the previous layer. The input layer consists of dummy neurons which perform no function other than to distribute the network inputs to the next layer. The neurons in the hidden layer have the general form shown in Fig.1(b) with an input-output relationship given by:

$$x = \sum_{i=1}^{n} w_i u_i + b, \qquad y = a(x) \tag{1}$$

The neuron activation function, a(.), is required to be smooth and differentiable and the most common choices are the sigmoid function and the hyperbolic tanh function:

$$a(x) = \frac{1}{1+\exp(-x)}, \qquad a(x) = \frac{1-\exp(-x)}{1+\exp(-x)} \tag{2}$$

The output layer neurons are normally chosen to be linear, with the activation function $a(x) = x$.

The overall input-output relation for such a multi-input, single-output, single hidden layer network can be expressed mathematically as:

$$y = g(\underline{u}, \underline{w}) = \sum_{j=1}^{n_h} c_j a\left(\sum_{i=1}^{n_i} w_{ij} u_i + b_j\right) + d \tag{3}$$

where $n_i$ and $n_h$ are the number of inputs and hidden layer neurons respectively, y is the network output and $u_i$ is the $i^{th}$ element of the input vector $\underline{u}$. The various weights which make up the overall weights vector $\underline{w}$ are:

$c_j$ = the weight between the $j^{th}$ neuron in the hidden layer and the output neuron.
$w_{ij}$ = the weight between the $i^{th}$ input and the $j^{th}$ hidden layer neuron.
$b_j$ = the bias on the $j^{th}$ hidden neuron.
d = the bias on the linear output neuron.

This is the form of the MLP which will be used in the following.

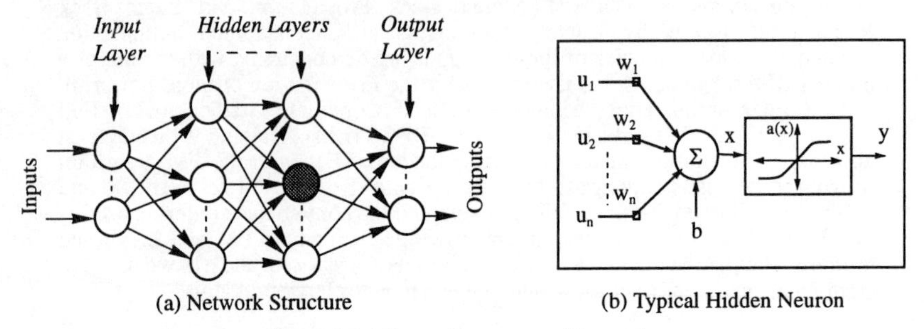


(a) Network Structure (b) Typical Hidden Neuron

**Fig.1.** Multilayer Perceptron Network

## 1.2 Training

When training an MLP to represent some mapping, $y_d = f(\underline{u})$ a training set is used, consisting of sample input vectors and desired outputs over the range of the function space to be learned. Each input vector is applied to the network and the resulting output is compared with the desired output to produce an error measure or cost function $E(\underline{w})$. This is normally chosen as the sum squared-error over the complete training set. Thus,

$$E(\underline{w}) = \frac{1}{2}\sum_{j=1}^{N_T}(y_j - y_{dj})^2 \tag{4}$$

where $N_T$ is the number of training vectors and $y_j$ and $y_{dj}$ are the actual and desired output of the network for the $j^{th}$ training vector. This can also be expressed as

$$E(\underline{w}) = \sum_{j=1}^{N_T} e_j(\underline{w}), \qquad e_j(\underline{w}) = \frac{1}{2}(y_j - y_{dj})^2 \tag{5}$$

and $e_j(\underline{w})$ is the squared error for the $j^{th}$ training vector.

The objective of training is to adjust the network weights iteratively according to some learning rule so that $E(\underline{w})$ is minimized and as such can be viewed as an unconstrained optimization problem.

## 1.3 Test Problems

Two simple test problems will be used throughout for the purpose of analysing the various training schemes under investigation.

*Problem 1*

Here an (1,5,1) MLP is trained to represent the quadratic function y=(x-2)(x+1) in the range [-3,+3]., The training set used consists of 25 points uniformly distributed over the range to be learned.

*Problem 2*

The simplest possible network is a single neuron network as shown in Fig.2 which has the advantage that, having only two weights, its error surface $E(\underline{w})$ can be plotted for a given training set.

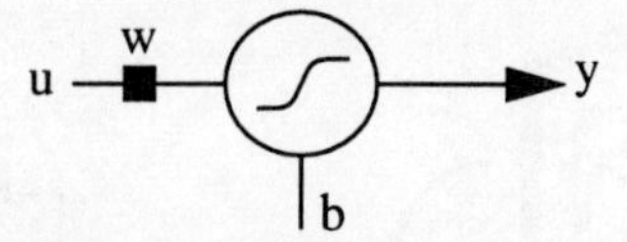


**Fig.2.** A Single Neuron Network

Assuming a tanh nonlinearity, E is then given by:

$$E(w, b) = \sum_{i=1}^{N_T} \left( \tanh\{w \cdot u_i + b\} - y_{d_i} \right)^2 \tag{6}$$

For the training set $u_i \, \varepsilon \, \{-0.5, -0.25, 0.0, 0.2, 0.5\}$, $y_{di}=u_i$ (i.e. the mapping y=x) the resulting error surface and the corresponding contour plot are given in Fig.3.

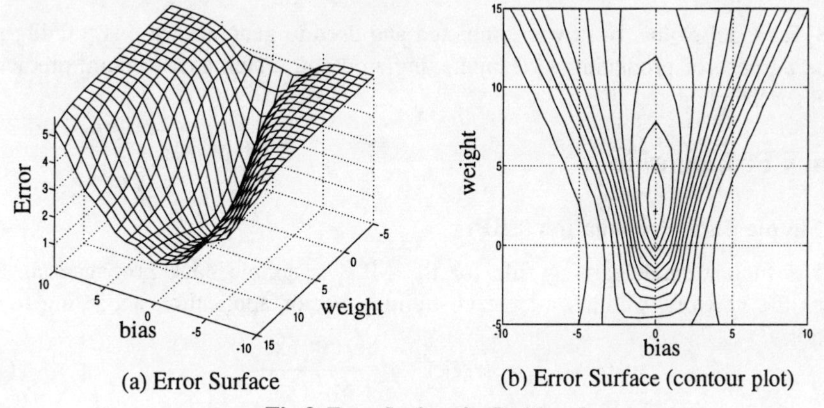


(a) Error Surface (b) Error Surface (contour plot)

**Fig.3.** Error Surface for Problem 2

This problem is particularly useful for gaining insight into the mechanisms of the various training algorithms since the trajectory of the weights during training can be plotted on the error profile.

*Problem 3*

The development of a neural predictive model for the Continuous Stirred Tank Reactor [1] will be used as a more realistic control case study. The Continuous Stirred Tank Reactor (CSTR) is a highly nonlinear plant and as such is a useful example for testing neural networks. The system is a single-input, single output one, the output being the concentration of a product compound and the input being the flow rate of a coolant. The reaction that takes place to produce the compound is exothermic, which raises the temperature and hence reduces the reaction rate. The introduction of a coolant allows the manipulation of the reaction temperature and hence the product concentration can be controlled. The reaction takes place in a container of fixed volume and the product flow rate, input concentration, temperature and output flow rate are assumed constant at their nominal values (Fig.4). The plant has a time delay (d) of 30 seconds, consisting of a transport delay at the output and a measurement delay when determining the product concentration. The equations for the CSTR plant can be found in [2].

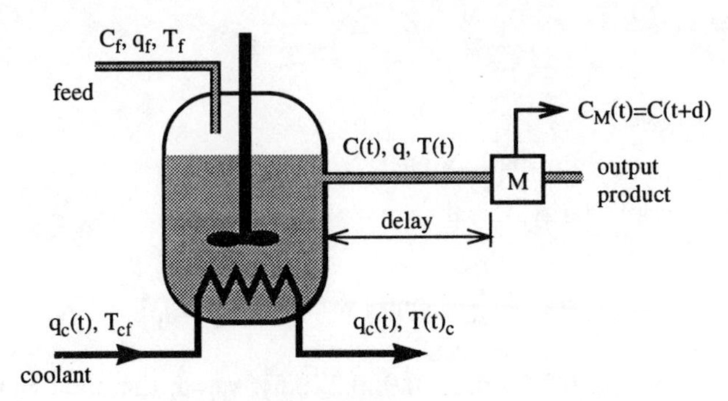


**Fig.4.** Continuous Stirred Tank Reactor

A CSTR model of this from was simulated and used to generate various training sets for the purpose of producing a 20 input, single output, 5-step-ahead neural predictive model for the plant.

# 2 Back Propagation

## 2.1 Simple Back Propagation (SBP)

The fundamental learning rule for the MLP is simple back propagation. This adjusts the network weights after each training vector application according to the equation:

$$w_i(k+1) = w_i(k) - \eta \frac{\delta e_j(w_i(k))}{\delta w_i(k)} \tag{7}$$

Here $w_i(k)$ is the $i^{th}$ element of the weights vector $\underline{w}$ at the $k^{th}$ iteration of the learning rule and $\eta$ is a small positive scalar referred to as the step size or learning rate. In vector format this can be written as:

$$\underline{w}(k+1) = \underline{w}(k) - \eta \underline{s}_j(k) \tag{8}$$

where

$$w = \begin{bmatrix} w_1 \; w_2 \; \ldots\ldots \; w_n \end{bmatrix}^T \tag{9}$$

and

$$\underline{s}_j = \begin{bmatrix} \frac{\delta e_j(w_1)}{\delta w_1} \; \frac{\delta e_j(w_2)}{\delta w_2} \; \ldots\ldots \; \frac{\delta e_j(w_n)}{\delta w_n} \end{bmatrix}^T \tag{10}$$

By considering $e_j(\underline{w})$ as a surface or error profile in n-dimensional space, equation (7) can be seen to be a simple gradient descent rule where the direction of steepest descent at a given point $\underline{w}(k)$ is given by the negative of the error gradient $\underline{s}_j$ at that point. Adjusting the weights by a small amount in this direction will therefore lead to a reduction in $e_j(\underline{w})$ .

Thus simple back propagation reduces the squared error with respect to each training vector ($e_j(\underline{w})$ ) in turn. Provided successive training vectors are uncorrelated, and the weight adjustments for each training vector are kept small (i.e. small $\eta$), the overall effect is a reduction in cost function $E(\underline{w})$ for each cycle through the training set. (Uncorrelated training vectors can be obtained by randomising the training set before commencing training).

To implement the learning rule the error gradients $\delta e_j/\delta w_i$, must be evaluated for all the weights in the network at each iteration. This was a major stumbling block in the past, particularly in relation to the hidden layer weights, and it was not until 1986 that Humelhart, Hinton and Williams derived a method for determining them [3]. The method, known as back propagation is derived by using chain rule expansions to obtain expressions for the partial derivatives $\delta e_j/\delta w_i$ and consists of two equations, one for the output layer weights ($w_{ij}$),

$$\frac{\delta e}{\delta w_{ij}} = (y_i - d_j) \cdot y_j \cdot a'(x_i) \tag{11}$$

and one for the hidden layer weights ($w_{jk}$),

$$\frac{\delta e}{\delta w_{jk}} = \sum_i \left( \frac{\delta e}{\delta x_i} \cdot w_{ij} \right) \cdot y_k \cdot a'(x_j) \tag{12}$$

The back propagating nature of the equations arises from the fact that the $\delta e/\delta x$'s, for the neurons in the ($j^{th}$) layer, is a function of the sum of the $\delta e/\delta x$'s of the ($j^{th}$+1) layer. Thus there is an inherent back propagation of the error from the output towards the inputs.

## 2.2 Batch Back Propagation (BBP)

This is a variation of the SBP algorithm where the weights are adjusted with respect to an overall cost function $E(\underline{w})$ using the learning rule:

$$\underline{w}_{k+1} = \underline{w}_k - \eta_k \underline{g}_k \tag{13}$$

where

$$\underline{g} = \left[ \frac{\delta E(w_1)}{\delta w_1} \ \frac{\delta E(w_2)}{\delta w_2} \ \ldots\ldots \ \frac{\delta E(w_n)}{\delta w_n} \right]^T \tag{14}$$

and $\eta(k)$ is the step size at the $k^{th}$ iteration.

Again this is a gradient descent rule, with $E(\underline{w})$ being viewed as an n-dimensional cost surface. Referring to equation (5), it can be seen that $\underline{g}$ may be expressed as the sum of the gradients of the individual training vector error profiles,

$$\underline{g} = \sum_{j=1}^{N_j} \left( \frac{\delta e_j(\underline{w})}{\delta \underline{w}} \right) = \sum_{j=1}^{N_j} \underline{s}_j \tag{15}$$

Therefore $\underline{g}$ can be obtained by applying each of the training vectors in turn to the MLP so that the individual gradients, $\underline{g}_j$ can be calculated using the back propagation routine. These gradients are then summed to produce the overall or 'batch' gradient.

Unlike the SBP algorithm, where the step size is normally kept constant, or initialised to a relativity large value and gradually decreased as training progresses, the optimum step size ($\eta_{opt}$) in the BBP algorithm is usually estimated at each iteration, using a line search along the direction indicated by the error gradient vector, that is:

$$\eta_{opt} = \min_{\eta}(E(\eta)) = \min_{\eta}(E(\underline{w} - \eta\underline{g})) \quad (16)$$

Common line search approaches include uniform sampling (single or multi-step), the bisection method, the golden section method and the quadratic line-search technique. Of these the latter is the most efficient, in that it converges to the minimum with the least number of function evaluations along the search direction.

A line search is termed exact if it accurately determines the minimum along the search direction. However, even the efficient quadratic line search requires several function evaluations. In an inexact line search on the other hand, the accuracy with which the minimum is evaluated is limited either by the number of function evaluations allowable or by stopping when the difference between successive approximations is less than some specified value. Determining points on the $E(\eta)$ profile is particularly computationally intensive for neural network problems. Consequently it is usually much more efficient to employ inexact line searches with MLP training algorithms such as BBP. In general, the effect of using inexact line searches, as opposed to exact ones, is to increase the number of iterations needed to reach a minimum in the cost function, while reducing the overall training time because each iteration requires less computation.

## 3 Second Order Methods

### 3.1 Taylor Series Analysis

The Taylor Series expansion for an n-dimensional function $E(\underline{w})$ about the point $\underline{w}_0$ has the form:

$$E(\underline{w}) = E(\underline{w}_0) + (\underline{w} - \underline{w}_0)^T \underline{g}_0 + (\underline{w} - \underline{w}_0)^T H_0 (\underline{w} - \underline{w}_0) + \ldots + \ldots \quad (17)$$

where $\underline{g}_0$ is the vector of gradients $\delta E/\delta w_i$ at the point $\underline{w}_0$ and $H_0$ is the Hessian matrix of second order derivatives, the elements of which are given by:

$$h_{ij} = \frac{\delta^2 E}{\delta w_i \delta w_j} \quad (18)$$

The steepest descent algorithm can be regarded as being based on a linear approximation to the actual error surface given by the first two terms in the Taylor series.

$$E(\underline{w}) \approx E(\underline{w}_0) + (\underline{w} - \underline{w}_0)^T \underline{g}_0 \quad (19)$$

This approximation to $E(\underline{w})$ is a hyperplane and as such does not have a minimum. However setting $\underline{w} = \underline{w}_0 - \eta\underline{g}_0$, where $\eta$ is a positive scalar, does guarantee a reduction in error. The step size $\eta$ could be made infinite for a hyperplane, but this approximation only holds for a very small region of $E(\underline{w})$ about $\underline{w}_0$. Hence E only reduces for a small range of $\eta$ values, the best of which can be determined using a line search technique.

BBP, a steepest descent algorithm, has very poor convergence properties leading to protracted training times. This can be regarded as being a direct consequence of the lack of accuracy of the linear approximation to $E(\underline{w})$. The next logical step to improving training performance is to derive algorithms which are based on a quadratic model of the error surface, that is approximate $E(\underline{w})$ by the first three terms of the Taylor series.

$$E(\underline{w}) \approx E(\underline{w}_0) + (\underline{w} - \underline{w}_0)^T \underline{g}_0 + (\underline{w} - \underline{w}_0)^T H_0 (\underline{w} - \underline{w}_0) \quad (20)$$

In this case the approximation is quadratic and has a global minimum at $\frac{\delta E}{\delta \underline{w}} = 0$.

$$\frac{\delta E}{\delta \underline{w}} = \underline{g}_0 + H_0 (\underline{w} - \underline{w}_0) = 0 \quad (21)$$

Solving for $\underline{w}$ gives:

$$\underline{w} = \underline{w}_0 - H^{-1} \underline{g}_0 \quad (22)$$

For a quadratic function this would yield the exact minimum. More generally it is only an approximation, in which case an iterative approach has to be adopted where $-H^{-1}\underline{g}_0$ is used as the search direction in a similar manner to $-\underline{g}_0$ in the steepest descent algorithm.

$$\underline{w}_{k+1} = \underline{w}_k - H_k^{-1} \underline{g}_k \quad (23)$$

The Hessian matrix is very difficult and time consuming to calculate and as such is seldom used directly as above. Various methods exist which avoid this calculation as follows.

### 3.2 Conjugate Gradient Training Algorithm (CG)

This is based on the fact that the minimum of any quadratic function of n variables can be found by searching along at most n independent directions which are mutually conjugate with respect to their Hessian matrix. (Two non-zero vectors $\underline{u}$ and $\underline{v}$ are said to be conjugate with respect to a non-singular matrix B if, and only if, $\underline{u}^T B \underline{v} = 0$). Various methods exist for calculating suitable conjugate vectors without knowing the Hessian matrix. One such method, requiring only a small modification to the steepest descent algorithm, is the Conjugate Gradient (CG) or Fletcher-Reeves method [4]. The algorithm is summarised in Box.1.

This is effectively batch back propagation with a momentum term, the momentum gain $\beta$ being chosen to produce conjugate vectors. The method is derived for a quadratic function and when applied to general problems conjugacy is not maintained. It therefore takes more than n iterations to reach the solution. For training an MLP, n is the number of weights in the network and is given by:

$$n = \sum_{i=0}^{L-1} N_{i+1}(N_i + 1) \quad (24)$$

where L is the number of layers in the network and $N_i$ is the number of neurons in the $i^{th}$ layer.

(1) Start with k=0 and $\underline{w}_0$.

(2) Calculate $\underline{g}_k$ for $\underline{w}_{(k)}$. If $|\underline{g}_k| < \varepsilon_1$ go to (9).

(3) Otherwise set the search direction $\underline{d}_k = -\underline{g}_k$.

(4) Find $\underline{w}_{k+1}$, the minimum point along $\underline{d}_k$.

(5) Calculate $\underline{g}_{k+1}$.

(6) If $|\underline{w}_{k+1} - \underline{w}_k| < \varepsilon_2$ or $|\underline{g}_{k+1}| < \varepsilon_1$ go to (9)

(7) Otherwise set $d_{k+1} = -\underline{g}_{k+1} + \beta_{k+1}\underline{d}_k$, where $\beta_{k+1} = \dfrac{[\underline{g}_{k+1}]^T[\underline{g}_{k+1}]}{[\underline{g}_k]^T[\underline{g}_k]}$

(8) Set k=k+1 and repeat from (4).

(9) Stop.

**Box.1.** Fletcher-Reeves Conjugate Gradient Algorithm

In practice, conjugacy deteriorates with the number of iterations carried out, as does the performance of the method. To counter this problem the algorithm is periodically reset to steepest descent, usually every n iterations. In terms of the procedure outlined in Box.1, this corresponds to going from step (8) to step (3) instead of (4) once every n iterations.

When using the CG technique the line search (step (4)) must be exact in order to obtain conjugate search directions. The algorithm performs very poorly if inexact line searches are used.

### 3.3 Analysis of CG Training.

*Problem1*

Fig.5 shows the learning curves obtained when using the CG technique with various reset values (r). CG with r=1 is simply BBP (steepest descent) while CG with $r = \infty$ corresponds to pure conjugate gradient (i.e no reset). CG with reset performs much better than pure conjugate gradient and CG in general is significantly better than BBP.

The performance varies considerably depending on the choice of reset (r). In this example r=50 produces the best result but, in general, there is no clear choice for the optimum reset value as it varies widely from problem to problem. However, small reset values ($r<n$) tend to give the best results in the early stages of training while larger values ($r>n$) do better in the later stages. Thus, rather than keeping the reset value fixed throughout training, it could be allowed to vary, starting small and increasing as training proceeds. Possible rules for determining r could take the form $r_{i+1} = f(r_i)$ where f(.) is a function of some or all of the following; the number of iterations; the number of weights (n); the sum squared error; and the magnitude of the gradient. Some basic variable reset rules, based on the first of these, have been tested and found to give better learning curves than those obtainable using a fixed reset parameter, as well as much greater consistency of performance from problem to problem.

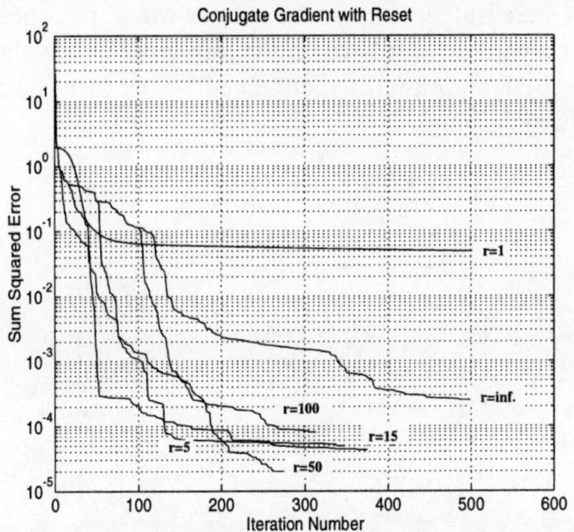


**Fig.5.** Performance of the CG Algorithm

The step sizes (η) obtained when using the CG technique are compared with those of the BBP algorithm in Fig.6. The oscillatory pattern of the BBP step sizes can be interpreted as indicating the inefficiency of this method because it contains information not utilised by the algorithm. In the case of the CG technique there does not appear to be any particular pattern in the neta values which are almost random. A second feature of these values its that they cover a much wider range than those of the BBP algorithm and are on average 10 times larger. This again is an indication that conjugate gradient is a much better method, that is, large step values occur because the search directions are much more accurate.

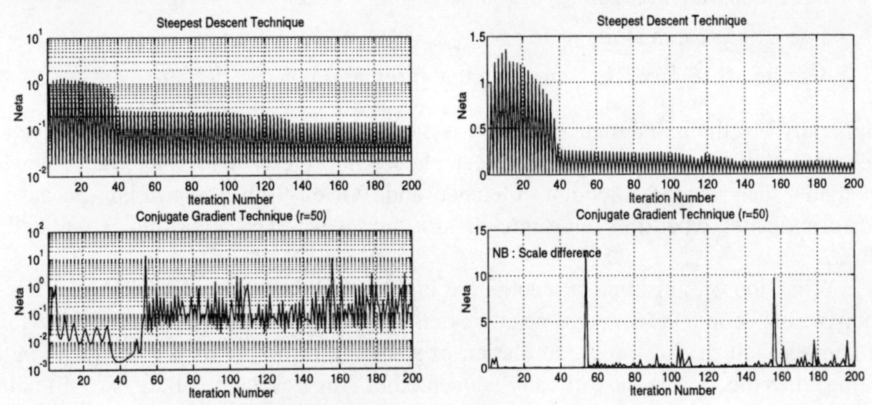


**Fig.6.** Comparison of Step Sizes for CG and BBP

*Problem 2*

The superiority of the conjugate gradient search directions is clearly evident in the contour plot for Problem 2 (Fig.7). Here the CG technique has almost reached the minimum after only two iterations, that is, after two searches. This is consistent with the fact that the error surface in this problem is approximately quadratic in the region of the minimum. If it were truly quadratic the contour would be concentric circles and the conjugate gradient algorithm would reach the minimum in exactly two iterations. The inefficiency of the BBP algorithm is highlighted by the fact that its weight trajectory zig-zags towards the minimum.

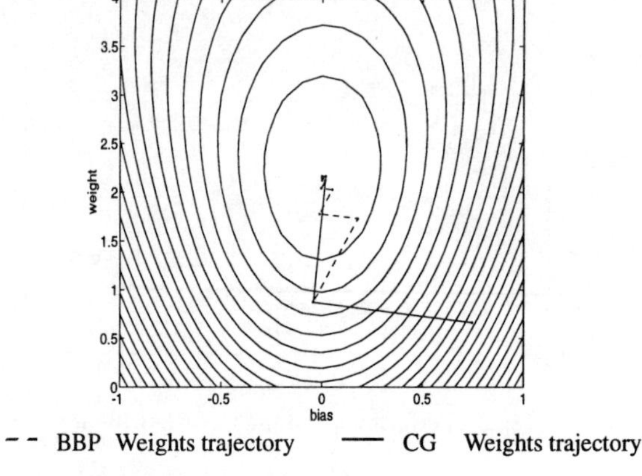


-- BBP Weights trajectory —— CG Weights trajectory

**Fig.7.** CG Weights Trajectory in Problem 2

## 3.4 Full Memory BFGS Training Algorithm (FM)

The BFGS (Broyden, Fletcher, Goldfarb and Shanno) algorithm belongs to a family of optimization methods referred to as Quasi-Newton methods [5] which are characterised by the fact that instead of calculating the Hessian matrix (Newton's methods) its inverse is approximated by a symmetric positive definite matrix $M_k$ which is updated iteratively. The basic computational structure is as follows.

- Set the search direction ($\underline{d}_k$) equal to $-M_k\underline{g}_0$, where $M_k = H_k^{-1}$.
- Use a line search along $\underline{d}_k$ to give $\underline{w}_{k+1} = \underline{w}_k + \eta_k . \underline{d}_k$.
- Update $M_k$ to give $M_{k+1}$, ensuring that it remains positive definite.

Generally $M_1$, the initial matrix is taken as the identity matrix corresponding to steepest descent since $\underline{d}_k = -\underline{g}_k$. BFGS has been chosen in preference to other Quasi-Newton methods such as the Davidon, Fletcher and Powell (DFP) algorithm because it performs better when the line searches are inexact [4]. The algorithm is outlined in Box.2.

When the optimisation problems are highly nonlinear (i.e. non quadratic) the M matrix can become indefinite or negative definite with the result that the algorithm fails. It can also fail due to the accumulation of rounding errors. If M is reset to its initial value when this occurs the difficulty is overcome. This Reset On Failure (R.O.F) policy is also applicable to the CG technique and the Limited Memory BFGS algorithm to be discussed in the next section. If the algorithms still fails after resetting then a local minimum has been reached.

(1) Select $\underline{w}_0$ and some arbitrary positive definite matrix $M_0$ (nxn) e.g. I.
Set $\varepsilon_1, \varepsilon_2$ to some small positive numbers. Calculate $\underline{g}_0$. Stop if $|\underline{g}_0| < \varepsilon_1$.
Set k=0.

(2) Calculate $\underline{d}_k = -M_k \underline{g}_k$.

(3) Use a line search along s(k) to find the minimum error along s(k) giving a new point $\underline{w}_{k+1} = \underline{w}_k + \eta_k . \underline{d}_k$.

(4) Determine $\delta_{k+1}$ where $\delta_{k+1} = \underline{w}_{k+1} - \underline{w}_k$. Stop if $|\delta_{k+1}| < \varepsilon_2$.

(5) Calculate $\underline{g}_{k+1}$. Stop if $|\underline{g}_{k+1}| < \varepsilon_1$.

(6) Determine $\lambda_{k+1}$ where $\lambda_{k+1} = \underline{g}_{k+1} - \underline{g}_k$.

(7) $A_{k+1} = \left(1 + \frac{\lambda_{k+1}^T M_k \lambda_{k+1}}{\delta_{k+1}^T \lambda_{k+1}}\right) \frac{\delta_{k+1} \delta_{k+1}^T}{\delta_{k+1}^T \lambda_{k+1}}$ $\quad B_{k+1} = -\frac{\delta_{k+1} \lambda_{k+1}^T M_k + M_k \lambda_{k+1} \delta_{k+1}^T}{\delta_{k+1}^T \lambda_{k+1}}$

**Box.2.** The Full Memory BFGS Training Algorithm

Resetting M periodically during training can lead to better performance in some problems. This is illustrated in Fig.8 (a) which shows the results obtained when the BFGS algorithm with different reset values was used to train a (1,5,1) MLP network with sigmoid hidden neurons to represent the 3-D mapping z=x.sin(x)+y.cos(y)-0.75. Here the training set consisted of 225 points drawn uniformly from x and y in the range [-2,+2]. In this example the performance with r=20 and r=50 is much better than when $r = \infty$ (i.e BFGS without reset). The best results were obtained when r was varied (r=comp.) throughout training as follows: r=5 for iteration1 to 5; r=10 for iteration 6 to 105; and $r = \infty$ for iteration 106 onwards. These values were chosen manually, but clearly the reset variable could be determined automatically using an iterative rule of the form described earlier for the CG method.

In many problems, however, the performance of the BFGS algorithm without reset is as good as that obtained with reset. Fig. 8(b), which shows the variation in performance with r for Problem 1 is a typical example.

The M matrix can be thought of as building up a quadratic model of the cost function over a number of iterations. In the earlier stages of training the algorithm may be operating in a region of the error space which is highly non-quadratic with the result that the validity of the model being built is very localised. Frequent resetting under these conditions is advantageous as it discards information which is only locally applicable, information which would distort the model in the later stages of training. As training moves the weights nearer to a minimum the quadratic model will be become more widely valid with the result that resetting should be employed less frequently or not at all. The effect of resting too frequently is clearly seen in Fig. 8(b).

A second interpretation of the effect of resetting is that it limits the accuracy of the quadratic model and as a result the search directions produced by the algorithm. Consequently the algorithm is less likely to be attracted towards a shallow minimum, which is of course advantageous. However, once in the basin of attraction of a minimum, resetting has a detrimental effect and should be phased out.

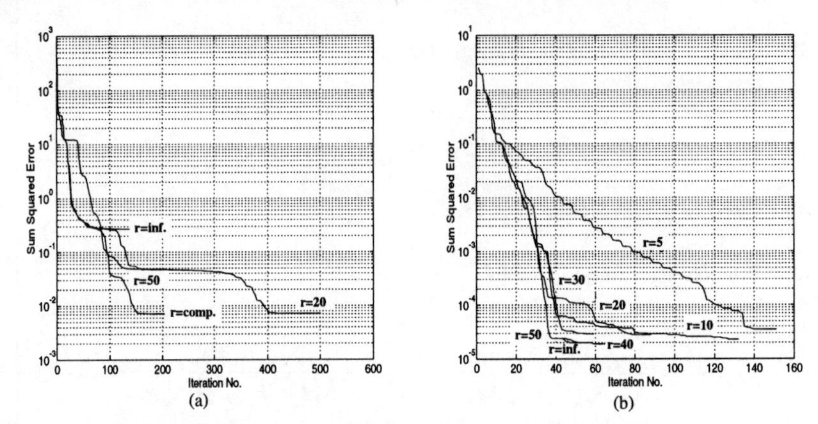


**Fig.8.** Performance of the BFGS algorithm for various reset values.

### 3.5 Limited Memory BFGS Training Algorithm (LM)

The Limited Memory BFGS algorithm [5] is simply the BFGS full memory algorithm with the M matrix reset to the identity matrix every iteration. Since $M_k$ is always I the algorithm can be simplified so that there is no need to calculate or store matrices. This leads to a large saving in memory, hence the name.

### 3.6 Analysis of the LM and FM Training

*Problem 1*

Fig.9(a) shows the learning curve obtained for Problem 1 with the FM (no reset) and LM algorithms. The best result obtained with the CG method has also been included for comparison. There is little difference between the methods over the first few iterations, but in the later stages the rate of convergence to the minimum of the FM technique is much faster than the other algorithms. In this example the CG result is better than that obtained using the LM algorithm. In general there is little difference between the methods, but the LM algorithm has one major advantage over CG. There are no parameters to be chosen when using LM approach, whereas with CG the reset value (r) has to be carefully selected to give good performance.

The superiority of the FM algorithm is also reflected in the step sizes obtained when using it. These are compared with the step sizes obtained using the CG method in Fig.9(b). On average the FM step sizes are much larger than those of the CG technique.

The mean values are 2.366 and 0.7104 respectively. This is an indication that the FM search directions are much more accurate than those of the CG algorithm. The LM algorithm step sizes, which have been omitted from Fig.9(b) for reasons of clarity, are similar to those obtained with the CG technique.

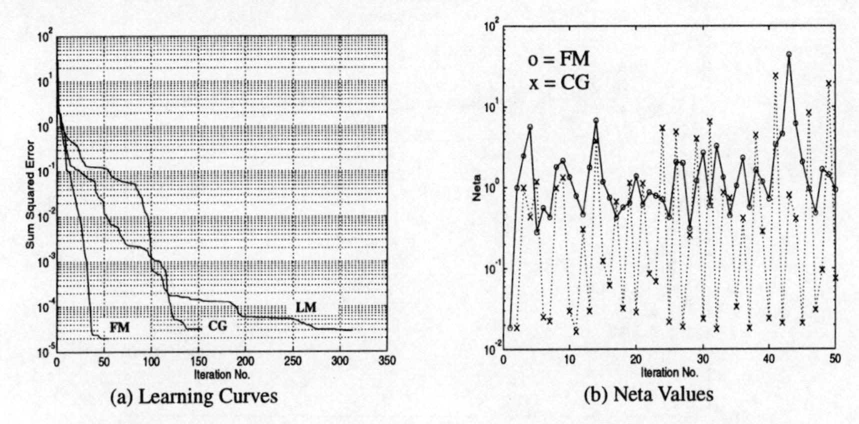


(a) Learning Curves (b) Neta Values

**Fig.9.** Comparison of the FM, LM and CG algorithms for Problem 1

*Problem 2*

The Full and Limited Memory BFGS algorithms produce results which are almost identical to those obtained with the CG algorithm when applied to Problem 2. This is as would be expected because the FM algorithm with exact line searches is equivalent to the Fletcher-Reeves conjugate gradient method when applied to a quadratic function [6].

### 3.7 Speed of Training

Thus far the performance of the various training algorithms considered has been assessed in terms of the rate of convergence to a minimum as a function of the number of iterations. This is useful when determining the power of an algorithm, but it does not take into account the algorithm iteration time and hence is not a true reflection of performance.

When comparing training methods the essential criterion is clearly the error reduction obtainable in a given amount of time. This can easily be assessed by plotting the learning curves as a function of time. Those obtained for Problem 1 are given in Fig.10 and clearly show that the FM algorithm is the best for this problem, followed by LM algorithm. Each of the three second-order methods stopped training after only a few seconds, having reached minima in the error surface. As can be seen, each ended up in a different local minimum with the FM algorithm finding a much deeper minimum than the other methods. This observation is true in general and appears to be related to the ability of the FM algorithm to find its way out of plateau type regions on the error surface.

The size of the network, the number of training vectors and the degree of accuracy with which the line searches are carried out all influence the iteration time. The effect of these parameters will be illustrated using a realistically sized problem in the next section.

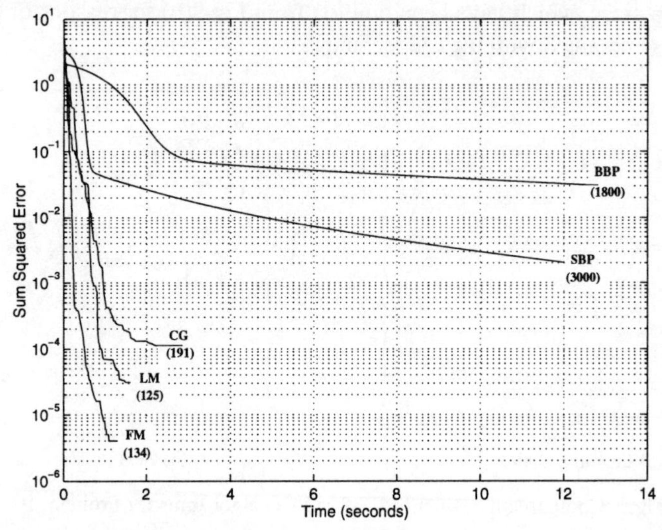


**Fig.10.** Training Algorithm Performance for Problem 1. The number of iterations performed by each algorithm is given in brackets. SBP ($\eta$=0.1, $\beta$ =0.1), CG (variable reset) and FM (no reset)

### 3.8 Training Performances for the CSTR Problem

Four different MLP network and training set combinations were investigated (Table 1). An inexact quadratic line search was used with each of the batch algorithms. Various different degrees of inexactness were tested with each of the algorithms and the best learning curves obtained in each case are given in Fig.11.

| Name | Network Structure | Training Set Size |
|---|---|---|
| Test 1 | (20,5,1) MLP | 400 vectors |
| Test 2 | (20,10,1) MLP | 400 vectors |
| Test 3 | (20,20,1) MLP | 400 vectors |
| Test 4 | (20,20,1)MLP | 800 vectors |

**Table 1:** Definition of training tests

| | Test 1 | Test 2 | Test 3 | Test 4 |
|---|---|---|---|---|
| SBP | 0.193 | 0.370 | 0.732 | 1.50 |
| BBP | 0.294 | 0.560 | 1.090 | 2.30 |
| CG | 0.430 | 0.866 | 1.632 | 2.49 |
| LM | 0.290 | 0.750 | 1.132 | 2.37 |
| FM | 0.300 | 0.660 | 1.580 | 2.73 |

**Table 2**: Iteration times (seconds)

Note that the CG algorithm requires significantly more function evaluations, due to the dependency of the method on accurate line searches. Note also that the FM training requires the least number of function evaluations, with the result that its iteration times are comparable to those of the other algorithms, even though its method for determining the search direction is much more computationally intensive (Table 2).

These results show that, even for relatively large problems, the FM training approach is very much superior to the others. CG and LM also perform reasonably well when compared to the BBP algorithm. The SBP algorithm also gives very good results for this problem though, in general, it tends to be inferior to the second-order methods.

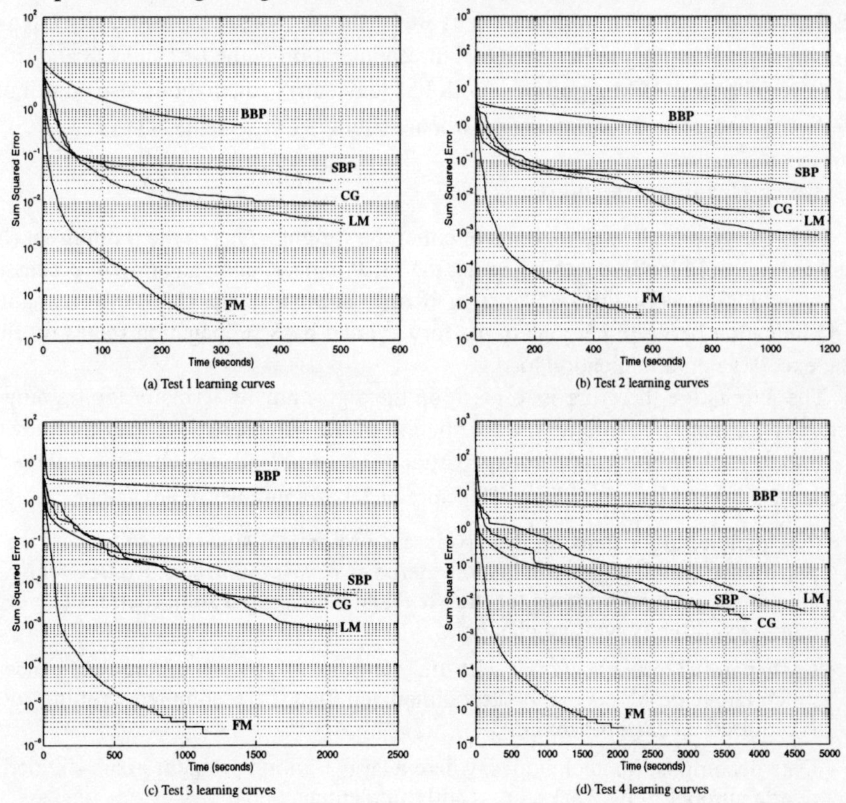


**Fig.11.** Learning curves for the CSTR neural network predictor. SBP ($\eta$=0.1 and $\beta$=0.1), CG (variable reset) and FM (no reset)

# 4 Choosing a Training Algorithm

While FM is clearly the best training algorithm it cannot always be used because of hardware limitations. It employs three (n x n) matrices, where n is the number of weights, with the result that memory requirements become extremely large as n increases. ($memory \propto n^2$). In the past this meant that only very small networks (n<100) could be trained using FM. For larger networks the CG or LM algorithm can be employed since the memory requirements are much less ($memory \propto n$). Modern computers have an increasing memory capacity with the result that FM can now be used to train relatively large problems (n < 5000).

A second reason why the FM algorithm may not be practical for very large networks is the fact that the amount of matrix computations increases as $n^2$. Thus the iteration time for the FM method will increase dramatically to the extent that the CG and LM algorithms will overtake it, especially in the earlier stages of learning.

## 5 Potential for Parallelisation

An analysis of LM, CG and FM methods reveals that the batch gradient calculation and the step-size calculation (line search) are by far the most computationally intensive components, amounting to 98 to 99% of the computation in the LM and CG algorithms and 80 to 95% in the FM algorithm. With FM, the matrix calculations also contribute a significant amount to the computation when the number of weights is large.

### 5.1 Batch Gradient Calculation

This can be parallelised in terms of either the weights ($N_W$) or the training set ($N_V$). The former would involve implementing the MLP in parallel on an array of processors. This would give very little advantage, if any, because there is a large amount of communication between neurons in the forward and back propagation stages resulting in an excessive communication load.

The alternative therefore is to partition the algorithms in terms of the training set vectors ($N_V$), that is, divide up the calculation of the batch gradient over a number of processors, each process calculating partial batch gradients which are combined to produce the overall batch gradient. This approach has a number of advantages:

- The parallel algorithms are relatively easy to create, since the subsets of the partitioned training data can be considered as new training sets. Consequently, existing sequential routines for the MLP and back propagation routines can be employed directly without change.
- Each parallel process works with only a portion of the training set, hence much larger problems can be accommodated than if the complete training set is required on each processor.
- Data decomposition techniques, where a large training problem is sub-divided over a number of networks, are readily implemented [7].

### 5.2 Step-Size Calculation

Here there are three choices for parallelisation, on the weights, on the training set or on the function evaluations ($N_C$). The comments made above on the first two possibilities, in relation to the batch gradient, also apply here.

Parallelisation in terms of the function evaluations cannot be done directly as the quadratic line-search approach is essentially sequential. However, if a bisection approach is used, where a number of points are evaluated uniformly over a given interval and the minimum chosen, then each evaluation can be assigned to a parallel processor [8]. To get the desired accuracy the bisection can be done twice, the second time using the interval around the lowest points obtained in the first interval. Thus if $T_c$ seconds is the duration of one evaluation over the training set then the bisection method when parallelised takes $2T_c$ seconds. Increasing the number of parallel processes simply increases the accuracy with which the step-size is calculated. For 6 processors a total of 12 evaluations would be carried out which would take $12T_c$ seconds if implemented sequentially. This method has a number of disadvantages:

- It requires the complete training set to be available on each processor.
- Increasing the number of processors does not improve speed-up.
- If there are a limited number of processors then the accuracy of step-size determination will be restricted.

Consequently parallelisation of the training set is favoured here also. In this instance the evaluation of the sum squared-error over all the training set is divided up over a number of parallel processors so that each one evaluates a sum squared-error for the portion of the training set assigned to it.

Adopting this approach the sequential quadratic line-search technique can be parallelised, without affecting the structure of the method. For n processors the speed-up will be of the order of $N_c$ x $T_c$/n while the sequential implementation takes approximately $N_c$ x $T_c$ seconds. For a quadratic line-search with $N_c$=8 (a value which would give an accuracy greater than the 12 point bisection method) the speed-up will be greater than that achievable with the bisection technique whenever n > 4. The advantages of partitioning the step-size calculation in this manner are:

- It is compatible with the batch gradient calculation in that only a portion of the training set is stored on each processor, hence memory saving.
- Speed up improves with the number of processors and accuracy is the same irrespective of the number of processors.
- The algorithm is numerically equivalent to the sequential version.

## 5.3 FM Matrix Calculations

Parallelising the matrix calculations in the FM algorithm is worth considering for two reasons:

- For large values of $N_w$, these become significant, especially when the number of training vectors is relatively small.
- The storage required for the M, Δ and B matrices becomes a limiting factor on the size of problem which can be handled.

Consider, for example, a T800 transputer with 2 MB of RAM. Taking into account reasonable program, variable and training set storage requirements, the memory available for the 3 arrays would be about 1.2 MB, corresponding to 300,000 floating point numbers. Each array thus can have at most 100,000 elements and the largest network that can be accommodated using full matrix storage has 310 weights. This would be equivalent to a (20,14,1) network.

However, the matrices are symmetric and this can be exploited to reduce both the memory and the computational requirements. In the case of an (mxm) symmetric matrix only m(m+1)/2 elements need to be stored. With this memory saving, matrices of dimension (445 x 445) can be used corresponding to a (20,20,1) network.

A good choice for parallelisation is to partition the M, B and Δ matrices by rows so that if the matrices are (r x r) then r/n rows will be placed on each of n slave processes. If this approach is taken matrices of dimensions up to ($310\sqrt{n}$ x $310\sqrt{n}$) can now be accommodated in the case of the T800 transputer example considered above. Note that n=2 gives a memory reduction which is equivalent to exploiting symmetry in the sequential implementation. Symmetry cannot be exploited in the parallelised matrix calculations.

While this parallelisation option is beneficial in terms of dealing with memory restrictions, the speed-up achievable is small compared to that obtainable by vector based parallelisation of the step size and batch gradient calculations. Consequently it will not be considered further here.

# 6 Hardware Implementation

In this section the implementation of the parallel training algorithms will be considered for a Sun network running as a Parallel Virtual Machine and for a transputer network consisting of T805 TRAMS with 2 MB of RAM.

## 6.1 PVM

PVM (Parallel Virtual Machine) is a software package that allows parallel programs to be run on a heterogeneous network of Unix computers[9]. It consists of two parts:

(a) A daemon process that resides on all the computers making up the virtual machine.

(b) A PVM library containing user callable C functions for message passing, spawning processes, coordinating tasks, and modifying the virtual machine.

The speed of communication between machines across a network is slow in comparison with that of dedicated concurrent processing hardware. It also varies considerably depending on the network and machine load. Consequently running parallel programs under PVM is only beneficial when problem granularity is very high. The major advantage offered by PVM is that it allows the processing power and large memory capacity of workstations to be exploited.

## 6.2 Transputers

A transputer network is a dedicated parallel programming architecture offering high speed communication between adjacent processors. Unlike the PVM system which allows direct communication between all processors, each transputer can transfer data directly to at most four other transputers. Therefore each processor has only 4 serial data links available to it. Consequently, when parallel programming on transputers a suitable network configuration has to be selected and data transfer mechanisms built into the parallel programs. Typical transputer configurations include the tree, pipe-line, ring and array. Those best suited to the parallel batch gradient and step size calculations are the pipeline and the ring (Fig.12). In the former information is passed along the pipeline to each processor. The results produced by each one are then passed back along the pipeline to the root transputer. The structure of the parallelised batch gradient and step-size calculations for this configuration are illustrated schematically in Fig. 13(a). The communication time ($T_c$) for this implementation of the parallel programs is given by:

$$T_c = n\tau(3N_w + 2N_c) \tag{25}$$

where n is the number of parallel processes and $\tau$ is the transmission time per float per process. An advantage of using this configuration is that, once the master process on the root transputer sends the search direction to the first transputer in the pipeline, it can continue with the rest of the program without having to wait for the information to reach the last transputer in the pipeline. This reduces the communication time given in equation (25) by $(n-1)N_w\tau$.

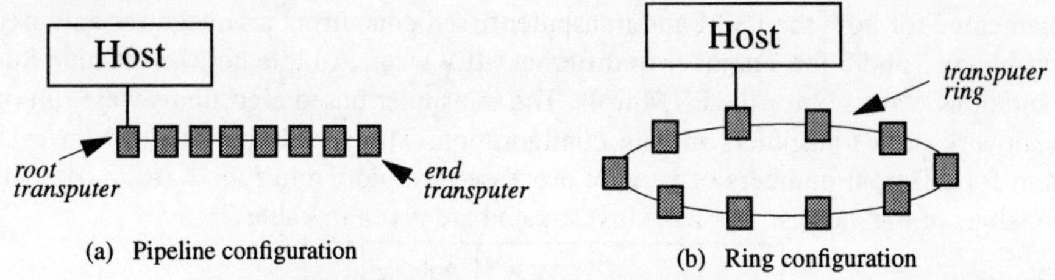


**Fig.12.** Transputer Configurations

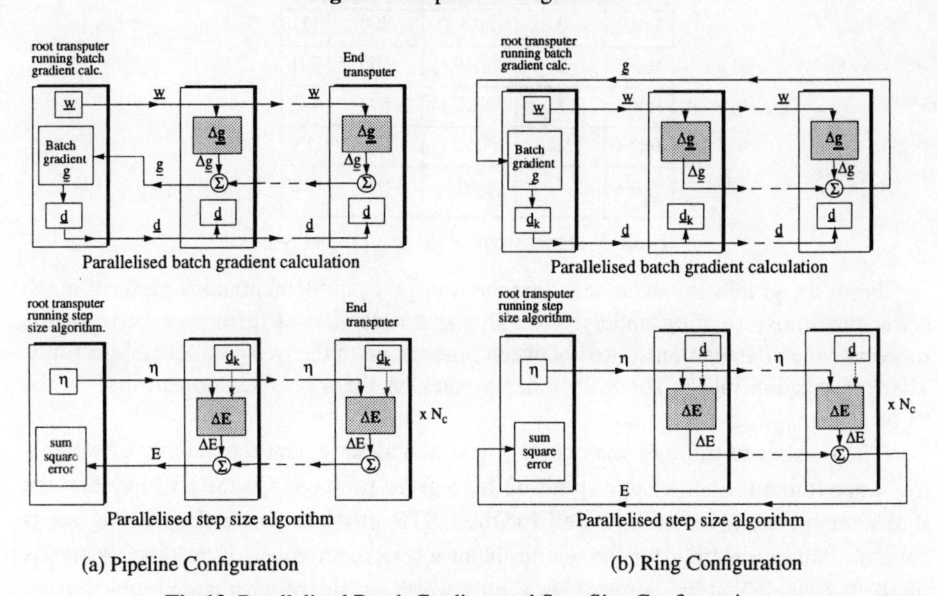


**Fig.13.** Parallelised Batch Gradient and Step Size Configurations

A ring architecture is simply a pipeline with the last transputer connected back to the first one to form a loop making it possible to have a unidirectional flow of information which can be advantageous. The structure for the parallelised batch gradient and step size calculations for the ring architecture are shown in Fig.13(b). In this case $T_c$ is given by:

$$T_c = n\tau(2N_w + N_c) + \tau(N_w + N_c) \qquad (26)$$

Again it is not necessary to wait until the search direction has gone the whole way round the ring and hence $T_c$ can be reduced by $(n-1)N_w\tau$.

Comparing equations (25) and (26) it can be seen that the ring configuration is preferable in this instance as it results in less communication within the parallel programs.

### 6.3 Parallel Algorithm Performances

Parallel versions of the LM and FM training algorithms, as described above, were implemented for both the PVM and transputer based concurrent architectures and used on problems 1 and 3 for a number of different MLP sizes. An idle network of nine Sun workstations was used for the PVM tests. The transputer based algorithms were run on a a network of 6 transputers in ring configuration. The speed-ups obtained on each system for different numbers of parallel processes are plotted in Fig.14 (a) to (d). The dimensions of the various problems investigated are given in Table 3.

| Name | Problem | MLP Structure | Training Set Size | $N_w$ |
|---|---|---|---|---|
| Test 1 | 3 | (20,5,1) | 800 | 111 |
| Test 2 | 3 | (20,10,1) | 800 | 221 |
| Test 3 | 1 | (20,20,1) | 800 | 441 |
| Test 4 | 1 | (1,5,1) | 25 | 16 |
| Test 5 | 1 | (1,10,1) | 25 | 31 |

**Table 3**: Dimensions of the test problems

From the graphs it can be seen that the transputer implementations perform much better than those running under PVM. This is a reflection of difference between the speed at which data is transmitted between processors on the two parallel architectures. The transmission rate is about 20 times greater on the transputer system that on the PVM system.

The number of training vectors ($N_v$), and to a lesser extent the number of weights ($N_w$), determine the achievable speed-up for a given problem. On the PVM system the algorithms perform reasonably well for the CSTR problem where the training set is large but fail completely for the test problem where the number of training vectors is small. Thus the PVM implementation is only worth considering for large problems.

The speed-ups achievable for the CSTR problem, using the transputer based training algorithms, approach the theoretical maximum for the number of parallel processes considered. The performance is also reasonable for problem 1. However, in the case of the problem with the largest memory requirement (Test 3), the FM algorithm could not be used because of insufficient memory on the transputers. The

PVM based FM algorithm performs reasonably well for this size of problem and therefore becomes a feasible alternative to transputer in these circumstances.

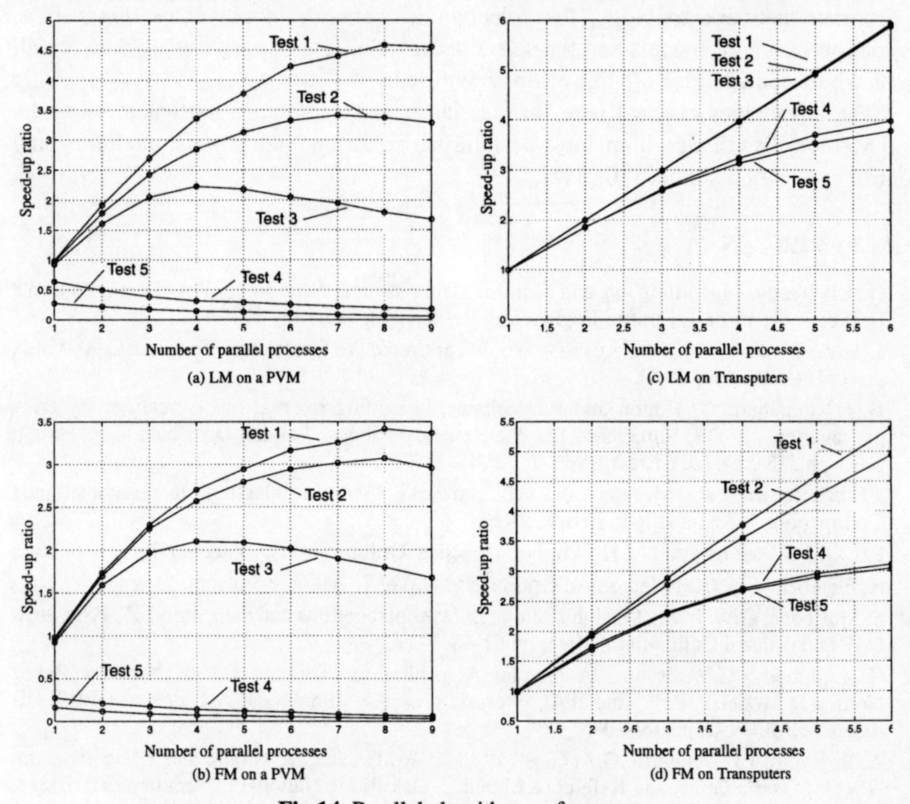


**Fig.14.** Parallel algorithm performance

# 7 Concluding Discussion

Several conclusions arise from this study of off-line, gradient based training of Multilayer Perceptrons:

- Second-order training algorithms are at least an order of magnitude faster than the standard BBP algorithm.
- Of the second-order methods examined the Full Memory BFGS algorithm is by far the superior, offering speed-ups of anything between 20 and 1000 over BBP depending on the severity of the error goal chosen and the size of the problem.
- FM requires a large amount of memory and as such cannot be used for very large problems. In these circumstances the LM or CG is preferred.
- Further reductions in training times are achievable through parallelisation of the training algorithms. This has been demonstrated for the LM and FM algorithms and shown to give significant speed-ups when implemented on a dedicated concurrent architecture such as a transputer network.

The availability of fast and powerful algorithms for off-line training of Multilayer Perceptrons is important in practical applications. For example, the determination of an appropriate network topology, the selection of network input vector dimensions, validation of neural models and the selection of training times with noisy data are all areas where considerable off-line effort is required.

The advantages claimed here for Hessian based training, in particular using the Full Memory BFGS algorithm, have been further confirmed in industrial modelling and control application studies [10], [11].

## 8 REFERENCES


1. G. Lightbody, "Identification and Control Using Neural Networks", PhD thesis, Queen's University of Belfast, Control Engineering Research Group, May 1993.
2. J.D. Morningred et al., "An Adaptive Nonlinear Predictive Controller", Proc. ACC 90, Vol.2, pp. 1614-1619, May 1990.
3. D.E. Rumelhart, G. Hinton and R. Williams, "Learning internal representations by error propagation", in D.E Rumelhart, J.L. McClelland, (editors), Parallel Distributed Processing, Vol.1 pp 318-364. MIT Press, 1986
4. J.J McKeown, D. Meegan and D. Sprevak, "An Introduction to Unconstrained Optimization", Adam Hilger, Bristol, 1990.
5. P.E. Gill, W. Murray and M.H. Wrights, "Practical Optimization", Academic Press, London.
6. R. Fletcher, "Practical Methods of Optimization", Vol.1, Wiley & Sons, pp.51.
7. S. McLoone, G.W. Irwin, "Insights into multilayer perceptrons and their training", Proc. Irish DSP and Control Colloquium, 1994, pp.61-68.
8. G. Lightbody, G.W. Irwin, "A parallel Algorithm for Training Neural Network Based Nonlinear Models", Proc. 2nd IFAC Workshop on Algorithms and Architectures for Real-time Control, 1992, pp. 99-104.
9. A. Beguelin, J.J. Dongarra, G.A. Geist, W.Jiang, R.Manchek, K. Moore and V.S. Sunderam, "PVM 3 User's Guide and Reference Manual", Oak Ridge National Laboratory, Oak Ridge, Tennessee 37831, 1993.
10. G. Lightbody, G.W. Irwin, A. Taylor, K. Kelly and J. McCormick, "Neural Network Modelling of a Polymerisation Reactor", Proc. IEE Int. Conf., Control '94, Vol.1, pp. 237-242.
11. M.D. Brown, G.W. Irwin, B.W. Hogg and E. Swidenbank, "Modelling and Control of Generating Units using Neural Network Techniques", 3rd IEEE Control Applications Conference, Glasgow, August 1994, Vol.1, pp. 735-740.

# Kohonen Network as a Classifier and Predictor for the Qualification of Metal-Oxide-Surfaces

Waltraud Kessler and Rudolf W. Kessler

Institut für Angewandte Forschung, Fachhochschule Reutlingen
Alteburgstr. 150, D-72762 Reutlingen, Germany

**Abstract.** The corrosion of metals and the paint adhesion on metals is a result of the superposition of complex reactions on the surface, which depend on surface oxide thickness, its porosity and chemical composition. By means of diffuse reflectance spectroscopy and evaluation of the spectra by a Kohonen self-organizing map, it is possible to predict the future corrosion behaviour of low carbon steel. Combining a Kohonen map and an interpolation method in the output layer allows to determine the layer thickness of conversion layers on aluminium from their interference spectra. This offers a fast, reliable and on-line applicable tool to calculate the thickness of transparent surface layers on aluminium or other metals even in the range below 100 nm.

## 1 Introduction

In order to maintain and improve the quality of its products, industry requires rapid techniques for on-line process control and monitoring. This need for rapid responses is associated with other requirements, such as the possibility to simultaneously determine many distinct factors with non invasive methods. In recent years, much research has therefore been done on the development of rapid quality control and analytical techniques for on-line and in situ analysis during processing. Classical methods for quality control e.g. off-line analytical tools or optical inspection systems are more and more replaced by spectroscopic methods such as Near Infrared, Infrared, Raman or diode array UV/VIS-detectors. These techniques have been implemented already in the food industry, pharmaceutical industry and chemical process industry. But little work has been done on the characterization of metal surfaces and on integrated optical and spectroscopic tools for process control during the production of surface conversion layers on metals. State of the art tools are image analysis systems, based on CCD-camera devices or e.g. Laser interferometric methods. However, these methods do not give sufficient information on complex surface properties, like quality with respect to corrosion or paint adhesion.

Spectroscopic measurements of surfaces show the superposition of many factors, which influence the wavelength dependent reflectance and absorption of electromagnetic waves. Direct correlation of the spectral features with macroscopic properties are therefore difficult to visualize. Ideal mirrors reflect the incident light according to the Fresnels equations. Powder as an ideal light scatter shows only diffuse reflectance. Surfaces of metals or conversion layers on metals, however, often exhibit the superposition of these ideal features depending on the surface roughness of the metal. Thus, the analysis and interpretation of the spectroscopic results need an analytical tool which is able to classify a complex behaviour. Due to non-linearities within the spectral features, neural networks seem to be ideally suited to manage this problem. The Kohonen self-organizing map offers the advantage of an unsupervised learning for classification. In addition, the weights of the Kohonen map reveal information on the detailed spectral features, which are important for the classification and therewith provide hints for a scientific interpretation. This is a major advantage of the Kohonen network.

This paper presents the general aspects of the use of spectroscopic measurements for on-line control in combination with a Kohonen self-organizing map. Two examples will be presented here: Improved prediction of the corrosion behaviour of car body steel by diffuse reflectance measurements and the control and prediction of surface conversion layer thicknesses by a specular reflectance measurement.

The combination of fast and reliable spectroscopic measurements in combination with a neural net, especially a Kohonen map, allows an intelligent processing of materials. That means, the process is directly controlled by the quality of the final product which thereby influences the process variables like temperature, concentration etc. Thus the design of the process can respond to different raw materials as well as to different product properties.

# 2 Neural Networks for Classification and Interpolation

## 2.1 Classification with a Kohonen Map

For classification we use a Kohonen self-organizing feature map [1]. This kind of neural net compresses a multidimensional input into a one- two- or three-dimensional output space. A very important concept of this neural net is, that it preserves the topology of the multidimensional input data within the map. That means: the topology of the map corresponds to the topology of the intrinsic data.

All neurons in the Kohonen layer receive the same input. The output of each neuron is not connected to all other neurons in this active layer, but only to a small number that are topological close to it. Such local feedback of possible corrections has the result, that topological close neurons behave similarly when similar signals are input. Figure 1 shows the architecture of a Kohonen self-organizing map.

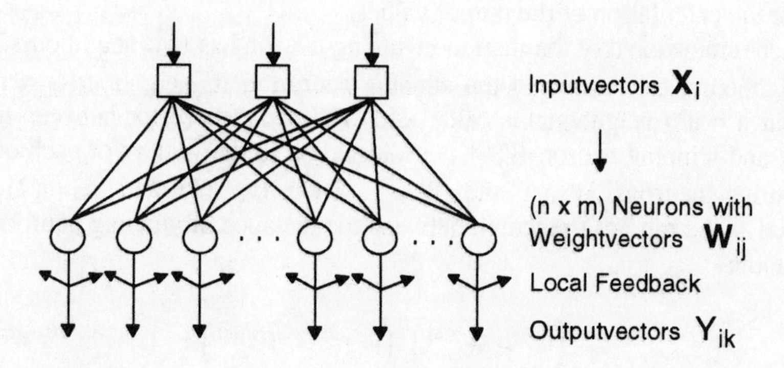


**Fig. 1.** Architecture of a Kohonen self-organizing map

The algorithm for one cycle of the Kohonen learning is as follows: An $m$-dimensional input vector $X_i$ enters the network. The Euclidean distance between the input vector $X_i$ and the weights $W_{ij}$ of the neurons is calculated for all neurons. The position $c$ of the neuron, whose weights are most similar to the input vector is found. The weights of neuron $c$ are corrected to improve its response for the same input $X_i$ on the next cycle. The weights of all neurons in the neighbourhood of the $c$-th neuron are corrected by an amount that decreases with increasing topological distance from $c$. This process repeats for all input vectors $X_i$ for several times. A trained Kohonen network maps similar signals to similar neuron positions.

We use a square neighbourhood with 4 nearest neighbour neurons and apply this Kohonen network to classify diffuse reflectance spectra according to the later corrosion behaviour of the car body steel samples.

## 2.2 Interpolation with a Kohonen Map

In this work we use the counter-propagation architecture [2] with a self-organizing map in the competition layer and add an interpolation method to the output layer [3]. Each neuron in the competition layer associates an input configuration with a corresponding output. The interpolation method takes into account the most important feature of the self-organizing map, namely to preserve topology [1].

In a standard Kohonen network each input vector is assigned to the winning neuron. Since neurons are discrete points of the function that is to be approximated, the output values are discrete, too. Input vectors, which are between the neurons of the map are approximated to discrete output values. The idea of topological interpolation is to use the winning neuron and its topological neighbours for the calculation of the output values.

In each dimension d of the map the winning neuron has two neighbours. For both neighbours, the ratio α of the winning neuron to its neighbour (*r* = right neighbour, *l* = left neighbour) is calculated. The distance vector between input vector $\boldsymbol{X}$ and winning neuron $\boldsymbol{W}_{\mathrm{w}}^{(\mathrm{in})}$ is projected on to the vectors of each of the neighbouring neurons $\boldsymbol{W}_{\mathrm{d,l}}^{(\mathrm{in})}$ and $\boldsymbol{W}_{\mathrm{d,l}}^{(\mathrm{in})}$. Then $\alpha_{d,l}$ (1) and $\alpha_{d,r}$ (2) are calculated as the ratio of the length between the distance of winning neuron and its neighbour.

$$a_{d,l} = \frac{\left(\boldsymbol{X} - \boldsymbol{W}_{w}^{(\mathrm{in})}\right)^{T}\left(\boldsymbol{W}_{d,l}^{(\mathrm{in})} - \boldsymbol{W}_{w}^{(\mathrm{in})}\right)}{\left(\boldsymbol{W}_{d,l}^{(\mathrm{in})} - \boldsymbol{W}_{w}^{(\mathrm{in})}\right)^{T}\left(\boldsymbol{W}_{d,l}^{(\mathrm{in})} - \boldsymbol{W}_{w}^{(\mathrm{in})}\right)} \tag{1}$$

$$a_{d,r} = \frac{\left(\boldsymbol{X} - \boldsymbol{W}_{w}^{(\mathrm{in})}\right)^{T}\left(\boldsymbol{W}_{d,r}^{(\mathrm{in})} - \boldsymbol{W}_{w}^{(\mathrm{in})}\right)}{\left(\boldsymbol{W}_{d,r}^{(\mathrm{in})} - \boldsymbol{W}_{w}^{(\mathrm{in})}\right)^{T}\left(\boldsymbol{W}_{d,r}^{(\mathrm{in})} - \boldsymbol{W}_{w}^{(\mathrm{in})}\right)} \tag{2}$$

These α values calculated in input space represent the interpolation parameters for the output vector $\boldsymbol{Y}$ (3) in output space. $K_d$ indicates the number of existing neighbours in dimension *d*. (In border regions $k_d = 1$).

$$\boldsymbol{Y} = \boldsymbol{W}_{w}^{(\mathrm{out})} \sum_{d=1}^{dimensions} \frac{a_{d,l}\left(\boldsymbol{W}_{d,l}^{(\mathrm{out})} - \boldsymbol{W}_{w}^{(\mathrm{out})}\right) + a_{d,r}\left(\boldsymbol{W}_{d,r}^{(\mathrm{out})} - \boldsymbol{W}_{w}^{(\mathrm{out})}\right)}{k_d} \tag{3}$$

It is important to emphasize, that this kind of interpolation needs maps with a topology that corresponds to the topology of the data set. Otherwise topological defects produce errors, and geometric interpolation [4] is the better method.

We apply this topological interpolation method to evaluate interference spectra of oxide layers on aluminium alloys.

## 2.3 Interpolation with Radial Basis Function Network

Another widely used technique for interpolation is the radial basis function network (RBF) [5]. The results obtained by the prediction with the Kohonen interpolation technique are therefore compared with the results obtained by the RBF network. The topology of this network is presented in Fig. 2. This type of network is able to approximate any real-valued continuous function within a certain accuracy [6].

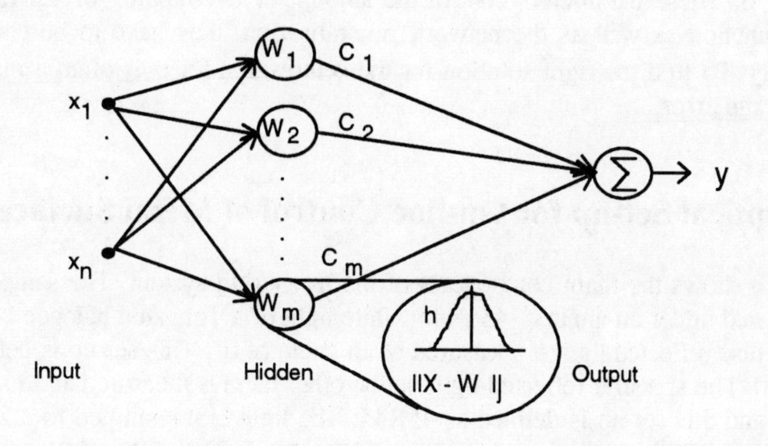


**Fig. 2.** Radial basis function network

Each neuron of the hidden layer has a radial activation function. A typical choice for this radial basis function $h_i$ is the Gaussian function (4) which has a peak at the centre **z** and decreases monotonically as the distance from the centre increases.

$$h(z) = e^{-(z/\sigma)^2} \qquad (\sigma > 0) \tag{4}$$

The training algorithm for the RBF net is as follows: First, the $m$ centres $\boldsymbol{W}_i$ for each hidden neuron are chosen. The simplest technique is to choose all the $m$ training data $\boldsymbol{X}_j$ as centres $\boldsymbol{W}_i$. And therewith is the function value $y_m$ defined for each centre. Then, a radial basis function $h_i$ is associated with each neuron. The input $z$ to this radial basis function $h_i$ is the Euclidean distance between the input vector $\boldsymbol{X}_j$ and the centre vector $\boldsymbol{W}_i$. The interpolation function is (5)

$$f(\boldsymbol{X}_j) = \sum_{i=1}^{m} c_i h_i \left( \left\| \boldsymbol{X}_j - \boldsymbol{W}_i \right\| \right) = y_j \tag{5}$$

The unknown weighting factors $c_i$ are calculated by solving the linear equation system for all $m$ centres $\boldsymbol{W}_i$ and all $m$ input vectors $\boldsymbol{X}_j$. The final output, that means the interpolated function value for an input vector $\boldsymbol{X}$ is the weighted sum (6) of the different radial basis functions of the hidden neurons.

$$f(X)=\sum_{i=1}^{m} c_i h_i(\|X - X_i\|) \tag{6}$$

Two parameters are associated with every RBF node, the 'centre' $\mathbf{z}_i$ and the 'width' $\sigma$. These parameters control the amount of overlapping of the radial basis function as well as the network generalisation, they have to be chosen correctly. To find the right solution for the centres and for $\sigma$ is often a matter of trial and error.

# 3 Optical Set-up for On-line Control of Metal Surfaces

Figure 3 shows the main components of the measuring system. The sample is illuminated under an angle of 45° with white light of a Tungsten halogen lamp. The diffuse reflected light is measured at an angle of 0°. This set-up is defined as 45R0. The specular reflected light on the other hand is measured at an angle of 45° and this set-up is defined as 45R45. The light is transmitted to a Zeiss MCS UV/VIS diode array spectrometer (Zeiss MCS 210) with a holographic grid. The spectrometer has 512 diodes and covers a wavelength range from 360 to 780 nm. The spectra are registered with a repetition rate of about 100 Hz (minimum sampling time of the spectrometer is 10 ms) and analysed by a standard intel computer (486 processor). During initialisation of the system the dark current spectrum is registered as well as the reference spectra. The diffuse reflectance spectra are measured relative to bariumsulfate ($BaSO_4$). The reference sample for the specular reflected spectra is a metal surface with no conversion layer.

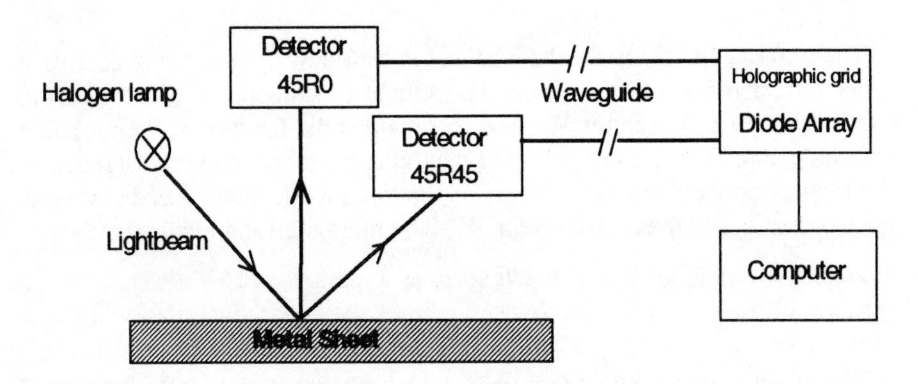


**Fig. 3.** The main components of the measuring system

# 4 Predicting the Thickness of Chromate Layers on Aluminium by Neural Networks

## 4.1 Materials: Conversion Layers on Aluminium

The aluminium sheets with conversion layers are taken from a standard production-line. The chromate conversion layers with a thickness between 20 and 300 nm are produced by varying the processing time of the chemical treatment [7]. The chromate layer thickness is determined by X-ray fluorescence analysis (XRF) which measures the chromium content of the surface and by an electron microscope. Figure 4 shows some measured reflectance spectra of these chromate layers.

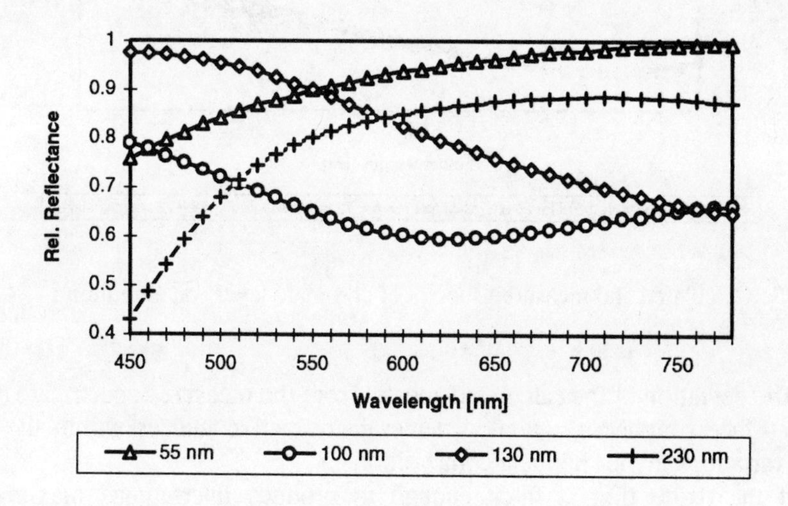


**Fig. 4.** Typical interference spectra of chromate conversion layers on aluminium

## 4.2 Physical background - Interference spectra

When a metal surface is covered by a transparent or weak absorbing thin film, interference effects occur when the surface is illuminated by white light. The interference effects depend on the thickness of the coating and the optical constants of the surface layer and the metal. The reflectance R of light is calculated from the complex refraction index ñ (n + ik), the reflection coefficients for perpendicular and parallel polarised light, the wavelength of the incident light and the angle $\alpha$ of incidence according to the Fresnel formulas [8, 9].

In practice it is more convenient to measure the intensity of the reflected light relative to a standard e.g. the surface without the coating, to avoid absolute measurements.

The optical constants for calculating the reflectance spectra of the produced chromate-oxide/aluminium layer system were determined by elipsometry. They are $n_1 = 1.58 - 0.03i$ for chromate and $n_2 = 1.65 - 5.65i$ for aluminium. A comparison of calculated spectra and measured spectra for two samples in the wavelength range between 450 and 800 nm can be seen in Fig. 5.

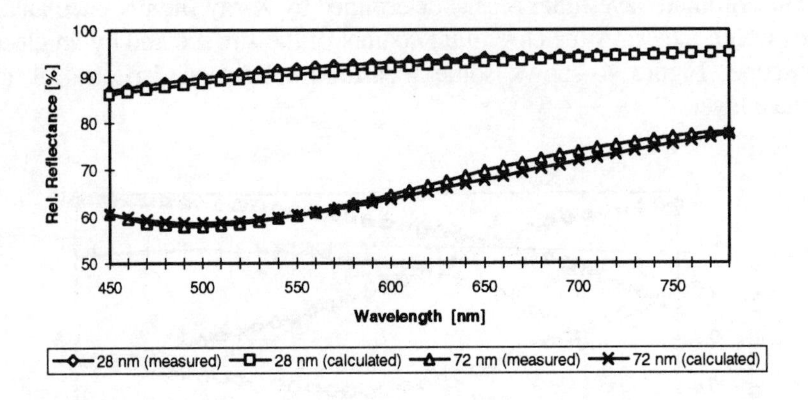


**Fig. 5.** Calculated and measured spectra of chromate layers on aluminium

The deviation of the calculated spectra from the measured spectra are due to the surface roughness, inhomogeneities in refractive indices within the layer and variations in the chemical composition.

If the oxide film is thick enough to produce interference maxima and minima, it is possible to calculate the layer thickness from these maxima and minima [10]. However, layers thinner than about 60 nm show no interference maxima or minima within the wavelength range of visible light, thus classical evaluation techniques fail. Evaluation of the spectra by means of neural nets offer the possibility to handle even this layer thickness range.

## 4.3 Predicting the Chromate Layer Thickness by a Kohonen-Topological-Interpolation Methods

### 4.3.1 Results from Calculated Spectra

In order to test the ability of the interpolation method with the Kohonen network, we train the Kohonen net with calculated spectra of chromate layers on aluminium and test the net with calculated spectra of different, but as well known, layer thicknesses. Since we want to compare these results with classical interpolation techniques and other neural nets, we apply the same data to a RBF network and a principal component regression method (PCR).

Four different network configurations are trained to examine the influence of the size of the network. The number of training spectra and the number of neurons in the Kohonen layer, respectively the hidden layer of the RBF net, is stepwise reduced from 10 to 3. The nets are tested with 20 spectra of layer thickness between 10 and 100 nm (10, 15, 20 .. 95, 100 nm). The predicted layer thickness of the 3 different approximation techniques are compared to the true layer thickness. The standard deviation in nm of predicted to measured layer thickness depending on the number of training patterns is shown in Fig. 6.

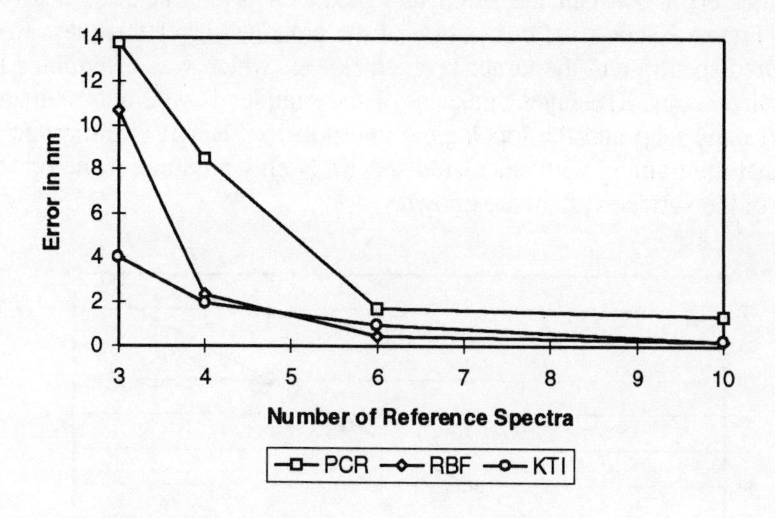


**Fig. 6.** Standard deviation (in nm) for predicted layer thickness of calculated spectra depending on the number of training spectra.
(KTI = Kohonen net with topological interpolation, RBF = Radial basis function network, PCR = Principal component regression)

When more than 6 training spectra are used the various methods show hardly any difference. But the multilinear approximation of the principal component regression yields for all configurations the worst result. The Kohonen net with topological interpolation allows to reduce the map size down to 3 neurons and the standard deviation remains smaller than 4 nm. Thus it is possible to cover the whole range between 10 and 100 nm layer thickness with only 3 spectra of known layer thickness. This is an important feature for industry since the production of reference material is a time consuming and costly procedure.

### 4.3.2 Results from Industrially Processed Samples

In the next evaluation phase, the same configurations are tested with 10 measured spectra within the thickness range of 20 to 95 nm. First, the training is done with calculated spectra and validation is carried out with measured spectra. We have 10 samples of different layer thickness. For each layer thickness there exist 10 spectra, taken at various points of the metal sheet. Since the measured spectra are not perfectly simulated by the calculated spectra (see Fig. 5), a greater error between measured and predicted layer thickness is to be expected. Figure 7 shows the correlation of the predicted layer thickness from the measured spectra and the target layer thickness, which was determined by electron microscopy. The layer thickness of the samples is well approximated by the Kohonen map and the topological interpolation. But it is important to mention that the training with calculated spectra is only possible, if the optical constants of the surface system are known.

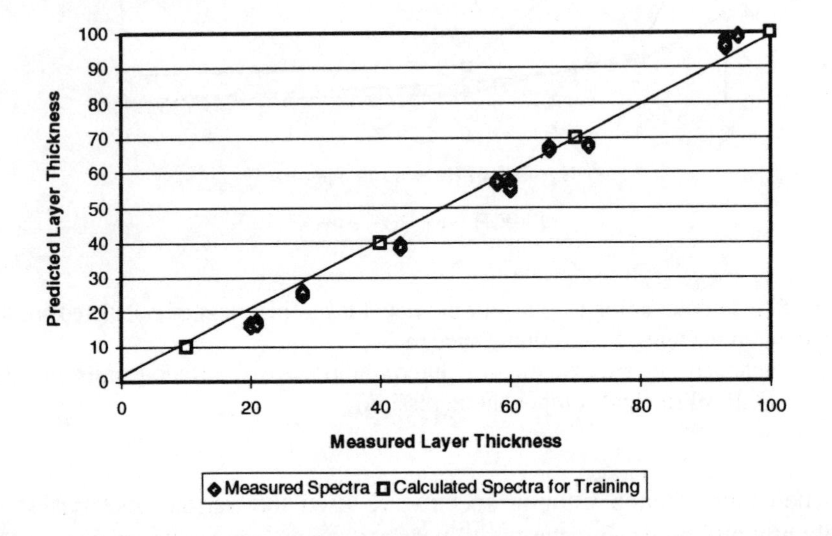


**Fig. 7.** Correlation between predicted layer thickness and measured layer thickness for industrially processed samples. Training spectra are calculated spectra.

For industrially processed oxide layers it is often not possible to determine the correct optical constants of the surface layer and the bulk material well enough to simulate the spectra with these data. It is therefore necessary to train a neural net with the measured spectra of a few reference samples, whose thickness is determined by other methods (e. g. electron microscopy, which is accurate enough but not suitable for on-line control). Since the preparation of reference samples is usually very costly and often time consuming, it is necessary to do an approximation with as few reference samples as possible. Therefore the Kohonen net and the RBF net are trained with only 3 measured spectra of samples of known layer thickness. In our case we take the spectra of layer thickness 20, 58, and 93 nm. The spectra for testing the nets are taken of the 10 different aluminium sheets of different oxide layer thicknesses as described before. Each sheet is measured at 10 different points to get an impression of the variance of the reflectance spectra for similar layer thicknesses. We test the nets with these 100 measured spectra. The correlation between predicted (interpolated by the net) and measured layer thickness is shown in Figure 8. The correlation coefficient is r = 0.98 for the Kohonen map with topological interpolation and r = 0.97 for the RBF net. One sample was neither interpolated correctly by the Kohonen nor by the RBF net. We believe this is due to the errors in the reference layer thickness.

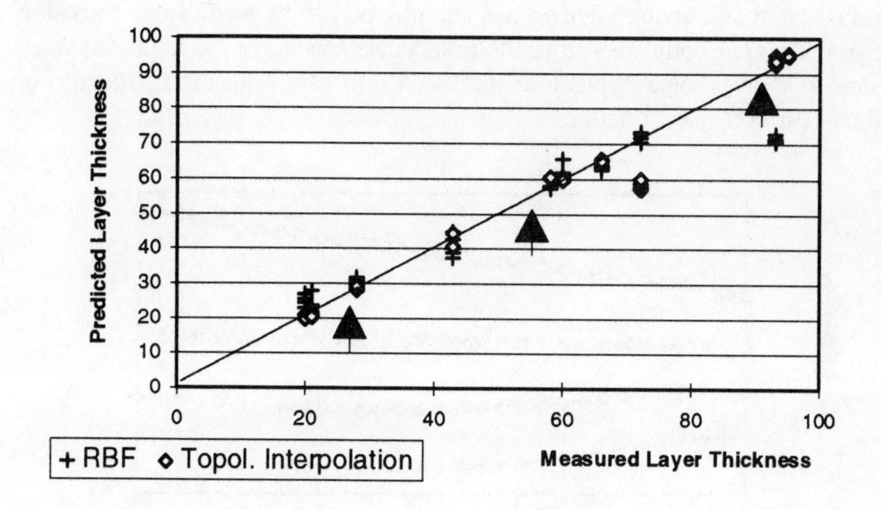


**Fig. 8.** Predicted versus reference layer thickness. Layer thickness of training vector is indicated by ↑.

The variance within the 10 spectra of the same sample is caused by inhomogeneities within the layer and the roughness of the surface.

The good correlation coefficient proves that it is possible to cover the layer thickness range from 10 to 100 nm with only 3 calibration samples in the case of aluminium, if the interpolation is done by a neural net. The Kohonen net with topological interpolation gives slightly better results than the RBF net. The major advantage of the Kohonen net is, that it doesn't need any trial and error to find the optimal configuration, whereas the RBF network demands a bit of skill to find the best solution.

Conventional diode array spectrometers measure a complete spectrum every 10 ms and since the data evaluation is very fast, this method is ideal for on-line control of thin conversion layers on metals. It has been applied successfully in aluminium industry for a test period of 6 month and it is about to be installed.

# 5 Prediction of the Corrosion Behaviour of Car Body Steel

## 5.1 Materials: Car Body Steel (Low Carbon Steel)

A diffuse reflectance spectrometer was installed at the production-line of a German car manufacturer during a 6 months period to predict the corrosion rate of coated car body steel from the reflectance spectra of the uncoated steel. Figure 9 shows some typical examples of diffuse reflectance spectra of different car body steel quality.

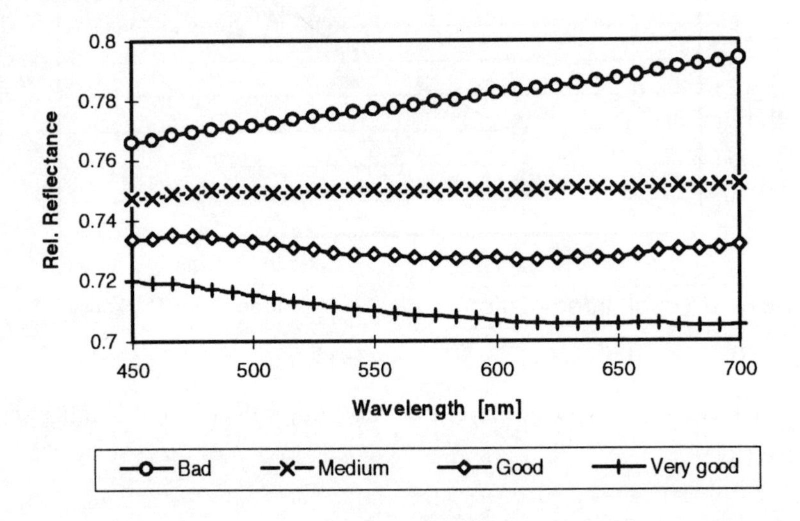


**Fig. 9.** Typical diffuse reflectance spectra of different car body steel samples

More than 140 samples of car body steel from 19 different coils, manufactured by 7 different European steel manufacturers were collected. The samples are characterized by their diffuse reflectance spectra and the optical corrosion index, calculated from the intensity of the reflected light at 400 and 700 nm. [11]. The spectra are measured with the optical set-up 45R0 as described in chapter 3 and shown in Fig. 3. The spectra of these individual samples are furthermore compared to spectra of the whole coil which were measured on-line during production to get an impression of the variance within the coil. The spectra of the complete coil didn't differ much and it became clear that an individual sample is a perfect substitute for the whole coil [12]. After full coating of the samples and a 10 weeks exposure to a salt spray test at different temperatures in a climatic chamber (VDA test 621-415), the corrosion under paint is determined in mm. Four major classes are distinguishable with a corrosion under paint from ≤ 0.7 mm (very good) ranging up to ≥ 1.0 mm (very bad).

## 5.2 Physical and Chemical Background

The surface of a metal is always covered by a thin naturally formed oxide film. Corrosion and coating adherence are mainly influenced by the texture of this oxide film. It is possible to measure the main features of this film like its thickness, inhomogeneities and distortions (pits and cracks) within the film by specular and diffuse reflectance spectroscopy. This naturally formed surface oxide on low carbon steel is only 10 - 50 nm thick. It is formed by Fe(II)-Fe(III)-oxides [13]. The ratio of Fe(II) and Fe(III) ions determines the quality of the oxide layer in respect to corrosion. A high proportion of Fe(II) in the mixed oxide results in low corrosion, whereas more Fe(III) in the mixed oxide causes high corrosion. Furthermore, a 'good' oxide layer in respect of future corrosion behaviour is homogeneous. A 'bad' oxide layer shows already many initial corrosion pits and is therefore heterogeneous. Mixed oxides with a major part of Fe(II) are called green rust. They absorb at higher wavelengths and therefore we measure more reflection at lower wavelengths. If there is more Fe(III) in the mixed oxide, it is called brown rust and it absorbs blue and green light, so we get reflection at higher wavelengths. If the surface is heterogeneous, with many pits, the light is scattered and diffuse scattering is measured. This function is inverse to the wavelength, that means for shorter wavelengths the reflected intensity of the scattered light is higher.

The diffuse reflectance spectra of the surface show the superposed inherent information of the chemical composition and the homogeneity of the surface oxide. The quality of this surface layer predetermines the quality of the phosphate layer and therefore the adherence of all successive paintings. Thus it is understandable, that there should be a correlation between the measured reflectance spectra, representing the 'quality' of the natural oxide film, and the corrosion under paint. In previous work [14] the correlation between the optical corrosion index of the spectrum in the wavelength range from 400 to 700 nm and the corrosion under paint is described in detail. A correlation of r = 0.9 was found between the predicted corrosion from the diffuse reflectance spectra and the corrosion under paint. This means, that approximately 80 % of the samples were classified correctly.

Corrosion under paint, measured in mm is still the standard method for testing corrosion resistance. Due to difficulties in the reproducibility of the testing, the error of the reference method is high and classification in quality classes is preferable to a traditional correlation analyses.

Another problem arises because of the uneven distribution of the sample set. This means, very good and very bad samples, which are important for discrimination, are not as often represented in the sample set as a mean quality material. It is possible to overcome this difficulty by using a Kohonen map for classification. We have therefore used the following procedure [15]:

1. Selecting the number of neurons to represent the different classes. This offers the chance for a robust classification and a good generalizability by the Kohonen map.
2. Unsupervised learning of the statistically distributed sample set. The unsupervised learned spectra, represented by the weights of the Kohonen map, offer the opportunity to extract the scientific background of the Kohonen classification.
3. Modification of the unsupervised learned Kohonen weights by pre-setting the weights, taking into account the scientific background of the samples. This pre-set Kohonen map integrates external know-how into the weights and is therefore able to balance the unequal distribution of the sample-set.

## 5.3 Classification with an Unsupervised Kohonen Map

The neural net to classify the reflectance spectra is a standard Kohonen net with 4 weight neurons. Each of the 4 weight neurons is connected to one neuron in the output layer, which contains the corresponding corrosion class (see Fig. 10). The corrosion under paint in mm is attributed to the 4 classes as follows: the class 'very good' is ≤ 0.7 mm corrosion under paint, the class 'good' is around 0.8 mm corrosion under paint, the class 'medium' is around 0.9 mm and the class 'bad' is ≥ 1 mm corrosion under paint.

From all diffuse reflectance spectra the mean value of each spectrum is subtracted from the spectral values to reduce the influence of the light source intensity and the roughness of the steel surface, before they are used for training the Kohonen net. Since the spectra are very smooth it is sufficient to reduce the spectra to 32 components (one value every 10 nm wavelength).

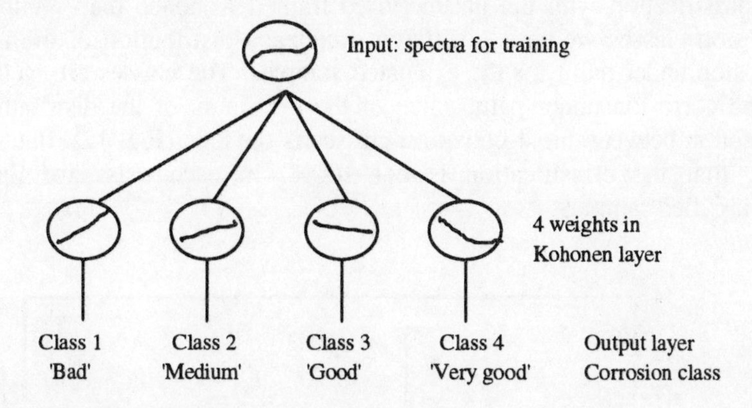


**Fig. 10.** Kohonen Map used for Classification of car body steel

A representative set of spectra from 50 samples is used for training the Kohonen net in order to learn from the weight vectors the physical background of the classification and to select the best matching representative spectra for successive pre-setting of the map. Figure 11 shows the spectra of the corresponding 4 weight neurons, which result from the unsupervised learning of the Kohonen map. These 4 spectra are supposed to be typical representatives of the 4 corrosion classes.

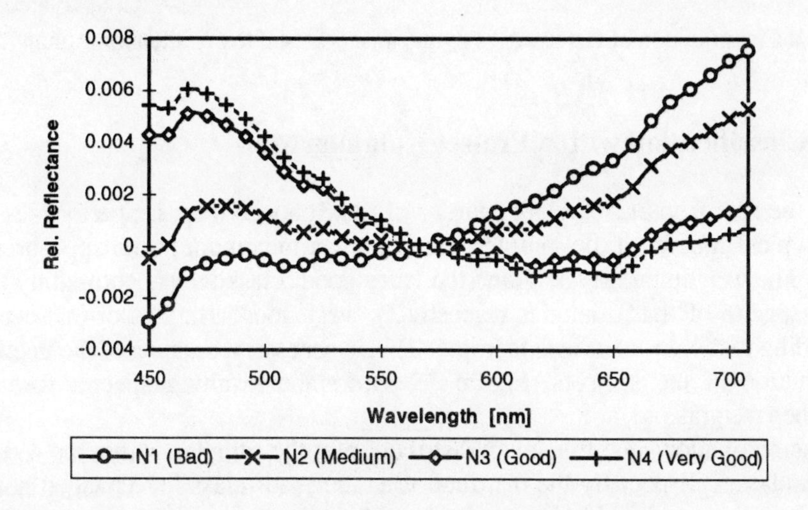


**Fig. 11.** Spectra (mean centred) trained to the Kohonen weights (N1 - N4) by the unsupervised learning

In the evaluation phase 90 unknown samples have to be classified by either the trained or later by the pre-set Kohonen map. Figure 12 shows the result of the classification with the unsupervised trained Kohonen map. Within the 4 corrosion classes we have a different frequency distribution of the measured corrosion under paint for the evaluated samples. The classes are well defined by the corrosion under paint value of the maximum of the distribution. The distinction between the 4 corrosion classes is obvious (Fig. 12), but it can be seen, that the classification is not 100%. In each class we find some misclassified samples.

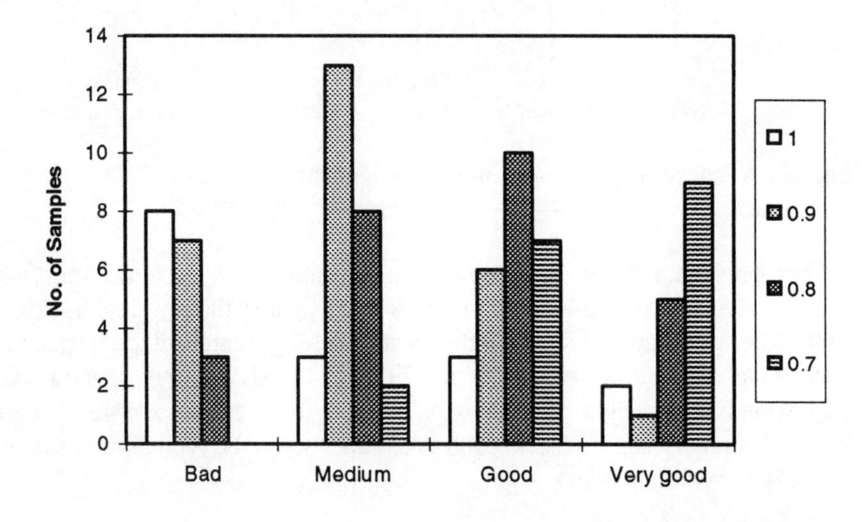


**Fig. 12.** Classification of test samples by the unsupervised learned Kohonen map

## 5.4 Classification with a Pre-set Kohonen Map

In the next step the 4 weight neurons of the Kohonen map are partly pre-set with typical spectra of the samples which show a maximum in the appropriate class. The weights of the 'bad' and the 'very good' class are substituted by the mean spectra of 'bad' samples respectively 'very good' samples of this class, according to the unsupervised learning. This is necessary because of the uneven distribution of the samples. Figure 13 shows the resulting spectra for the Kohonen weights.

Figure 14 shows the frequency distribution of the samples within the 4 corrosion classes. Especially the distribution in the 'bad'-class has changed notably. Even the 'medium' and 'good' class show a more distinct maximum of samples of the 'true' classification. Only the 'very good' class has not chanced significantly. Due to this result, a detailed analysis of the optical set up revealed, that this is not a problem of the classification by the neural net, but by the measuring system. To solve this deficiency an improved optical set-up is being built.

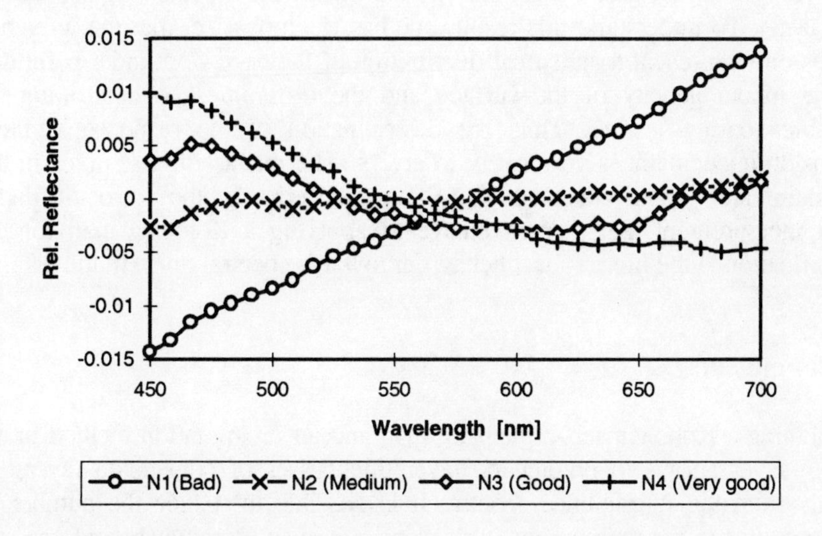


**Fig. 13.** Modified spectra (mean centred) for Kohonen weights (N1 - N4)

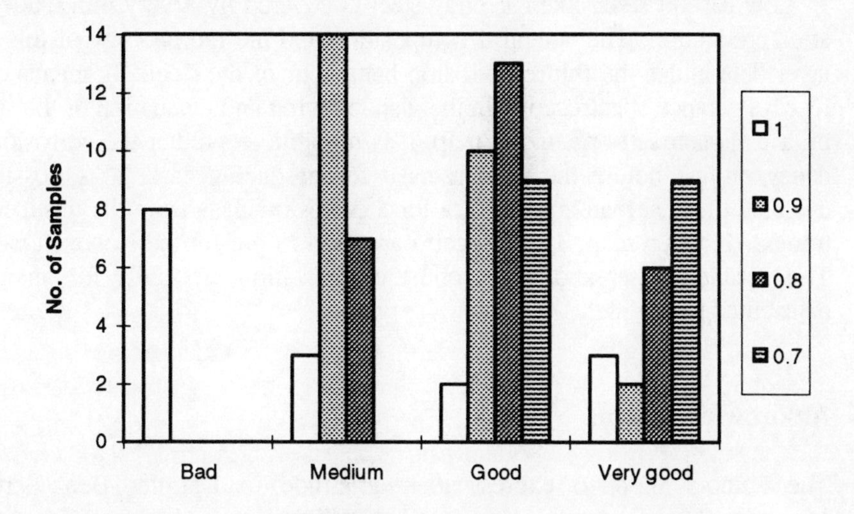


**Fig. 14.** Classification of the test samples by the pre-set Kohonen map

The spectroscopic corrosion index as described in [14] correctly classifies 80% of the 90 samples. The unsupervised trained Kohonen network finds the same classification of these samples but additionally classifies another 8% of the samples correctly. If the spectra are evaluated with the pre-set Kohonen net it is even possible to categorise 94 % of the samples. Another 6% of the samples remain unclassified not only by the traditional method but also by the neural net. To understand this result, one has to emphasise, that the reference values only represent a statistical distribution of the corrosion under paint due to the inhomogeneity of the surface and the difficulties of controlling all variables during testing. Thus the determination of the response variable corrosion under paint is very likely to errors. The unclassified samples in the 'medium' and 'good' class, probably show the error of the corrosion under paint measurement. Another advantage of applying a Kohonen map for the classification is the more robust behaviour towards spectral noise in the data.

## 6. Conclusion

Combining a Kohonen self-organising map and an interpolation method in the output layer allows to predict the layer thicknesses of conversion layers on metals from the interference spectra. It is possible to reduce the number of training spectra for the neurons in the Kohonen map enormously and one still obtains satisfactory results. This offers a fast, reliable and on-line applicable tool to calculate the thickness of transparent surface layers on aluminium or other metals even in the range below 100 nm.

Low carbon steel, like car body steel is covered by a very thin oxide layer after production. The chemical composition and the morphology of this oxide layer determines the future corrosion behaviour of the steel. By means of diffuse reflectance spectroscopy in the visible region and evaluation of the spectra by a Kohonen self-organizing map, it is possible to predict the corrosion tendency on-line before the steel is used for producing cars. It is possible to evaluate the best matching spectra for a corrosion class from the unsupervised trained Kohonen map. These spectra are used to pre-set the Kohonen weights. This enables better and more robust classification especially for an uneven distributed sample set.

## Acknowledgement

The authors wish to express their gratitude to Daimler Benz/Germany, Mercedes Benz/Germany, Alusuisse-Lonza/Suisse and Alusingen/Germany for providing samples for this work and their co-operation. Part of the work was supported by the Ministry of Science and Research, Baden-Württemberg, Germany.

## References

1. Kohonen, T.: Self-organization and associative memory. Third Edition, Springer Verlag, Berlin, Germany (1989)

2. Hecht-Nielsen, R.,: Counterpropagation Networks. Appl. Optics **26** (1987) 4979-4984

3. Kessler, W., Göppert, J., Kessler, R. W.: Prediction of oxide layer thickness by a topology preserving interpolation method in a self-organizing map. Proccedings of the 7th International Conference on Systems Research, Informatics and Cybernetics, Baden-Baden (1994) 99-104

4. Göppert, J., Rosenstiel, W.,: Self-organizing maps vs. backpropagation: An experimental study. Proceedings of Design Methodologies for Microelectronics and Signal Processing, Giwice, Poland (1993) 153-162

5. Poggio,T., Girosi, F.: A theory of networks for approximation and learning. A.I. Memo No. 1140, MIT (1989)

6. Broomhead, D. S. and Lowe, D.: Multivariable functional interpolation and adaptive network. Complex Systems, **2** (1988) 321-355

7. Ende, D., Kessler, W., Oelkrug, D., Fuchs, R.: Characterization of chromate- phosphate conversion layers on Al-alloys by electrochemical impedance spectroscopy (EIS) and optical measurements. Electrochimica Acta, **38** (17) (1993) 2577-2580

8. Hecht, E.: Optik. Addison-Wesley, Bonn (1989)

9. Vasicek, A.: Optics of thin Films. North-Holland Publishing Company (1960)

10. Gauglitz, G., Brecht, A., Kraus, G., Nahm, W.: Chemical and biochemical sensors based on interferometry at thin (multi-)layers. Sensors and Actuators B, **11** (1993) 21-27

11. Kessler, R. W., Böttcher, E., Füllemann, R., Oelkrug, D.: In situ characterisation of electrochemical formed oxide films on low carbon steel by diffuse reflectance spectroscopy. Fresenius Z. Anal. Chem **319** (1984) 695-700

12. Kessler, R. W., Brögeler, M., Tubach, M., Degen, W. Zwick W.: Determination of the corrosion behaviour of car-body steel by optical methods. Werkstoffe und Korrosion **40** (1989) 539-544

13. Kessler, R.W., Kessler, W., Quint, B., Kraus, M: C.Jochum (Ed.) Multivariate Analysis of unstructured diffuse reflectance spectra from car body steel surfaces. Software Development in Chemistry **8**, Gesellschaft Deutscher Chemiker (1994) 91-98

14. Kessler, R.W, Degen, W., Zwick, W.: Optical on-line sensor to determine the corrosion behaviour of low carbon steel. Deutscher Verband für Materialforschung und -prüfung, Tagungsband Werkstoffprüfung, Bad Nauheim, Germany (1989) 329ff

15. Kessler, W. Kessler, R. W., Kraus, M. Kübler, R., Weinberger, K.: Improved prediction of the corrosion behaviour of car body steel using a Kohonen self-organizing map. Colloquium Digest IEE Colloquium on Advances on Neural Networks for Control and Systems, Berlin (1994)

# Analysis and Classification of Energy Requirement Situations Using Kohonen Feature Maps within a Forecasting System

Steffen Heine and Ingo Neumann

Best Data Engineering GmbH Berlin, Storkower Straße 113, 10407 Berlin, Germany
e-mail: heine@hp2.rz.htw-dresden.de

**Abstract.** Improving the accuracy of the electrical energy demand forecasting in the short term range up to seven days can decrease the operation costs of the energy system significantly by the following optimising of the energy management. Because of the different objectives of energy management in different energy systems a considerable amount of engineering power is necessary to build a forecast model for each application case. The efficiency of using the Kohonen Feature Map for a stepwise automatisation of this process is discussed by means of two application cases. Finally, the embedding of the new analysis capability in a modular-constructed forecasting system is presented.

## 1 Load Forecast within Electrical Energy Systems

### 1.1 Overview

Because of the limited scope for storage of electrical energy the knowledge about the load demand in the future is a necessary information for the operation of electrical energy systems (EES). Different modern requirements within the energy supply industry as for example

- Introduction of energy trading and competitive energy procurement processes,
- Decrease of the operation costs, increase of the return of investment of each separate company,
- Improvement of the quality of energy supply (reliability, frequency, voltage) because of the high developed technical equipment of the customers,

lead to a need of sophisticated load forecasting solutions in the short term range up to one week. Regardless of extensive research and development in the past statistical load forecasting approaches (multiple regression, time series based or state space models) were used in only a few load control centres due to their

- Lack of accuracy also in comparison to the forecast "by hand" from the experience of the operator and
- Complicated model structure and absence of a user-friendly interface with explanation and evaluation facilities.

By introduction of AI based techniques, especially artificial neural networks (ANN) and fuzzy-rule based systems, to a hybrid system combined with the development of the computer hard- and software aspects a comprehensive forecast solution can overcome this situation with the following essential new qualities:

1. Reducing the expenditure of data analysing and modelling by using self adapting systems,
2. Possibilities of putting the operators experience into the forecast model by using rule based modelling especially for "special days" or "special situations" and
3. User-friendly graphical interface that enables the integration of the necessary explanation and evaluation facilities into the prediction system.

In the next section some aspects of the load model building process are discussed, followed by the presentation of a data analysis capability based on self-organising ANN and their integration into the system structure of a prediction system.

### 1.2 Short Term Load Forecast Modelling

To point out the problem of short term load forecast (STLF) the following facts should be mentioned:

1. There are many grid companies, electric utilities and big industrial customers with different dependencies and relationships of the load.
2. The load is a non-stationary process with a daily, weekly and yearly periodicity.
3. Time, weather and economic factors as well as random disturbances (TV, strikes) influence the load. They are, together with the past behaviour of the load, potential parameter for modelling.

It is not worth trying to consider all the different situations of an EES within a unique model because such a model can have a limited accuracy only. It is therefore better to separate different situations of load demand (for example by using the similarity of load shapes as a criterion for clustering) and to create "specialist" models for those. In the short term range up to one day or one week this classification and model structuring process is based on the daily load shape. Following the previous load data analysis is a necessary prerequisite for building a forecast tool with a high accuracy. An additional objective of the data analysis is to get information not only for the load forecast but also for short, medium and long term planing tasks (Fig. 1).

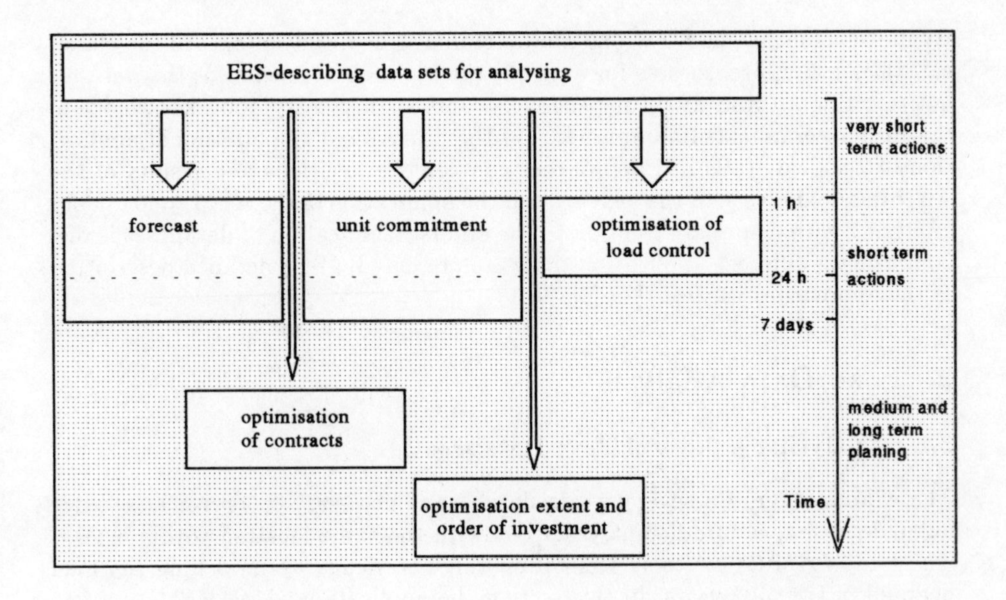


**Fig. 1.** Support of planing activities by load data analysis

Because of the clustering and dimensionality reduction abilities the Kohonen feature map (KFM) [6], a self-organising ANN, can be the preferable tool for deriving knowledge about dependencies of the load consumption in EES. There are two promising ways of using the KFM for data pre-processing purposes concerning STLF:

1. *Especially for the very STLF in an online manner*
   Presenting the load shape at the moment of forecast to a responsible KFM; Determining the most similar load shapes from the past on the basis of the neighbourhood; Building a model for example by training a feedforward ANN with these vectors extended by temperature information and the load in the next time steps and getting the forecast model [3].
2. *For the STLF in a preparatory manner*
   Tool for analysing extensive load databases and, together with additional interpretation and analysis methods, for separate the database into clusters according to the similarity of daily load shapes (or parts of them).
3. *For the direct load forecasting*
   Training the KFM with data from the past; disconnecting some inputs and presenting the current data to the KFM; determining a winner neuron and following the forecast (the weights of the disconnected inputs to the winner neuron). By using the KFM in the so called auto-associative mode one can get only a poor accuracy because there is a restricted number of possible values for the forecast [7]. Therefore a following compensation of some additional factors influencing the load is necessary

By means of using the KFM one can overcome some disadvantages of the classical statistical approaches as for example the high requirements for the data for not rejecting a thesis about dependencies and the necessary good knowledge of the rather complicated approaches. On the other hand the additional application of classical statistical approaches can confirm decisions and results of the ANN-based analysis. In the next section the application of the KFM to load data analysing with a special attention to the offline classification of databases is discussed. Two different application cases are presented. For a detailed description of the approaches used see [1], [6], [9].

# 2 Load Data Analysis

## 2.1 Analysing a Several Year Database

The examination of a several years database is necessary to derive knowledge about load shapes of different years but with the same location within a year. The building of STLF tools sets up on this knowledge by modelling the load demand of the past years. In this section the application of the KFM to a two years database of a grid company is discussed. The load of each day is described as a vector of 24 values (load in steps of 1 hour). The vectors were normalised by using (1) to equalise differences in the load level to reflect differences in the load shape, only. The vectors were normalised a second time using (2) to obtain vectors of unity length.

$$V_n(i) = \frac{V(i) - V_{\min}}{V_{\max} - V_{\min}} \tag{1}$$

with $i = 1..24$, $V_{\max}, V_{\min}$ = minimum and maximum value of the vector

$$V' = \frac{V_n}{||V_n||} \quad \text{with } V' = \text{ normalised load vector} \tag{2}$$

Figure 2 shows a part of an organised KFM based on the daily load shape. The thick lines are boundaries between clusters and results of a first rough classification. A fine clustering should based on the following information:

1. On the basis of presenting properties of the vectors (temperature, in the case of Fig. 2 the date, ...) one can recognise the distribution of these properties. An additional filter (e.g. the temperature above a determined value) supports this process.
2. Free neurons connected are hints of boundaries between sections in the input space.
3. A suitable presentation of the distance (e.g. Euclidean) between neighbouring neurons (Fig. 4, [9]) subdivides the competitive layer into different categories.
4. By application of pure clustering algorithms (e.g. Fuzzy-C-Means [1]) to the raw (load shape) or the weight vectors and mapping the results to the KFM additional information about classes recognised is available.

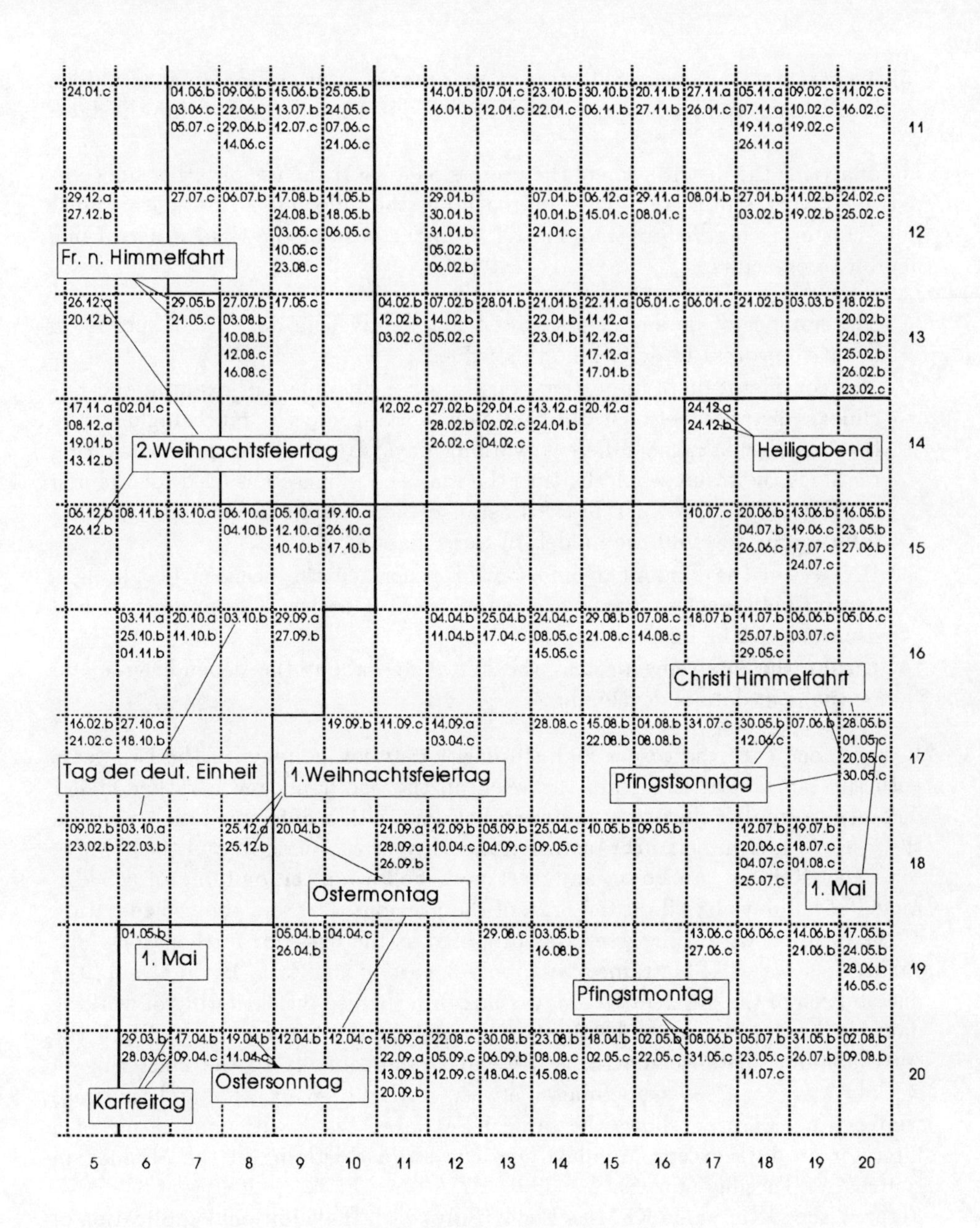


**Fig. 2.** Part of an organised 20x20-KFM, grid company, 14.09.91–13.09.93, date of the vectors; a = 91, b = 92, c = 93

5. The separate analysis of clusters of load shapes after getting an general idea is necessary and leads to new knowledge about dependencies within the data.

For analysing the distribution of the vectors on a KFM the use of rather big competitive layers is preferred. This approach has the following advantages against pure clustering algorithms and the KFM in the case of the rigid order of one neuron to one class:

1. The number of possible classes need not be given in advance. There is an iterative process to determine this value.
2. The consideration of important constraints is possible. For example the Euclidean distance between different non working days is much higher than the distance between different working days. By applying pure clustering methods the result would be that the majority of classes is used for the non working days. Due to this circumstance there would be not enough data information to build the models of these classes.
3. Because of the amount of information generated the decision of adding a neuron between two classes to one of them is easier. The number of bad decisions can be decreased.
4. During the analysing process the knowledge about the dependencies and relations in data is developing.

In Fig. 2 one can recognise a high similarity between holidays of the two years with the same location within the week on the one hand and a rather higher distance (a smaller similarity) between holidays with a different location within the week (e.g. Himmelfahrt–Ascension is each year a Thursday, and the 1. May–May Day Holiday can be on any weekday). Following the building of a STLF model of holidays based on the data of the previous year and the consideration of the location within the week was proposed as the basis for further work [8].
Figure 3 shows a KFM trained with only a part of the data. By analysing the distribution of the different weekdays one can recognise the clustering of working days in the upper part and non working days in the lower part. In Fig. 4 the corresponding distance matrix is presented. Additionally to the clustering of weekdays one can see a boundary between daylight saving time and (normal) winter time. Figure 5 shows the same KFM as in Fig. 3 with application of a filter for all days except Monday. One can see a clustering of the Monday in contrast to Tuesday (Fig. 6).
Figure 7 shows the same KFM as Fig. 2 (fully) with the additional application of the Fuzzy-C-Means clustering algorithm to the weight vectors of the organised KFM (highest membership values). At first sight it seems to make no sense to apply a second classification algorithm. But the KFM is used here in the meaning of a simple visualisation tool of high-dimensional data. The clustering algorithm supports the classification decisions of ordering a neuron to a class. In Figure 8 the membership values of a single category within the KFM are presented.
By a stepwise refined analysing of the database one can get the realisations necessary for a practical determination of day types for the respective supply

company and so the basis for the creation of specialist forecast models. In the next section the application of the KFM to a rather small database is discussed.

| | 1 | 2 | 3 | 4 | 5 | 6 | 7 | 8 | 9 | 10 | 11 | 12 | 13 | 14 | 15 | 16 | 17 | 18 | 19 | 20 |
|---|---|---|---|---|---|---|---|---|---|---|---|---|---|---|---|---|---|---|---|---|
| 1 | DI 0 | DO 0 | FR 0 | FR 0 | MI 0 | DO 0 | MO 0 | | FR 0 | DO 0 | | DO 0<br>FR 0 | DI 0 | | FR 0 | | FR 0<br>FR 0 | DI 0 | FR 0<br>FR 0 | FR 0 |
| 2 | FR 0 | | | DI 0 | DO 0 | | MI 1 | | DI 0 | MI 0 | | MI 0 | DO 0 | FR 0 | FR 0 | | DO 0 | DI 0 | MI 0 | FR 0 |
| 3 | | MI 0 | DI 0<br>DO 0 | | DI 0 | MO 0 | MO 0 | | DO 0 | | DI 0 | | MI 1 | | DI 0 | | MI 1<br>DI 0 | | | FR 0<br>DO 0 |
| 4 | FR 0 | | | FR 0 | | MO 0 | | | MI 0 | | MO 0 | MO 0 | | MI 0 | | FR 0<br>DI 0 | DI 0 | DO 0 | FR 0 | DO 0 |
| 5 | DI 0<br>DO 0 | FR 0 | DI 0 | DO 0 | MI 0 | MO 0 | | FR 0 | | DI 0 | | DI 0<br>MO 0 | | MI 0<br>FR 0 | | DI 0<br>MI 0 | | MI 1<br>MI 0 | | MI 0<br>DI 0<br>MI 0 |
| 6 | FR 0 | MI 0 | | MI 0 | | DO 0 | | | | FR 0 | DI 0 | | | | | DO 0 | MI 0 | DI 0<br>MI 0 | | DO 0 |
| 7 | FR 0 | | | DO 0 | | DI 0 | MI 1 | | DO 0 | DO 0 | MO 0 | MI 0 | DI 0<br>DO 0 | MI 0 | DO 0 | DO 0 | | | MI 0 | DI 0 |
| 8 | DO 0<br>MI 0 | DO 0 | MI 0 | FR 0 | DI 0 | | MO 0 | | | | | MO 0 | | DO 0 | DI 0 | FR 0 | | MO 0 | MO 0<br>MO 0 | DO 0 |
| 9 | DO 0 | | | | | MO 0 | MO 0 | | MI 0 | | MO 0 | MO 0 | | MI 0<br>MI 1 | DI 0 | DI 0 | | | | MO 0<br>MO 0 |
| 10 | | MI 0<br>FR 0 | DI 0 | DI 0<br>MI 1 | | | | | | | MO 0<br>MO 0 | | DO 0 | | | DO 0 | | MO 0 | DI 0<br>MO 0<br>MO 0 | |
| 11 | FR 0 | FR 0 | | | MO 0 | MO 0 | MO 0 | | MO 0<br>MO 0 | | DI 0 | MO 0 | | | DO 0<br>DI 0 | DO 0 | MI 0 | | | MO 0 |
| 12 | DI 0 | | | DO 0<br>DI 0 | | | MO 0 | | | FR 2 | | | MO 0<br>MO 0 | | FR 0<br>FR 0 | FR 0 | | DI 0<br>MI 1<br>MI 0 | DO 0 | DO 0 |
| 13 | | MI 0 | | | | | | | | | SA 0 | | | | | | FR 0 | | | |
| 14 | | | | SA 0 | SA 0<br>SA 0 | | SA 0 | SA 0 | | | | SA 0 | | SA 0 | SA 0 | | | | SA 0 | |
| 15 | | SA 0<br>SA 0 | SA 0 | | | | | SA 0 | | SA 0<br>SA 0 | SA 0 | | SA 0 | | SA 0 | SA 0 | SA 0 | SA 0 | | SA 0 |
| 16 | SA 0 | | | SA 0 | | SA 0<br>SA 0 | | | | | | | SA 0 | SA 0 | SA 0 | | SA 0 | SA 0 | | |
| 17 | SO 0 | | SO 0 | | | | | SO 0 | | | SA 2 | | | | | | | SA 0 | | SA 3<br>DO 3<br>SO 3 |
| 18 | | SO 0 | | SO 0 | | SO 0 | | SO 0 | | MO 3 | | SO 0 | SO 0 | SO 0 | SO 0 | | SA 0 | SA 0 | | |
| 19 | | SO 0 | | SO 0 | SO 0 | | | | | SO 0 | | SO 0 | | SO 0 | | | | | SO 0<br>SO 0 | SO 0 |
| 20 | FR 3 | | | SO 0 | SO 0 | | SO 3 | FR 3 | SO 0 | | SO 0 | SO 0<br>SO 0 | SO 0 | SO 0 | MO 3 | SO 0<br>SO 0 | SO 0 | SO 0 | SO 0 | SO 0<br>SO 0 |

**Fig. 3.** Organised 20x20-KFM, grid company, 01.01.93–13.09.93, weekdays; 3 = holiday, 2 = adjacent day to a holiday, 1 = third Wednesday of the month

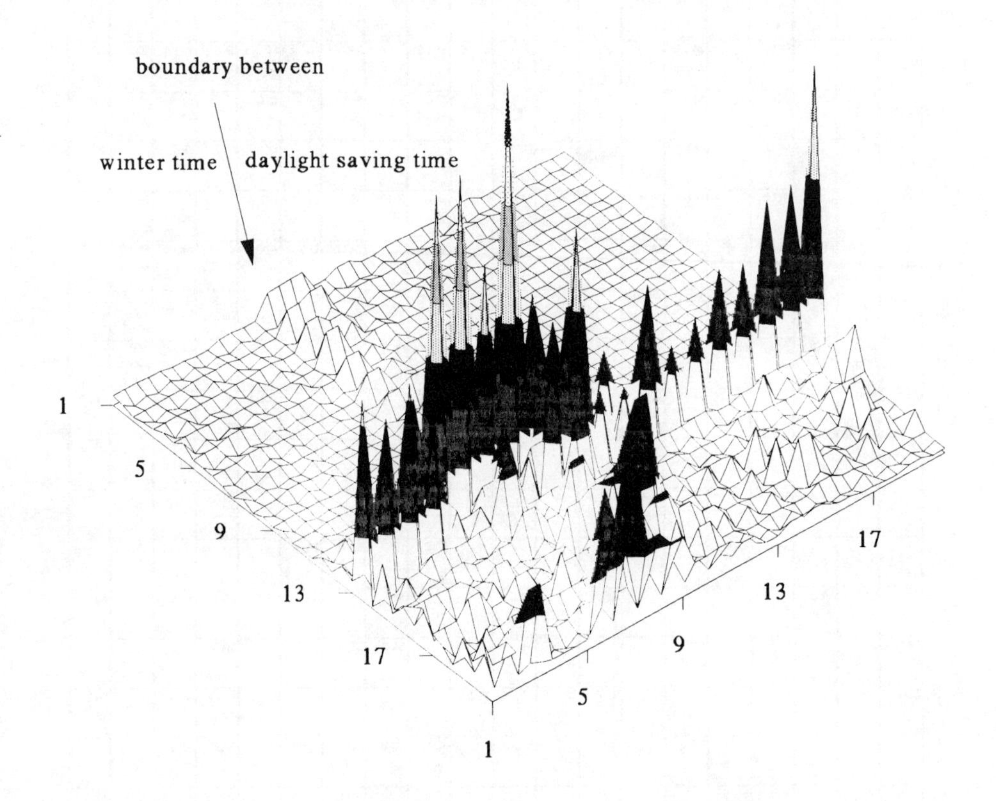


**Fig. 4.** Unified distance matrix of the KFM of Figure 3

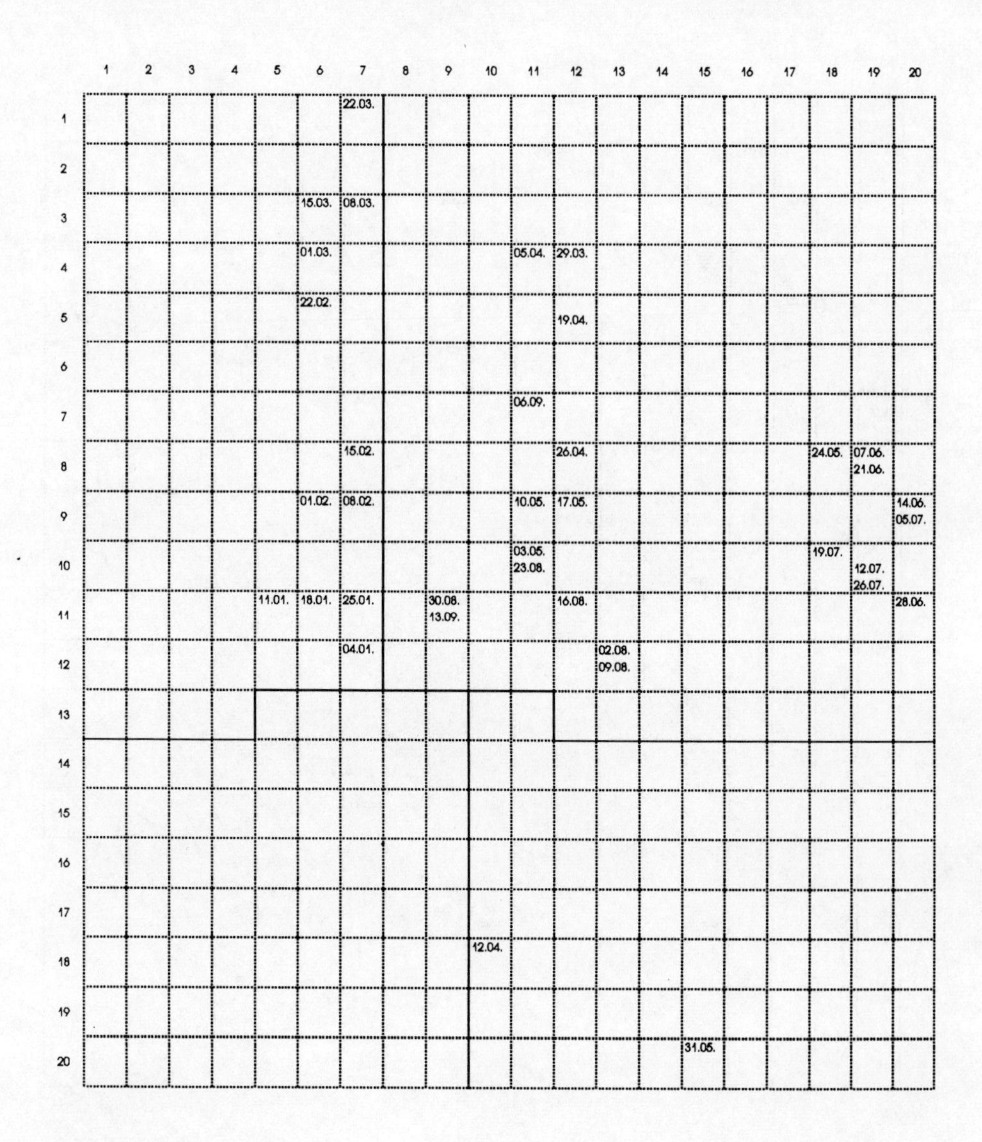


**Fig. 5.** Organised 20x20-KFM, grid company, 01.01.93–13.09.93, date of the vectors, Monday only

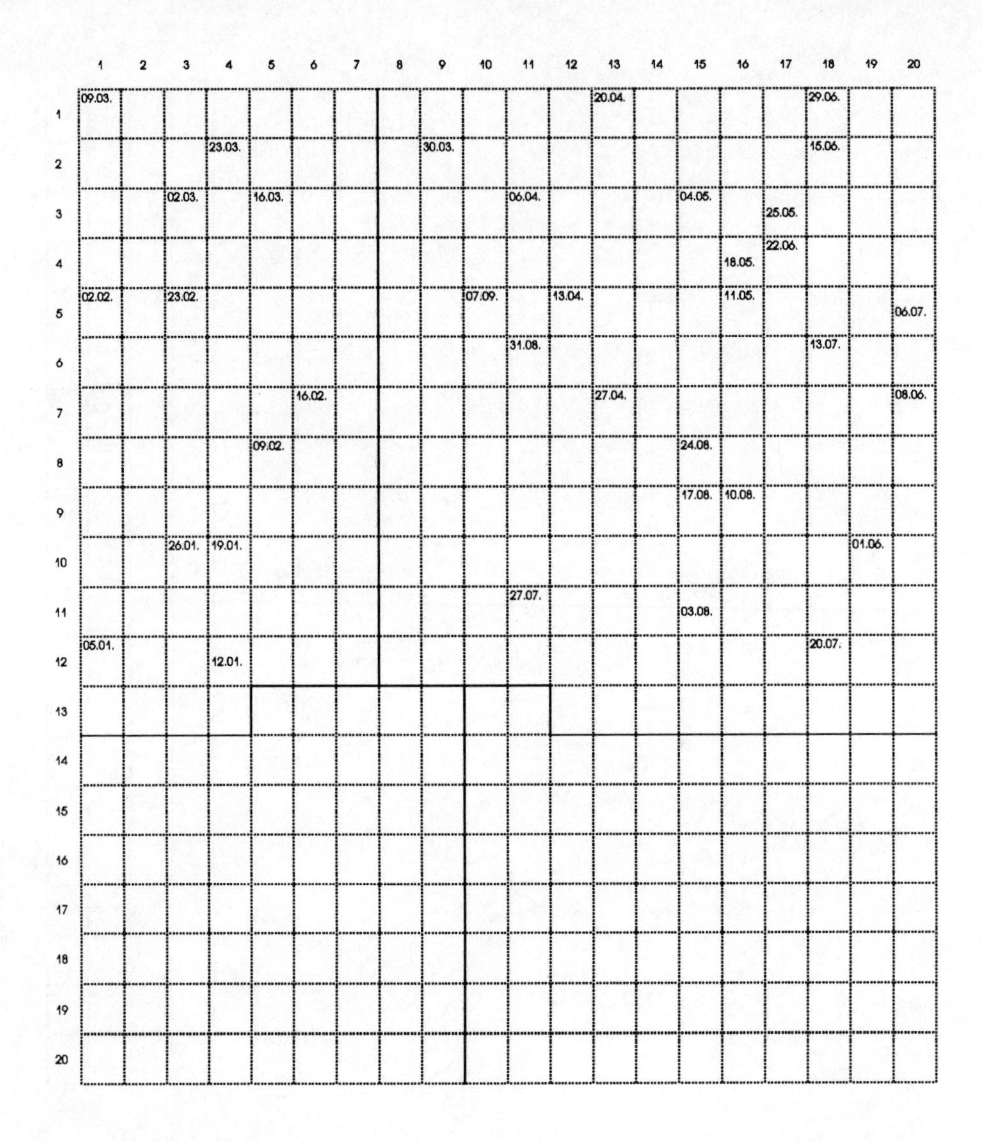


**Fig. 6.** Organised 20x20-KFM, grid company, 01.01.93–13.09.93, date of the vectors, Tuesday only

| | 1 | 2 | 3 | 4 | 5 | 6 | 7 | 8 | 9 | 10 | 11 | 12 | 13 | 14 | 15 | 16 | 17 | 18 | 19 | 20 |
|---|---|---|---|---|---|---|---|---|---|---|---|---|---|---|---|---|---|---|---|---|
| 1 | 4 | 4 | 4 | 4 | 7 | 4 | 4 | 4 | 4 | 4 | 4 | 7 | 7 | 7 | 7 | 7 | 7 | 7 | 7 | 7 |
| 2 | 1 | 4 | 4 | 4 | 7 | 4 | 4 | 4 | 4 | 4 | 4 | 4 | 7 | 7 | 7 | 7 | 7 | 7 | 7 | 7 |
| 3 | 1 | 4 | 1 | 1 | 1 | 4 | 4 | 4 | 4 | 4 | 4 | 4 | 7 | 7 | 7 | 7 | 7 | 7 | 7 | 7 |
| 4 | 1 | 1 | 1 | 1 | 1 | 1 | 4 | 4 | 4 | 4 | 4 | 7 | 7 | 7 | 7 | 7 | 7 | 7 | 7 | 7 |
| 5 | 1 | 1 | 1 | 1 | 1 | 4 | 7 | 4 | 4 | 4 | 4 | 7 | 7 | 7 | 7 | 7 | 7 | 7 | 7 | 7 |
| 6 | 1 | 1 | 1 | 1 | 1 | 4 | 7 | 7 | 7 | 7 | 7 | 7 | 7 | 7 | 7 | 7 | 7 | 7 | 7 | 7 |
| 7 | 1 | 1 | 1 | 1 | 1 | 7 | 7 | 7 | 7 | 7 | 7 | 7 | 7 | 7 | 7 | 7 | 7 | 7 | 7 | 1 |
| 8 | 1 | 1 | 1 | 1 | 1 | 7 | 7 | 7 | 7 | 7 | 7 | 1 | 1 | 1 | 1 | 1 | 1 | 1 | 1 | 1 |
| 9 | 3 | 3 | 1 | 1 | 7 | 7 | 7 | 7 | 7 | 7 | 7 | 1 | 1 | 1 | 1 | 1 | 1 | 1 | 1 | 1 |
| 10 | 3 | 3 | 3 | 3 | 3 | 7 | 7 | 7 | 7 | 7 | 7 | 1 | 1 | 1 | 1 | 1 | 1 | 1 | 1 | 1 |
| 11 | 3 | 3 | 3 | 3 | 3 | 7 | 7 | 7 | 7 | 7 | 7 | 1 | 1 | 1 | 1 | 1 | 1 | 1 | 1 | 1 |
| 12 | 3 | 3 | 3 | 3 | 3 | 3 | 7 | 7 | 7 | 7 | 1 | 1 | 1 | 1 | 1 | 1 | 1 | 1 | 1 | 1 |
| 13 | 3 | 3 | 3 | 3 | 3 | 3 | 7 | 7 | 7 | 7 | 1 | 1 | 1 | 1 | 1 | 1 | 1 | 1 | 4 | 1 |
| 14 | 3 | 3 | 3 | 3 | 3 | 3 | 7 | 7 | 7 | 1 | 1 | 1 | 1 | 1 | 1 | 1 | 2 | 2 | 2 | 7 |
| 15 | 3 | 3 | 3 | 3 | 3 | 3 | 3 | 6 | 6 | 3 | 1 | 4 | 2 | 2 | 2 | 2 | 2 | 2 | 2 | 2 |
| 16 | 3 | 6 | 6 | 3 | 3 | 3 | 3 | 6 | 6 | 6 | 5 | 5 | 5 | 2 | 2 | 2 | 2 | 2 | 2 | 2 |
| 17 | 6 | 6 | 6 | 6 | 6 | 6 | 6 | 6 | 6 | 2 | 2 | 5 | 2 | 2 | 2 | 2 | 2 | 2 | 2 | 2 |
| 18 | 6 | 6 | 6 | 6 | 6 | 6 | 6 | 6 | 5 | 5 | 5 | 5 | 2 | 5 | 2 | 2 | 2 | 2 | 2 | 2 |
| 19 | 6 | 6 | 6 | 6 | 6 | 5 | 5 | 5 | 5 | 5 | 5 | 5 | 5 | 5 | 2 | 2 | 2 | 2 | 2 | 2 |
| 20 | 6 | 6 | 6 | 6 | 6 | 5 | 5 | 5 | 5 | 5 | 5 | 5 | 5 | 5 | 2 | 2 | 2 | 2 | 2 | 2 |

**Fig. 7.** Membership of the neurons of the KFM of Figure 2 to categories on the basis of Fuzzy-C-Means, 7 possible categories, random initialising, m = 1,7 (factor for determining fuzziness)

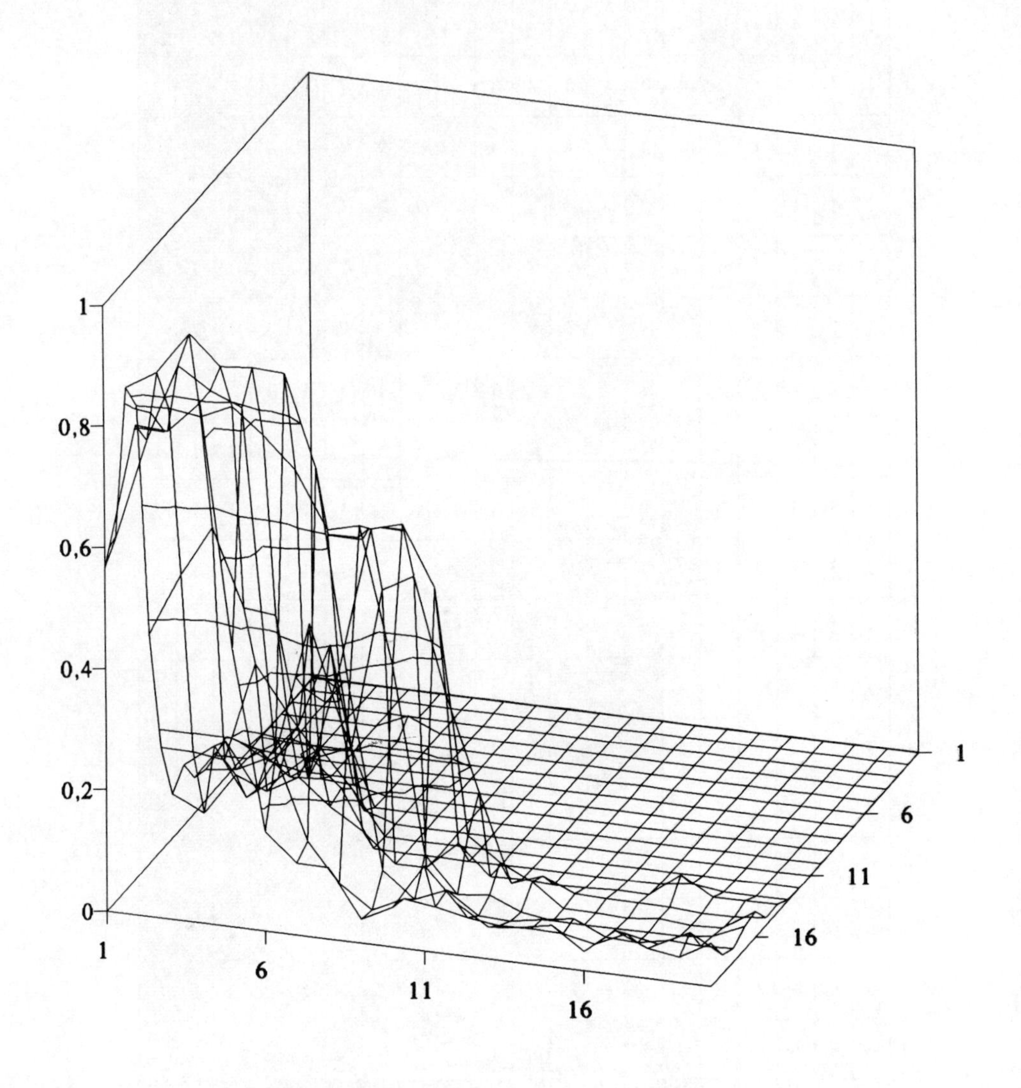


**Fig. 8.** Membership values of the neurons of the KFM of Figure 2 to category 6 according to Figure 7

## 2.2 Analysing a Four Weeks Database

The objective of a clustering of relative small databases is to get information about suitable input data vectors for a more flat model (i.e. to consider the time series nature and to use more current data). The data used here is from an island grid company. The load of each day is described as a vector of 48 values (load in steps of half an hour). The vectors were normalised by using the corresponding equation to (1) and by using (2). Figure 9 shows the KFM for presenting 17 input vectors (load shapes). It suggests the creation of a separate model for Friday because the Friday can be clearly clustered. Another recognisable feature is that the similarity of the load shapes decreases with increasing time range between two patterns.

An interesting result is obtained from Fig. 11 and Fig. 12. The Friday can't be clustered if only the first twelve hours of the day are taken into consideration. This suggests the creation of a separate model for Friday afternoon/evening only. Due to the KFM learning algorithm the order of the winner nodes in the two-dimensional matrix tends to a nearly even distribution, and therefore the distance (in the input space) between two adjacent nodes in one part of a trained KFM need not to be the same as in an other part of the KFM. A good possibility to get a distribution of the nodes according to the input space and a sharper separation of the winner nodes is to add some randomly generated patterns to the training patterns. The input space of the generated patterns should be the same as the input space of the actual training patterns. The order of the winner nodes in Fig. 10 suggests that the model learning should based on the last two weeks. It should be mentioned that the number of the additional patterns randomly generated can be determined in an iterative procedure only.

| | | | | |
|---|---|---|---|---|
| Fr4 | | Fr3 | | Fr1, Fr2 |
| | Tu3, We3<br>Th3 | | | |
| Th4 | | | | |
| Tu4 | | | Th1, Th2 | We2 |
| We4 | Mo4 | Mo3 | | Mo2, Tu2 |

**Fig. 9.** Organised 5x5 KFM, island grid company, 12.09.–04.10., load shape of 24 hours

| Mo4, Tu4,<br>We4, Th4 | Mo3 | | | |
|---|---|---|---|---|
| | We3, Th3,<br>Fr4 | | | Tu3 |
| | | | | |
| | | | Fr3 | Th1, Th2 |
| | | | | Fr1, Fr2,<br>Mo2, Tu2, We2 |

**Fig. 10.** Organised 5x5 KFM, island grid company, 12.09.–04.10., load shape of 24 hours; same vectors as in Figure 9 with additional learning of 63 randomly generated vectors (patterns)

| Mo2, Tu2 | We2 | Th1, Fr1,<br>Th2 | | Tu3, We3 |
|---|---|---|---|---|
| | | | Th3 | |
| Fr2 | | Fr3 | | Th4 |
| | | Tu4, Mo5 | | Fr4 |
| Mo3 | Mo4 | | We4 | |

**Fig. 11.** Organised 5x5 KFM, island grid company, 12.09.–04.10., load shape 01.00 a.m.–12.00 p.m.

## 3 Embedding the Analysing Capability into a Forecasting System

The requirements on the efficiency and accuracy of a STLF system can be fulfilled only by using powerful analysing, optimising and modelling techniques and implementing these into a modular-constructed frame and a suitable hardware. Figure 13 shows an overview of the whole developed procedure. The following explanations should be mentioned:

| | Fr4 | | We4 | Tu4 |
|---|---|---|---|---|
| Fr3 | | Mo3 | Mo4 | |
| | Th1 | Tu3 | | Th4 |
| Fr1, Fr2 | | Th2 | We3 | |
| | Mo2, Tu2 | We2 | | Th3 |

**Fig. 12.** Organised 5x5 KFM, island grid company, 12.09.–04.10., load shape 01.00 p.m.–12.00 a.m.

- By selecting the necessary data from the supervisory control and data acquisition (SCADA) system the necessary input vectors were committed for an offline model structuring and optimising process on the one hand (left side) and the online forecasting process on the other hand (right side).
- *Offline path:* The decision of building "specialist" models is based on the KFM (see Sec. 2). The functional inputs of the models and the model structure (in the case of feedforward ANN the network topology) can be optimised by an evolutionary tool [5]. The training of the models results in the adapted "specialist" models for the forecast.
- *Online path:* After the identification of the respective situation the responsible "specialist" model is activated and delivers the forecast vector. This information supports the unit commitment and the listing of a switch plan for the controllable load [2].

The described procedure was implemented on a UNIX-workstation under X-Windows. Figure Figure 14 shows the realised module structure, which corresponds with Figure 13. Some practice was made during an online test on a control centre [4], and at present we set the stage for the permanent insert into a load management system.

## 4 Conclusion

A new approach of analysing load consumption databases has been presented. The application of the KFM supports the deriving of knowledge about dependencies in data and the clustering of load shapes for the following creation of specialist forecast models. Finally, the embedding of the analysis capability into a modular prediction system is discussed.

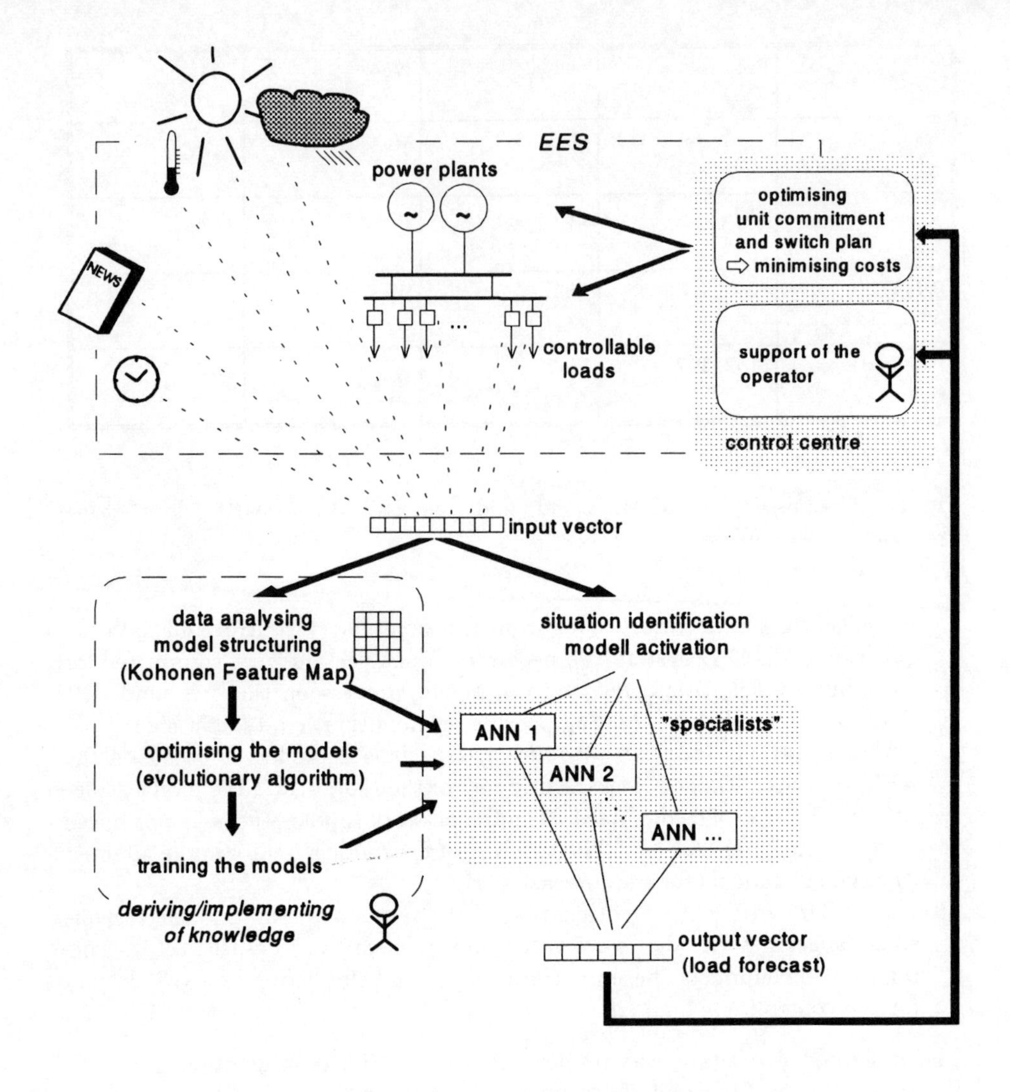


**Fig. 13.** Overview of the developed forecast procedure

## References


1. Bezdek, J. C.: Pattern Recognition with Fuzzy Objective Function Algorithms. Plenum, New York, 1981
2. Gottschalk, H., Heine, S., Fox, B., Neumann, I.: Economic Operation of a Power System with a Significant Amount of Controllable Load. Proceedings of 29th Universities Power Engineering Conference UPEC, University of Galway, 673–675, September 1994

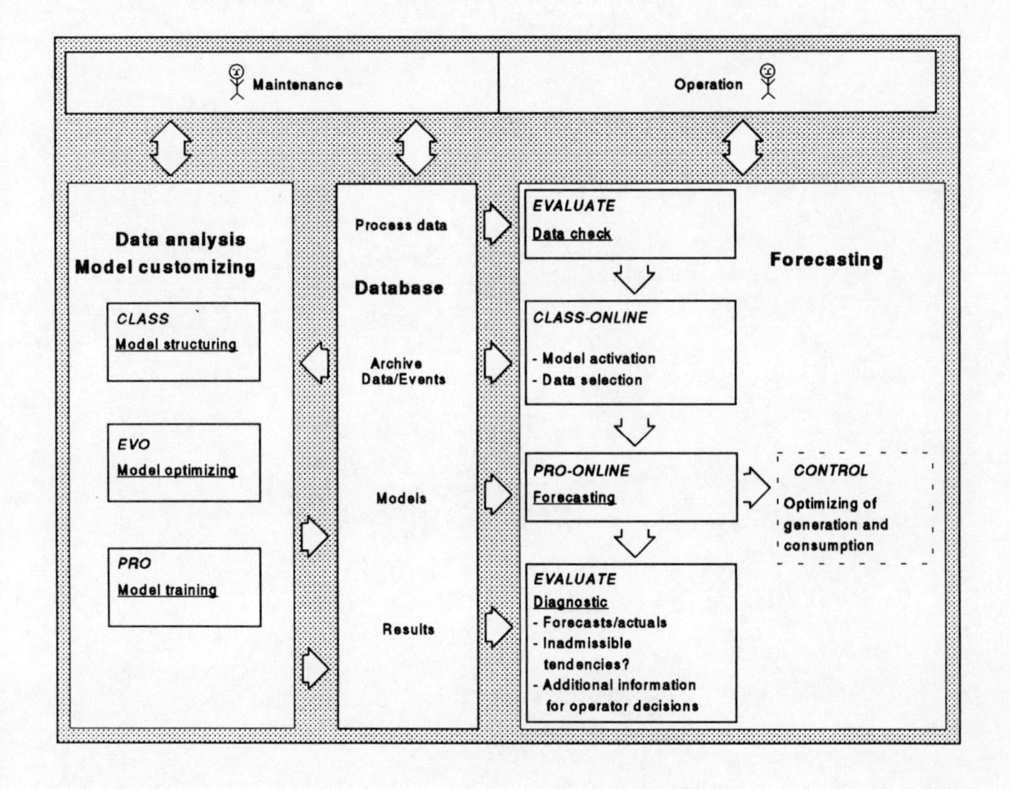


**Fig. 14.** Forecasting software components


3. Heine, S., Neumann, I.: Information Systems for Load-Data Analysis and Load Forecast by Means of Specialised Neural Nets. Proceedings of 28th Universities Power Engineering Conference UPEC, Staffordshire University, 279–282, Sept. 1993
4. Heine, S., Wilde, S.: Online-Prognose des Elektroenergieverbrauches eines sächsischen Stadtwerkes. Report, TH Leipzig, August 1994
5. Heine, S., Neumann, I.: Optimizing Load Forecast Models Using an Evolutionary Algorithm. Second European Congress on Intelligent Techniques and Soft Computing (EUFIT), 1690–1694, Aachen, September 1994
6. Kohonen, T.: Learning Vector Quantisation and the Self Organising Map. Theory and Applications of Neural Networks, 235–242, Springer Verlag, London, Berlin, Heidelberg, 1989
7. Macabrey, N., Baumann, T., Germond, A. J.: Prévision de charge dans un réseau électrique à laide du réseau de neurones de Kohonen. Bulletin SEV/VSE 83 (1992) 5, 13–19
8. Schreiber, H., Heine, S.: Einsatz von Neuro-Fuzzy-Technologien für die Prognose des Elektroenergieverbrauches an "besonderen" Tagen. Proceedings of the 4th Fuzzy Days, Universität Dortmund, 274–281, Juni 1994
9. Ultsch, A.: Konnektionistische Modelle und ihre Integration mit wissensbasierten Systemen. Research Report Nr. 396, Universität Dortmund, January 1991

# A Radial Basis Function Network Model for the Adaptive Control of Drying Oven Temperature

Olivier Dubois, Jean-Louis Nicolas, Alain Billat

Université de Reims Champagne Ardenne,
Laboratoire d'Applications de la Microélectronique, BP 347,
F-51062 Reims Cedex, France

**Abstract.** Artificial neural networks are new modelling tools for process control, especially in non-linear dynamic systems. They have been shown to successfully approximate non-linear relationships. This paper describes a neural control scheme for the temperature in a drying oven. The control strategy used is internal model control, in which the plant is modelled by a radial basis function network. The process was identified in an off-line phase —training the network while determining its optimal structure— and an on-line phase —adapting the neural model to any change in the process dynamics. The control strategy was tested experimentally in a series of trials with the oven empty and loaded.

## 1 Introduction

Artificial neural networks are new modelling tools for process control, especially in non-linear dynamic systems. First they can be used as black-box models. They can model any complex system without prior knowledge of its internal working. The only thing needed is a learning set containing input-output vectors generated by experiments and representing the system to model. This modelling can be global, on the whole process working space, so that approximations around working points are not necessary, even in non-linear domain. This learning can be done on-line, allowing for on-line adaptation of the model as new data for the system become available. A network that has been correctly trained also has the ability to generalize. It provides the right outputs in response to inputs not appearing in the learning set. Lastly the main advantage of neural networks that justifies their use for identifying non-linear dynamic systems is the ability of some of them to approximate non-linear relationships. It has been shown by several authors (Hornik et al. 1989; Funahashi 1989) that multilayer feedforward networks with as few as one hidden layer of non-linear units are capable of approximating any continuous function to any desired degree of accuracy. This accuracy becomes better as the number of hidden units is greater. Although this last property is also valid for multilayer recurrent networks, we limit the choice of the neural model to multilayer feedforward networks because their training is simpler and requires less computation time. Neural models are then particularly suitable for identifying unknown non-linear dynamic systems for which only input-output data are available and whose system dynamics are likely to vary.

The process control problem describes in this paper has the above characteristics. The process to be controlled was a drying oven. This system had two inputs, one for heating and one for cooling. The output to be controlled was the internal temperature.

The control strategy and the neural model used for drying oven identification had to satisfy two main objectives. First, the control strategy should be quite insensible to random disturbances (air leaks, door openings, etc.), second the neural model should be able to fit any change in the plant dynamics during its lifetime (change in internal load, heating system, etc.). The internal model control was used as control strategy, together with a radial basis function (RBF) network to identify the plant. The process identification had two phases, an off-line one and an on-line one. The off-line part consisted in training the network while determining its internal structure using an initial training set. The on-line phase was the adaptive part of the control scheme.

This paper first summarizes the principles of the identification and control of dynamic systems using multilayer feedforward networks. Section 3 introduces the RBF network model used, while the last section describes the procedure followed to control the drying oven temperature.

# 2 Identification and Control of Dynamic Systems Using Multilayer Feedforward Networks

## 2.1 Identification

Identification of a system consists of finding a model that approximates the system input-output relationship. Consider the non-linear discrete-time single-input single-output (SISO) system whose dynamics are described by the following difference equation:

$$y(k) = f\left[y(k-1),\dots,y(k-n_y),u(k-d),\dots,u(k-d-n_u+1)\right] \qquad (1)$$

where $u(k)$ and $y(k)$ are the measured input and output at time $k$, $n_y$, $n_u$ and $d$ represent the system orders and the time delay, and $f(.)$ is a non-linear continuous function that is characteristic of the system dynamics. Assume that nothing is known about this system, neither $f(.)$ nor $n_y$, $n_u$ and $d$. Identification then consists of determining the system orders and time delay, and approximating the unknown function $f(.)$ by a neural network model, using a set of measured input and output data (Fig. 1).

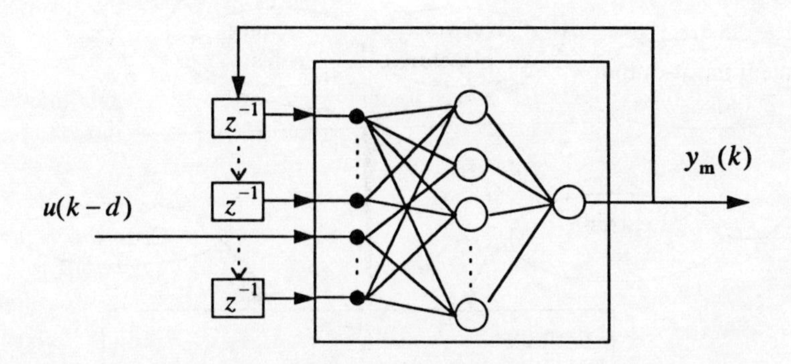


**Fig. 1.** Multilayer feedforward network model of a SISO dynamic system

The complete identification procedure we propose is summarized Fig. 2. This procedure is inspired by the traditional identification scheme employed with linear models (Ljung 1987; Landau 1993). The procedure begins with the choice of a neural model. The model is defined by its architecture and an associated learning rule. This choice can be made to satisfy particular criterion, such as real-time constraints, limited memory size or on-line adaptation necessity. If only three-layer feedforward networks are used for the model architecture, only the learning rule has to be defined. There are no precise rules for choosing a learning algorithm for a given application. The main tools are trial-and-error and user experience.

Once the neural model is chosen, learning can begin. Different structures are trained and compared using a learning set and a criterion. The optimal structure is the one having the fewest units and for which the criterion is minimum. This structure is determined stepwise by varying one parameter at a time. This estimation has two phases. The first concerns the input layer, whose number of units is fixed by the delays and the orders of the system, the second phase concerns the number of hidden units. One can notice that the weights adaptation is realised while the model structure is determined. The structure is then validated by correlation testing (Billings et al. 1992). The ability of the model to generalize is also evaluated by computing the criterion with testing sets that are different from the learning set. If the validation tests are satisfactory, the model is accepted, otherwise, another model is selected. This procedure is repeated until a valid model is found.

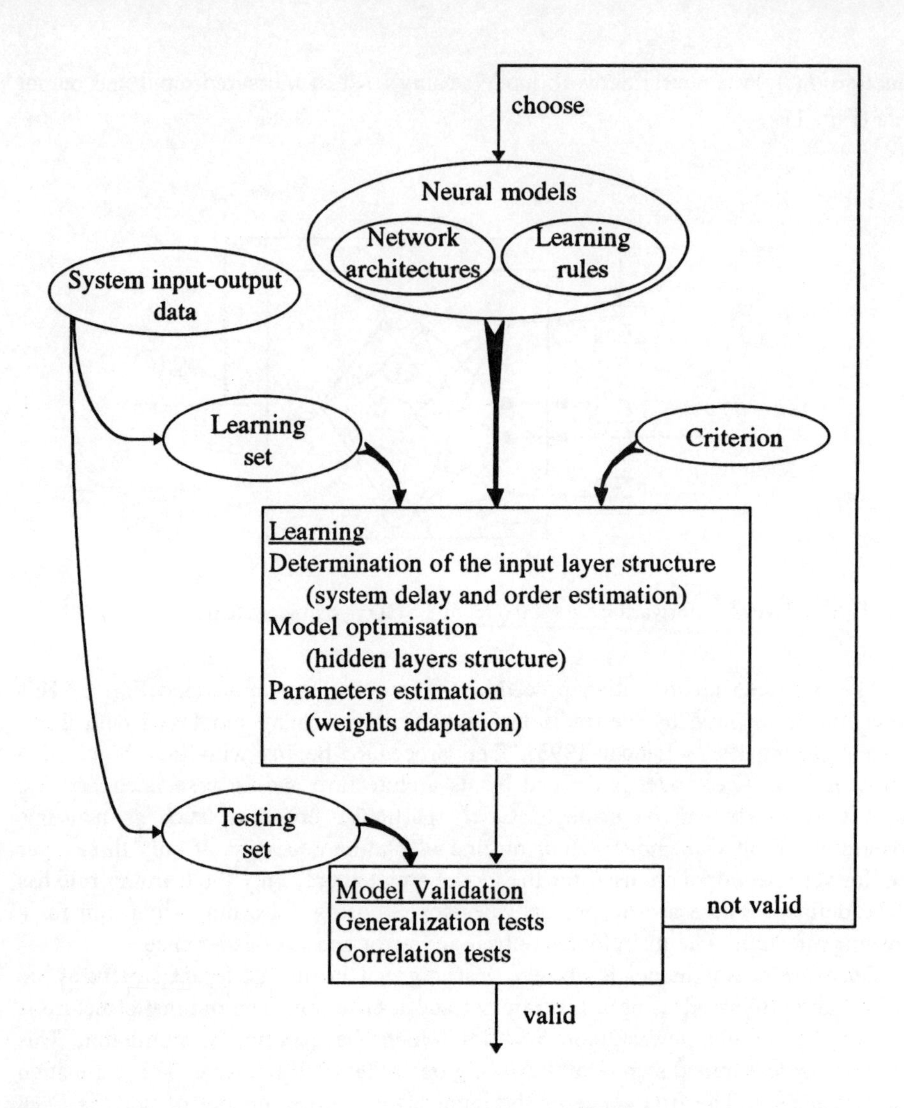


**Fig. 2.** General scheme for unknown dynamic system identification using neural network models

## 2.2 Neural Model Based Control

The neural model can be used in traditional control strategies that require a global model of the system forward or inverse dynamics, for example model predictive control (Saint-Donat et al. 1991; Song and Park 1993), or internal model control (Hunt and Sbarbaro 1991). All possible neural control strategies will not be described in detail; the reader should refer to Narendra and Parthasarathy (1990) or to Hunt et al. (1992) for a more complete survey of the subject.

# 3 The RBF Network Model

The RBF network relies on multivariable interpolation theory (Powell 1987). Its architecture is a feedforward multilayer type network, so it has good approximation capabilities (Park and Sanberg 1991).

## 3.1 Network Architecture and Propagation Rule

The RBF network architecture is shown in Fig. 3. This network is composed of three layers, an input layer, one non-linear hidden layer and a linear output layer.

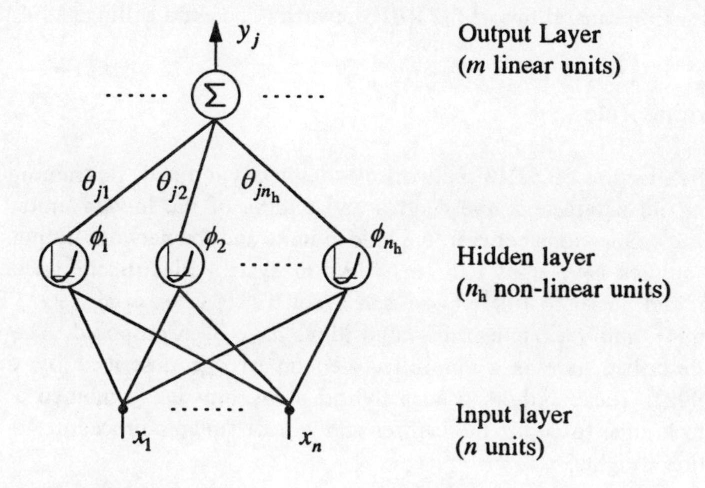


**Fig. 3.** RBF network architecture

The output $\phi_i$ of each of the $n_h$ hidden units is a non-linear function of the Euclidean distance between the input vector $x = [x_1, \cdots, x_n]$ and the centre vector $C_i = [c_{i1}, \ldots, c_{in}]$ of the considered hidden unit,

$$\phi_i = \phi(\|x - C_i\|, \rho_i), \qquad i = 1, \ldots, n_h \tag{2}$$

where $\rho_i$ are the widths and $\|.\|$ denotes the Euclidean norm. The network outputs $y_j$, which can be regarded as non-linear functions $NL_j(.)$ of the input vector, are simply weighted sums of the hidden unit outputs,

$$y_j = NL_j(x) = \sum_{i=1}^{n_h} \theta_{ji}\phi_i, \quad j = 1,\dots,m \tag{3}$$

where $\theta_{ji}$ is the weight of the connection linking the hidden unit $i$ and the output unit $j$; $m$ is the number of network outputs.

Many choices are possible for $\phi(.,\rho)$. The Gaussian function

$$\phi(z,\rho) = \exp(-z^2 / \rho^2) \tag{4}$$

or the thin-plate-spline function

$$\phi(z) = z^2 \log(z) \tag{5}$$

are often used. The choice of the non-linearity $\phi(.)$ does not seem to be crucial for the approximation capabilities of the RBF network (Chen and Billings 1992).

### 3.2 Learning Rule

When the architecture of a RBF network is completely defined, its training consists in estimating its parameters: the centres and widths of the hidden units, and the weights of the connections between the hidden units and the network outputs.

Several studies have used RBF networks in system identification and control fields (Hunt and Sbarbaro 1991; Lowe and Webb 1991; Chen et al. 1992, Elanayar and Shin 1994), and various learning algorithms have been proposed. The learning algorithm described here is a simplified version of one described by Chen and Billings (1992). These authors used a hybrid algorithm that combined a k-means clustering technique to adjust the centres and a least squares procedure to estimate the connection weights.

In the algorithm we propose, the centres are randomly distributed over the input space and are fixed during learning. By choosing the non-linearity (5) for the hidden units, we need not to determine the widths. Therefore, only the connection weights must be estimated. Since the network output values are linear with respect to the weights, they can be updated using a recursive least squares algorithm.

At each moment, the output vector of the hidden layer

$$\Phi(k) = \left[\phi_1(k),\dots,\phi_{n_h}(k)\right]^T \tag{6}$$

is computed after presentation of the training set input vector $x(k)$ to the network. If $o_j(k)$ is the $j$th desired output at the time $k$, the weight vector

$$\theta_j(k) = \left[\theta_{j1}(k), \dots, \theta_{jn_h}(k)\right]^T \tag{7}$$

of the $j$th network output is updated according to the following relation:

$$\theta_j(k) = \theta_j(k-1) + P(k)\Phi(k)\varepsilon_j(k), \quad 1 \le j \le m, \tag{8}$$

with

$$\varepsilon_j(k) = o_j(k) - \Phi^T(k)\theta_j(k-1), \quad 1 \le j \le m, \tag{9}$$

and

$$P(k) = P(k-1) - \frac{P(k-1)\Phi(k)\Phi^T(k)P(k-1)}{1 + \Phi^T(k)P(k-1)\Phi(k)}. \tag{10}$$

The use of a linear learning rule, such as the least squares algorithm that has a fast convergence rate, allows fast learning of the RBF network (Chen and Billings 1992).

## 4 Identification and Control of the Drying Oven

### 4.1 Description of the Drying Oven

The temperature and humidity of the drying oven could be controlled, but only temperature was controlled in this study. A drying oven can be used for testing the behaviour of a given product under specific temperature and humidity conditions. Therefore, the working drying oven always contains something representing a certain load that can alter its dynamics (Fig. 4).

Temperature control was made via two antagonist actuators, a heating electrical resistor circuit supplied by a static relay dimmer, and a refrigeration system whose fluid circulation was controlled by a valve. The inputs of the plant were bounded by 0 % and 100 % and correspond, for each actuator, to the fraction of the sampling period during which they were on. The temperature was measured by a probe placed inside the oven. The temperature range was 0 °C-100 °C.

Data acquisition and the generation of command signals were taken care of, by a PC linked to the oven. The temperature was measured 2 seconds before the end of each 12-second sampling period. The command values had to be computed, using the control algorithm, during this period and applied at the beginning of the next sampling period.

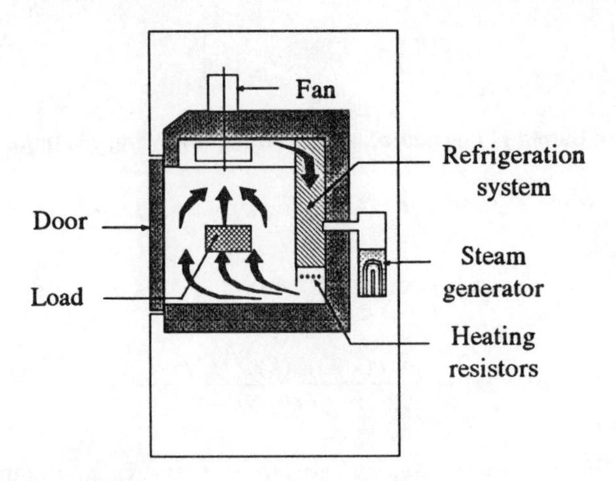


**Fig. 4.** Drying oven cutaway to show actuators location and air circulation

### 4.2 Neural Network Model

The drying oven non-linear dynamics, can be described by the following difference equation:

$$\begin{aligned} y_{\mathrm{p}}(k) = NL_{\mathrm{p}}\big[ & y_{\mathrm{p}}(k-1),\ldots,y_{\mathrm{p}}(k-n_y), \\ & u_{\mathrm{h}}(k-d_{\mathrm{h}}),\ldots,u_{\mathrm{h}}(k-d_{\mathrm{h}}-n_{u_{\mathrm{h}}}+1), \\ & u_{\mathrm{c}}(k-d_{\mathrm{c}}),\ldots,u_{\mathrm{c}}(k-d_{\mathrm{c}}-n_{u_{\mathrm{c}}}+1)\big] \end{aligned} \tag{11}$$

where $y_{\mathrm{p}}(k)$ is the value of the oven temperature at time $k$, and where $u_{\mathrm{h}}(k)$ and $u_{\mathrm{c}}(k)$ are the heating and cooling commands at the same time. The system output can then be considered as a non-linear function $NL_{\mathrm{p}}$ of the $n_{\mathrm{y}}$ lagged outputs, the $n_{u_{\mathrm{h}}}$ lagged heating commands and the $n_{u_{\mathrm{c}}}$ cooling commands ($n_{u_{\mathrm{h}}} \geq 1$ and $n_{u_{\mathrm{c}}} \geq 1$). $d_{\mathrm{h}}$ and $d_{\mathrm{c}}$ are respectively heating and cooling commands delays ($d_{\mathrm{h}} \geq 1$ and $d_{\mathrm{c}} \geq 1$). We used a RBF network to approximate this non-linear relationship. The topology of this network is shown in Fig. 5. The learning rule used to train this network was the one described Section 3.2. Let

$$\begin{aligned} x(k) = \big[ & y_{\mathrm{p}}(k-1),\ldots,y_{\mathrm{p}}(k-n_y), \\ & u_{\mathrm{h}}(k-d_{\mathrm{h}}),\ldots,u_{\mathrm{h}}(k-d_{\mathrm{h}}-n_{u_{\mathrm{h}}}+1), \\ & u_{\mathrm{c}}(k-d_{\mathrm{c}}),\ldots,u_{\mathrm{c}}(k-d_{\mathrm{c}}-n_{u_{\mathrm{c}}}+1)\big], \end{aligned} \tag{12}$$

the input-output relationship of the RBF model is then given by:

$$y_m(k) = NL_m(x(k)) = \sum_{i=1}^{n_h} \theta_i \phi_i = \sum_{i=1}^{n_h} \theta_i \phi\left(\left\|x(k) - C_i\right\|\right), \tag{13}$$

with

$$\phi(z) = z^2 \log(z) \tag{14}$$

This network model was selected for its fast convergence rate, as this allowed on-line learning, and hence, on-line adaptation.

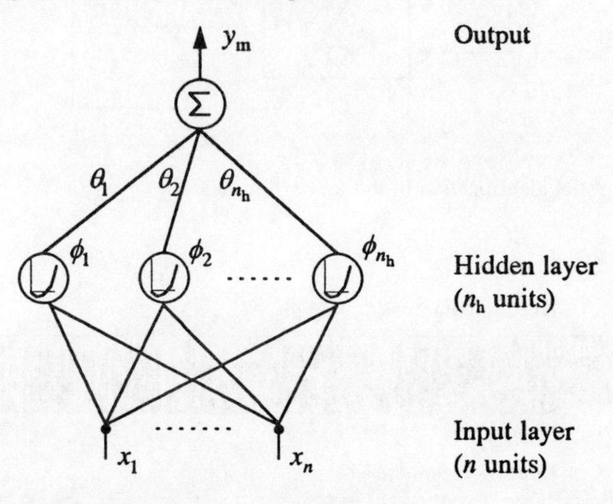


**Fig. 5.** RBF network model of the drying oven

### 4.3 Identification of the Forward Model

The plant was identified using a series-parallel scheme. The model output was computed from lagged inputs and outputs of the system (Fig. 6).

The data sets used for identification, were made up of the temperature response of the empty oven to a random input sequence (Fig. 7). This temperature response covered a space slightly bigger than the oven working space (from $-10\,^\circ\mathrm{C}$ to $+110\,^\circ\mathrm{C}$). The initial training set contained 800 samples of input-output data randomly picked from the above signals. The system was identified by the procedure described Section 2.1. The optimal structure of the network model was determined using the following criterion

$$J = \frac{1}{N} \sum_{k=1}^{N} (y_d(k) - y_m(k))^2 \tag{15}$$

which represents the output mean square error between the desired output $y_d$ and the model output $y_m$ at time $k$. The optimal structure is the one for which, after learning, $J$ is minimal over all the N samples of the training set.

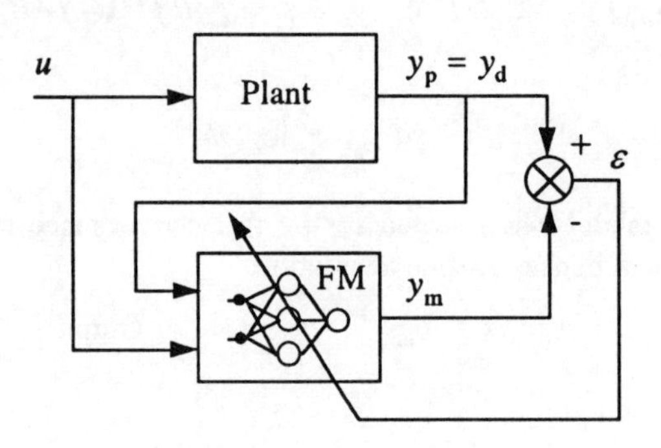


**Fig. 6.** Series-parallel identification of the forward model

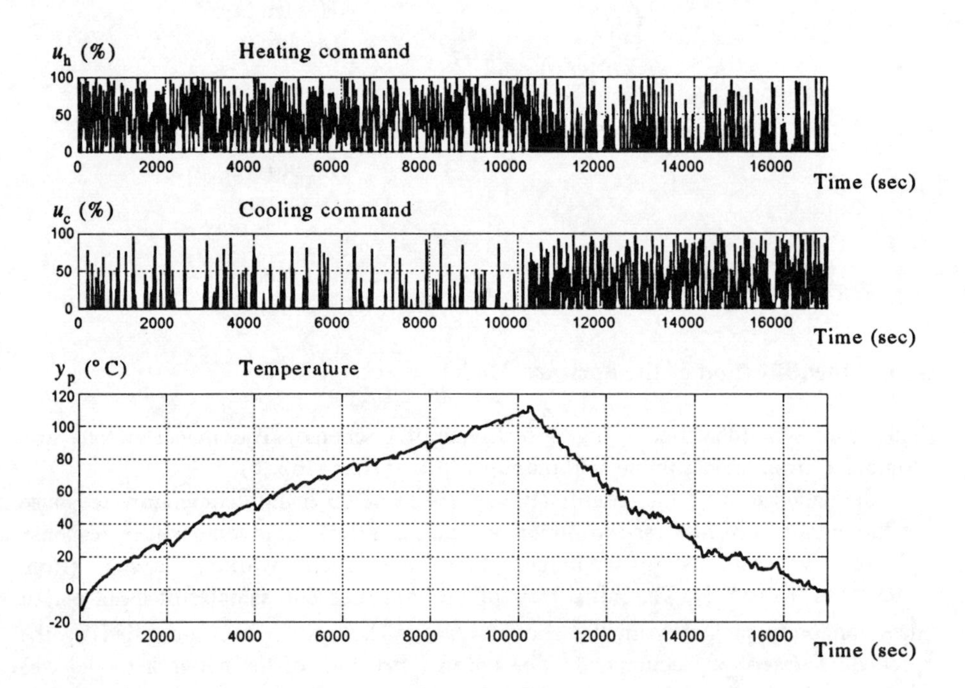


**Fig. 7.** Command sequences and temperature response used to build the learning and testing sets.

The network input number $n$ was determined by estimating the orders of the plant, because $n$ is given by:

$$n = n_y + n_{u_h} + n_{u_c} \tag{16}$$

Networks were trained for all possible combinations of delays $d_h$ and $d_c$ and orders $n_y$, $n_{u_h}$ and $n_{u_c}$, and the corresponding values of $J$ were computed (Fig. 8, Fig. 9 and Table 1). These figures only show the results for delays and orders that were identical for both heating and cooling commands ($d_h = d_c = d$ and $n_{u_h} = n_{u_c} = n_u$). These input layer configurations were the ones that gave the best results. The same procedure was used to determine the optimal number of network hidden units $n_h$ (Fig. 10).

The model became less accurate as the delay $d$ increased, but the accuracy increased as the number of lagged inputs $n_y$ increased. Finally, the mean square error increased as the order $n_u$ increased after passing through a minimum for $n_u = 2$, whatever the value of $n_y$. The network structure finally chosen had 9 inputs ($d_h = d_c = 1$, $n_y = 5$ and $n_{u_h} = n_{u_c} = 2$), 80 hidden units and one output. These parameter values were a compromise between good model accuracy and the smallest number of neurons.

The generalization capacity of the model was then evaluated by computing the criterion $J$ with testing sets. We have also performed correlation tests. All those results allowed us to validate the model.

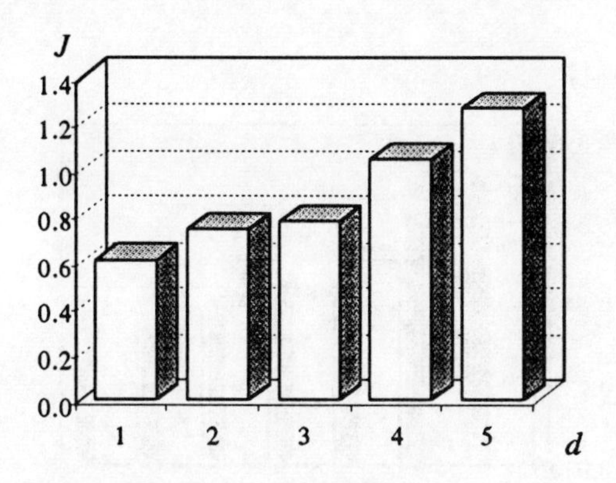


**Fig. 8.** Change in $J$ with respect to the delay $d$

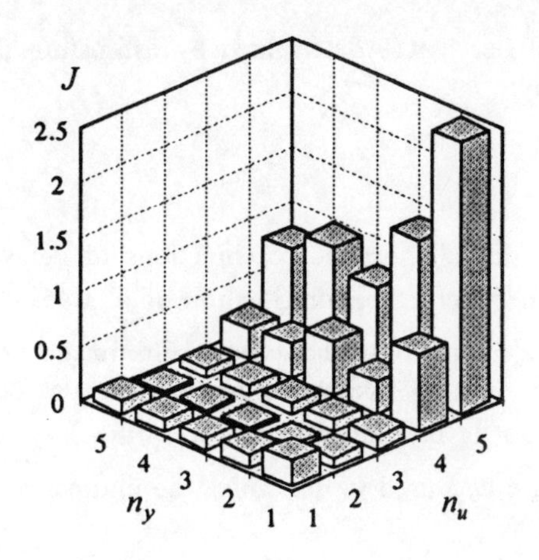


**Fig. 9.** Change in $J$ with respect to the orders $n_y$ and $n_u$

**Table 1.** Change in $J$ with respect to the orders $n_y$ and $n_u$

| $n_u$ \ $n_y$ | 1 | 2 | 3 | 4 | 5 |
|---|---|---|---|---|---|
| 1 | 0.229 | 0.13 | 0.113 | 0.112 | 0.113 |
| 2 | 0.093 | 0.04 | 0.036 | 0.034 | 0.033 |
| 3 | 0.121 | 0.104 | 0.099 | 0.099 | 0.079 |
| 4 | 0.695 | 0.29 | 0.494 | 0.309 | 0.29 |
| 5 | 2.484 | 1.41 | 0.812 | 1.026 | 0.839 |

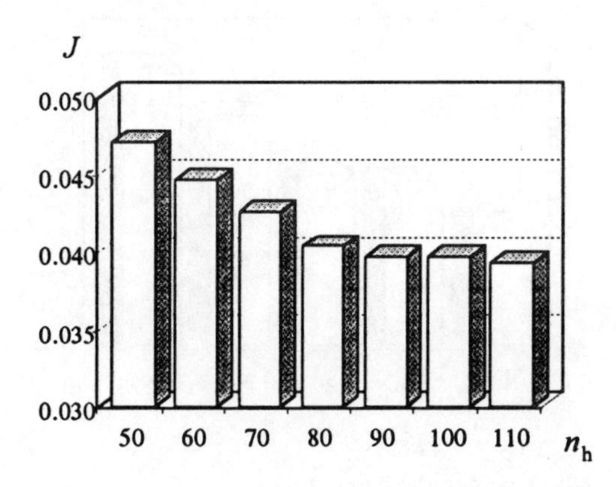


**Fig. 10.** Change in $J$ with respect to the number of hidden units

### 4.4 On-line Adaptation of the Model

The model adaptation was the on-line estimation of the weight vector so that it fitted any change in the plant dynamics. This was done by updating the contents of the initial training set. At each sampling time, the new measured data vector, which was characteristic of the process state, replaced its nearest training set vector, depending on the Euclidean distance. This new set could not be used to train the network again during one sampling period. We therefore developed a method to cancel the replaced vector effects on the network, before taking into account the new one, in one iteration.

Let $c$ be the index of the data vector that will be replaced, among the $N$ vectors of the training set. The weight vector $\theta(N-1)$ estimated without the $c$ index vector is computed by:

$$\theta(N-1) = P(N-1)\sum_{i=1}^{N-1} y_{\mathrm{d}}(i)\Phi(i), \tag{17}$$

with

$$P(N-1) = P(N) - \frac{P(N)\Phi(c)\Phi^T(c)P(N)}{\Phi^T(c)P(N)\Phi(c)-1}. \tag{18}$$

The new estimated weight vector $\theta(N)$, taking into account the new data vector, was then computed by the usual recursive least squares procedure.

This adaptation method consists of working with a fixed size learning set, containing observed data that are representative of the whole working space, at any time. Therefore, the network always keeps a global behavioural model of the drying oven, even after long periods of being regulated at a fixed temperature. This is a major advantage over recursive techniques based on the gradual forgetting of the past data.

### 4.5 Internal Model Control

In the internal model control (Fig. 11), the forward neural model FM is in parallel with the plant. One of the requirements for a maximal performance control, is that the controller C represents the inverse dynamics of the system (Hunt and Sbarbaro 1991).

We identified inverse dynamics of the drying oven using a RBF network and the same procedure used for system identification. This trial did not give an accurate enough model, we therefore generated plant commands from the forward model. At each sampling period, the command vector to be presented to the system input was computed by linear interpolation between the desired temperature and the temperature responses of the model to extreme command values ($u_{\mathrm{h}} = 0$ % and $u_{\mathrm{c}} = 100$ %; $u_{\mathrm{h}} = 100$ % and $u_{\mathrm{c}} = 0$ %; $u_{\mathrm{h}} = u_{\mathrm{c}} = 0$ %).

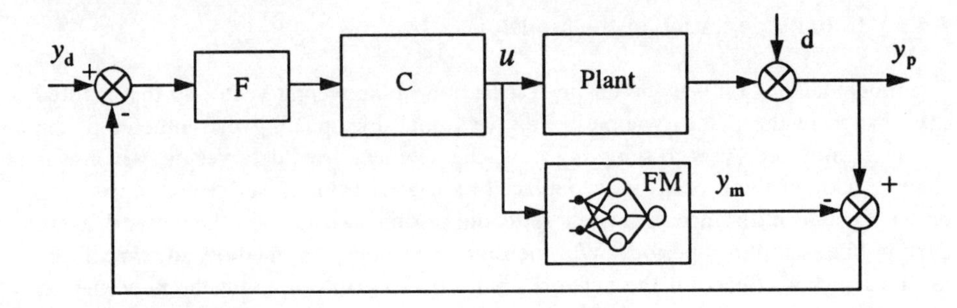


**Fig. 11.** Internal model control structure

## 4.6 Control Experiments

The control strategy performance was evaluated using a complex reference signal that covered the whole drying oven working space (from 0 °C to 100 °C), and passing through various levels (0 °C, 40 °C, 60 °C, 100 °C, 80 °C, 50 °C, 20 °C and 0 °C). The values of the slopes were ±2 °C/min. These values were the limits indicated by the drying oven maker. Figure 12 shows the temperature response of the empty oven. During this experiment, the drying oven door has been opened for 48 seconds (4 sampling periods).

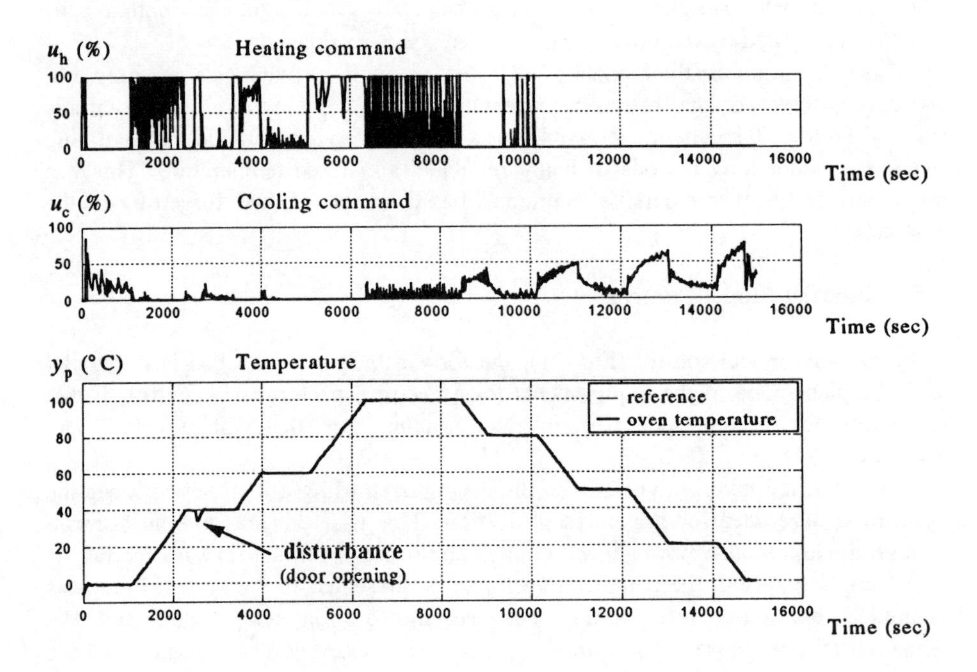


**Fig. 12.** Command signals and temperature response of the empty drying oven

We also tested the adaptive part of the control scheme with the oven loaded. The load was simulated by placing 50 kg of metal inside the drying oven (Fig. 13). The difference between desired and real temperature, during the last increasing temperature period, was not due to a control algorithm fault but to the insufficient heating power. Apart from that problem the drying oven temperature fitted the reference quite well, although the initial model was developed for the empty drying oven.

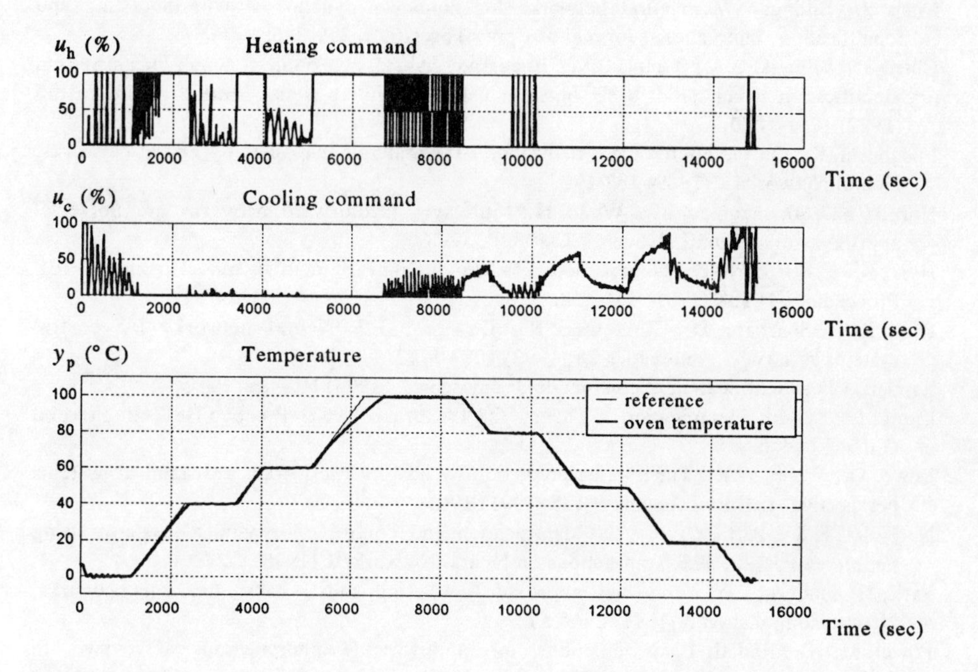


**Fig. 13.** Command signals and temperature response of the loaded drying oven

# 5 Conclusion

This study has tested the modelling abilities of radial basis function networks on a real plant. This type of neural network satisfied the real-time constraints and the on-line adaptation necessity for the control application. The results of control experiments were satisfactory, despite the controller inaccuracy.

The next step will be to control not only the temperature, but also the humidity inside the drying oven, using radial basis function network models.

## References


Elanayar, S.V.T., Shin, Y.C.: Radial basis function neural network for approximation and estimation of nonlinear stochastic dynamic systems. IEEE Transactions on Neural Networks **5** (1994) 594-603.

Billings, S.A., Jamaluddin, H.B., Chen, S.: Properties of neural networks with applications to modelling non-linear dynamical systems. International Journal of Control **55** (1992) 193-224.

Chen, S., Billings, S.A.: Neural networks for nonlinear dynamic system modelling and identification. International Journal of Control **56** (1992) 319-346.

Chen, S., Billings, S.A., Grant, P.M.: Recursive hybrid algorithm for non-linear system identification using radial basis function networks. International Journal of control **55** (1992) 1051-1070.

Funahashi, K.: On the approximate realization of continuous mappings by neural networks. Neural Networks **2** (1989) 183-192.

Hornik, K., Stinchcombe, M., White, H.: Multilayer feedforward networks are universal approximators. Neural Networks **2** (1989) 359-366.

Hunt, K.J., Sbarbaro, D.: Neural networks for nonlinear internal model control. IEE Proceedings-D **138** (1991) 431-438.

Hunt, K.J., Sbarbaro, D., Zbikowski, R., Gawthrop, P.J.: Neural networks for control systems-A survey. Automatica **28** (1992) 1083-1112.

Landau, I.D.: Identification et commande des systemes. (1993) Hermes, Paris.

Ljung, L.: System identification - Theory for the user. (1987) Prentice-Hall, Englewood Cliffs, London.

Lowe, D., Webb, A.R.: Time series prediction by adaptive networks: a dynamical systems perspective. IEE Proceedings-F **138** (1991) 17-24.

Narendra, K.S., Parthasarathy, K.: Identification and control of dynamical systems using neural networks. IEEE Transactions on Neural Networks **1** (1990) 4-27.

Park, J, Sanberg, I.W.: Universal approximation using radial basis function networks. Neural Computation **3** (1991) 246-257.

Powel, M.J.D.: Radial Basis functions for multivariable interpolation: a review. In J.C. Mason and M.G. Cox (Eds) Algorithms for Approximation (1987) 143-167 Clarendon Press, Oxford.

Saint-Donat, J., Bhat, N., McAvoy, T.J.: Neural net based model predictive control. International Journal of Control **54** (1991) 1453-1468.

Song, J.J., Park, S.: Neural model predictive control for nonlinear chemical processes. Journal of Chemical Engineering of Japan **26** (1993) 347-354.

# Hierarchical Competitive Net Architecture

Theresa W. Long[1] and Emil L. Hanzevack[2]

[1] NeuroDyne Inc., 108 Brooks St., Williamsburg, VA, USA 23185
[2] University of South Carolina, Columbia, SC, USA 29208

**Abstract**

Development of hypersonic aircraft requires a high degree of system integration. Design tools are needed to provide rapid, accurate calculations of complex fluid flow patterns. This project demonstrates that neural networks can be successfully applied to calculation of fluid flow distribution and heat transfer in a six leg heat exchanger panel, typical of the type envisioned for use in hypersonic aircraft.

We used training data generated from fluid flow and heat transfer equations in explicit form to train a neural net to solve the associated inverse dynamics problem. We developed an improved competitive net architecture. Finally, we successfully implemented a hierarchical neural network scheme to link multiple heat exchanger panels together. This method is direct, fast, and accurate.

## 1 Improved Competitive Net Architecture

Most currently popular algorithms for building mathematical models of unknown systems, or supervised learning neural networks, may be broadly described as global or local learning methods. Global neural network learning uses completely distributed weight representations, thus requiring update of all network weights per training pattern per epoch. Local neural network learning uses locally distributed weights, thus requiring update of small subsets of all network weights. Each approach has its advantages and disadvantages. However, we believe that local learning methods may be more appropriate for many practical engineering applications.

When a function is highly nonlinear, or when its dynamic characteristics vary in different operating regimes, it is difficult to fit the input-output data with a single global function (or system model) that faithfully describes the system behavior. The piecewise approach of function approximation is more suitable for this type of system. Perhaps the best known function approximation using piecewise fitting for a 2 dimensional problem is the cubic spline. A large curve can be decomposed into segments of local curves, and each local curve may be fitted with simple functions.

As a practical engineering example, a data set may contain some cases in which the heat load over a heat exchanger panel is relatively uniform, mixed with some cases in which one heat exchanger segment has a shock heat load several times greater than the other components. Thus there are two distinct heat load regimes in the data set. Another example would be the loss factor curves used in the pressure drop equations, in which there are several distinct regimes, depending on parameters such as Reynolds number. For these types of application, a competitive net architecture is preferable.

A neural network is used to approximate each local curve. We call these local-fitting neural nets the expert nets, to be consistent with the terminology coined by Jacobs et al.[1]. A competitive net is used to classify the input data and send it to the appropriate expert net. Previously suggested competitive nets are winner-take-all networks. Their convergence is extremely sensitive to correct selection of the weights[2]. Additionally, the classification is hard cut. For dynamic systems, a data point may lie within several probability distributions simultaneously. A hard cut classification does not provide details of this situation. Jacobs et al. proposed a competitive learning architecture where a gating net performs classification of the input data by assigning a weighting factor for each class represented by the expert net. The weights sum to one, so each expert net contributes to the function in varying degrees, and the classification is a soft cut.

Jacobs and Jordan[3] interpreted the gating net output as a priori probability, and used mixture density maximum likelihood as the objective function. The update rule for the gating net aims at moving the a priori probability towards the a posteriori probability. A slightly different way of interpreting the weights is that the gating net is learning the fuzzy set membership functions, and the weights are membership values. Pao[4] has indicated that the membership function values can help to provide an estimate of missing or incomplete knowledge. The generalization property of the neural net can be decomposed into two elements: the generalization of the expert net and the generalization of the membership function.

We have developed an improved version of a competitive network architecture (Figure 1). The number of expert nets determines the number of the gating net output whose activation function is the softmax[5]. The expert nets use backpropagation with one hidden layer. This same architecture can be used as a module at the bottom level of a cascade structure[6], with a higher level gating net regulating the output of the modules. We have made two modifications to the basic module architecture used by Jacobs and Jordan[3].

First, we added a hidden layer to the gating net. This hidden layer is necessary to provide additional connections to ensure that the classification problem is solvable. Consider, for example, a gating net with 2 inputs and 4 outputs, and for simplicity, assume that the output nodes are binary. Since this is not a winner-take-all net, all output nodes of the gating net may be activated at the same time. Therefore, there are $2^4$ output vector combinations, but there are only 8 weights connecting the input and output. This is equivalent to having 8 unknowns and 16 equations. It would be problematic to solve for the weights in this case. By adding a hidden layer, we provide the necessary number of additional weights needed for solution. But it is important not to add too many hidden nodes, or the gating net will overfit the membership function and lose generalization ability.

**Figure 1**
**Competitive Learning Nets**

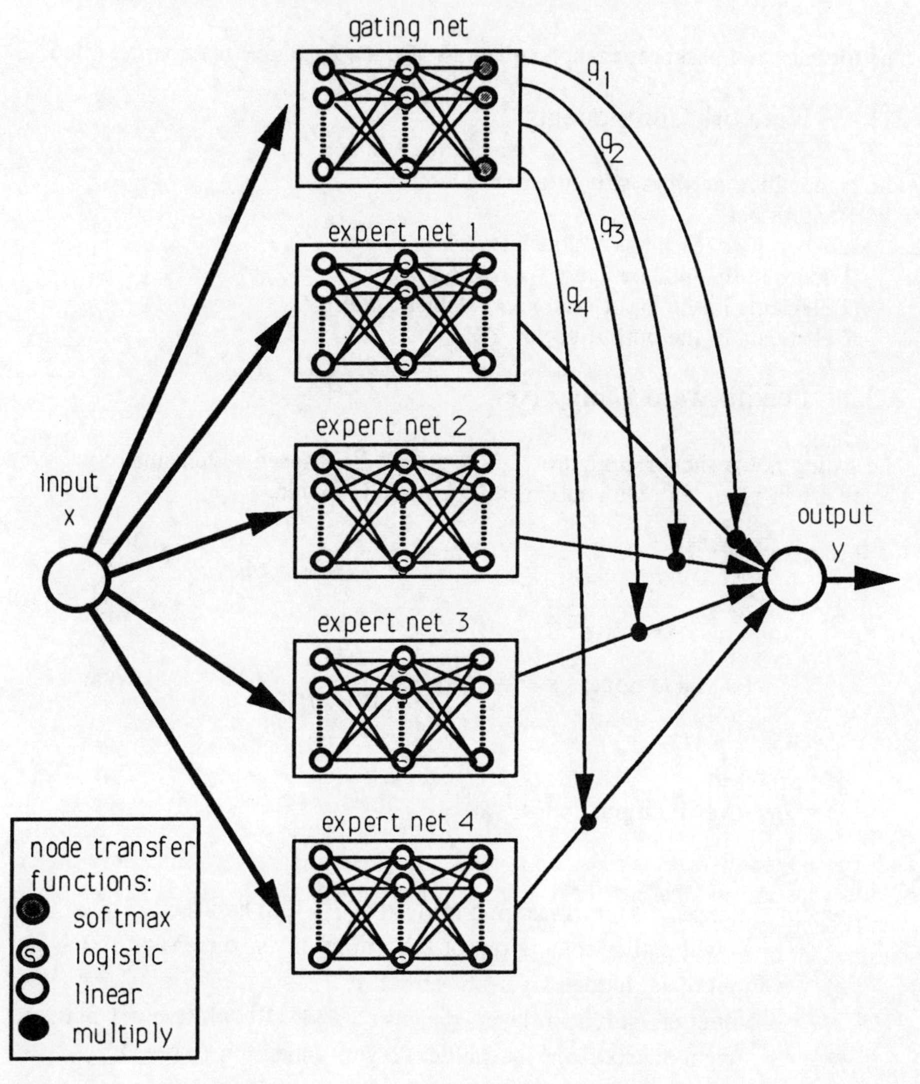


The second modification is that we used the inner product of the error vector (weighted by the gating net output), instead of the mixture density maximum likelihood, as the objective function for each expert net. For simple cases, we obtained similar results using either our objective function or the mixture density objective function. However, if we use a competitive net structure in which a cascade of gating nets is used to regulate the expert net output[6], our objective function is easier to track in backpropagation through the cascade structure. Using the mixture density objective function, the various levels of the posterior probability would require careful bookkeeping. As the complexity of system is increased, the cascaded gating net architecture should prove advantageous.

# 2 Development of the Competitive Net

The forward and backpropagation equations of our competitive net are presented.

## 2.1 Network Components

The competitive network structure has:

1 gating net
E expert nets (each has 1 hidden layer)
P nodes in the hidden layer, + 1 bias node
N elements in the input vector x, + 1 bias node
R elements in the output vector Y.

## 2.2 Feedforward Gating Net

The gating net is shown in Figure 2. Vectors are bold-faced. Elements of a vector are in plain text, with subscript j for node j omitted.

$$\mathbf{s}^{(1)} = \mathbf{w}^{(0)} \bullet \quad (1)$$

$$x^{(1)} = f(s^{(1)}) = \frac{1}{1+\exp(-s^{(1)})} \quad (2)$$

$$\text{for Elliott node:} \quad x^{(1)} = f(s^{(1)}) = \frac{s^{(1)}}{1+|s^{(1)}|} \quad (2a)$$

$$\mathbf{s}^{(2)} = \mathbf{w}^{(1)}\mathbf{s}^{(1)} \quad (3)$$

$$y = \exp(s^{(2)}) \quad (4)$$

$D = \sum y_i$ over E output nodes

$$g_i = \frac{y_i}{D} \quad (5)$$

where $\mathbf{x}$ = input vector, dimension N+1, x[N+1]=1, to be used for bias
$\mathbf{w}^{(0)}$ = weight matrix coming out of 0th (input) layer, dimension Px(N+1),
$\mathbf{s}^{(1)}$ = input of 1st hidden layer, dimension P.
$\mathbf{x}^{(1)}$ = output of 1st hidden layer, dimension P+1, (P+1)th term =1 bias,
$\mathbf{w}^{(1)}$ = weight matrix from 1st (hidden) layer, dimension Ex(P+1),
$\mathbf{s}^{(2)}$ = input to nodes of the output layer, dimension E
$\mathbf{g}$ = output vector of the gating net, dimension E.

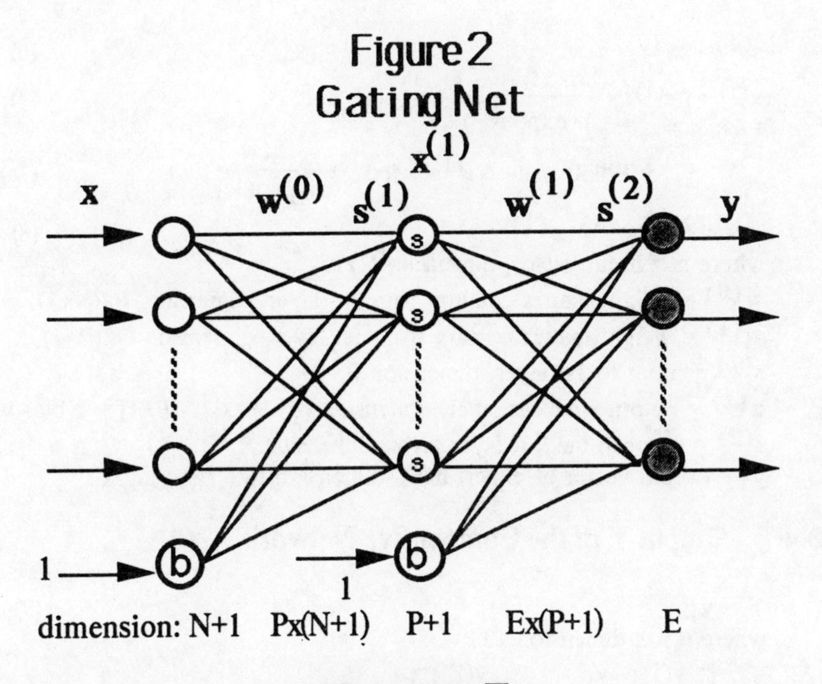


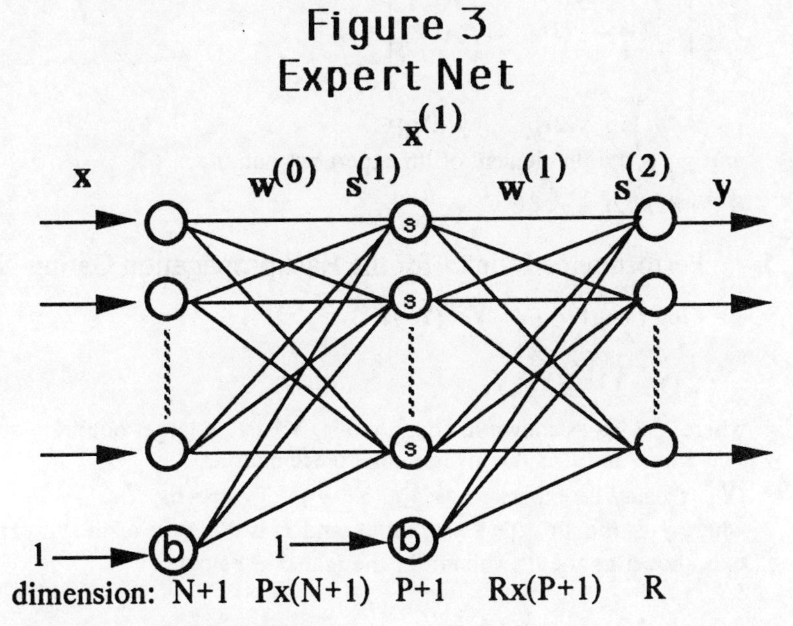


## 2.3 Feedforward Expert Net

An expert net is shown in Figure 3. We show only 1 expert net. The other expert nets are analogous. The input layer is labeled as the 0th layer, the hidden layer is the 1st layer, the output layer is the 2nd layer. The dimension P is not the same P as in the gating net.

$$\mathbf{s}^{(1)} = \mathbf{w}^{(0)} \bullet \mathbf{x} \tag{6}$$

$$x^{(1)} = f(s^{(1)}) = \frac{1}{1+\exp(-s^{(1)})} \tag{7}$$

for Elliott node: $$x^{(1)} = f(s^{(1)}) = \frac{s^{(1)}}{1+|s^{(1)}|} \tag{7a}$$

$$\mathbf{y} = \mathbf{s}^{(2)} = \mathbf{w}^{(1)} \bullet \mathbf{x}^{(1)} \tag{8}$$

where $\mathbf{x}$ = input vector, dimension N+1,

$\mathbf{w}^{(0)}$ = weight matrix coming from 0th layer, dimension Px(N+1),

$\mathbf{w}^{(1)}$ = weight matrix coming from 1st layer, dimension Qx(P+1),

$\mathbf{s}^{(1)}$ = input to 1st layer, dimension P,

$\mathbf{x}^{(1)}$ = output from 1st layer, dimension (P+1), $x^{(1)}$ [P+1] = 1 bias weight.

$\mathbf{s}^{(2)}$ = input to the last layer layer, dimension R,

$\mathbf{y_i}$ = output vector of expert net i, dimension R.

## 2.4 Output Y of the Competitive Network

$$\mathbf{Y} = \mathbf{y}\mathbf{g} \tag{9}$$

where y has dimension RxE

$$\mathbf{y} = \begin{bmatrix} y(1)_1 & y(2)_1 & \cdots & y(E)_1 \\ y(1)_2 & y(2)_2 & \cdots & y(E)_2 \\ \cdot & \cdot & \cdot & \cdot \\ \cdot & \cdot & \cdot & \cdot \\ y(1)_R & y(2)_R & \cdots & y(E)_R \end{bmatrix}$$

and $yi_j$ is the jth element of ith expert net output.

$$\mathbf{g} = [\, g_1, g_2, \ldots\ldots, g_E \,]^T$$

## 2.5 Performance Index J for the Backpropagation Gating Net

$$J = -\log\left[\, \exp\{-\frac{1}{2}(\mathbf{Y}^*-\mathbf{Y})^T(\mathbf{Y}^*-\mathbf{Y})\}\,\right]$$

$$= \frac{1}{2}(\mathbf{Y}^*-\mathbf{Y})^T C (\mathbf{Y}^*-\mathbf{Y}) \tag{10}$$

where $\mathbf{Y} = \mathbf{y}\mathbf{g}$ as shown in equation (7), $\mathbf{Y}^*$ is the target output.
and C is the inverse covariance matrix.

$(\mathbf{Y}^*-\mathbf{Y})$ may be expressed as: $\Sigma g_i(\mathbf{Y}^*-\mathbf{y}_i) = \Sigma g_i \mathbf{e_i} = \mathbf{e}\mathbf{g}$

where $\mathbf{y_i}$ is the ith expert net output, and $\mathbf{e_i}$ is the error of the ith expert net. $\mathbf{e_i}$ is shown as the ith column in the matrix e below.

$$\mathbf{e} = \begin{bmatrix} Y_1^*-y(1)_1 & Y_1^*-y(2)_1 & \cdots & Y_1^*-y(E)_1 \\ Y_2^*-y(1)_2 & Y_2^*-y(2)_2 & \cdots & Y_2^*-y(E)_2 \\ \cdot & \cdot & \cdot & \cdot \\ \cdot & \cdot & \cdot & \cdot \\ Y_R^*-y(1)_R & Y_R^*-y(2)_R & \cdots & Y_R^*-y(E)_R \end{bmatrix} \tag{11}$$

## 2.6 Backpropagation for the Gating Net

$$\frac{\partial J}{\partial w} = \frac{\partial g}{\partial s} e^T C(Y^*-Y) x^T \tag{12}$$

$$\frac{\partial g}{\partial s} = \begin{bmatrix} g_1-g_1^2 & -g_1 g_2 & \cdots & -g_1 g_E \\ -g_2 g_1 & g_2-g_2^2 & \cdots & -g_2 g_E \\ \cdot & \cdot & \cdot & \cdot \\ \cdot & \cdot & \cdot & \cdot \\ -g_E g_1 & -g_E g_2 & \cdots & g_E-g_E^2 \end{bmatrix} \tag{13}$$

$$d_j^{(2)} = -\frac{\partial g}{\partial s} e^T C(Y^*-Y) \tag{14}$$

$$\Delta w^{(1)} = m d_j^{(2)} x^{(1)T} + a\, \Delta w^{(1,old)} \tag{15}$$

$$w^{(1)} = w^{(1)} + \Delta w^{(1)} \tag{16}$$

where $w^{(1)}$ = weights connected to the output nodes, dimension Ex(P+1)
$\Delta w$ = new delta weights, dimension mxn
$\Delta w^{[\cdot,old]}$ = delta weights used in the previous backpropagation iteration
m = learning rate, use $x^T x$ as 1st try value, and a = momentum coefficient.

We can replace $e^T$ with $y^T$ in equation (10) and (11), but the sign in front must be switched. If we use the inverse covariance matrix C, replace $Y^*-Y$ by $C(Y^*-Y)$.

next layer back ($w^{(0)}$):

$$\frac{\partial x^{(1)}}{\partial s^{(1)}} = x^{(1)} - \{x^{(1)}\}^2 \tag{17}$$

P diagonal terms

or for Elliott node: $\frac{\partial x^{(1)}}{\partial s^{(1)}} = (1 - |x^{(1)}|)^2$ (17a)

P diagonal terms

$$d^{(1)} = \frac{\partial x^{(1)}}{\partial s^{(1)}} \bullet \{w^{(1)}\}^T\, d^{(2)} \tag{18}$$

$$\Delta w^{(0)} = -m d^{(1)}\, x^T + a\, \Delta w^{(0)[last]} \tag{19}$$

$$w^{(0)} = w^{(0)[last]} + \Delta w^{(0)}, \tag{20}$$

## 2.7 Performance Index for Expert Net i

$$J = -\log\left[\exp\{-\frac{1}{2}(g_i Y^* - g_i y_i)^T C (g_i Y^* - g_i y_i)\}\right]$$

$$= \frac{1}{2}(g_i Y^* - g_i y_i)^T C (g_i Y^* - g_i y_i)$$

$$d_i^{(2)} = -(g_i)^2 C(Y^* - y_i) \tag{21}$$

The rest is analogous to equations (15)-(20)

## 3 Successful Application to Flow Calculation

Much research is being done in the area of neural networks, and industry is actively seeking successful application to real world problems. We describe here a successful application. We have used a hierarchical neural network architecture to model complex coolant flow patterns, such as those encountered in design of hypersonic aircraft. This method is direct, fast, and accurate.

Development of hypersonic aircraft requires a high degree of system integration. A hypersonic engine may rely on fast reacting hydrogen for fuel. Since it is stored at cryogenic temperatures, this fuel can also be used as a coolant for aerodynamically heated surfaces. This dual service can potentially lead to very efficient design. Design tools are needed that can provide rapid, accurate calculations of complex fluid flow patterns. Existing methods, while reasonably accurate, are much too slow to allow multiple simulations (spanning a wide range of conditions) needed for efficient design[7].

The basic element of the coolant system is a heat exchanger panel. We have chosen a typical 6 leg panel (Figure 4) as the base case. Multiple sets of panels would then be connected in parallel and/or series to cover the outer surface of the vehicle. The basic engineering problem being addressed is as follows: in a gas/gas heat exchanger panel, given desired inlet pressure (p1) and temperature (T1), outlet pressure (p2), and heat load (Q), what mass flow rate of coolant (m1) is required? At first glance, this appears to be a relatively straightforward heat transfer calculation, except that there are two complications in this application.

The wide range of operating conditions, high pressure, high heat load, and gaseous nature of the coolant result in variable properties (density, viscosity, and specific heat) throughout each leg as well as across the panel. Thus, the heat exchanger must be broken down into small individual components (straight line, bends, splits, and merges) for accurate calculations. Properties are then assumed to be approximately constant in each small component. We also need to calculate the split flow fraction at each split component in the heat exchanger (e.g., 5 different split fractions in a typical 6 leg panel).

The previous approach was to write a fluid flow equation, equation of state, loss factor equation, pressure drop equation, and heat transfer equation for each panel component, and then solve this system of simultaneous equations (Figure 5). Thus there are five equations for each component, so for a typical 6 leg panel, with 3 straight segments per leg along with associated bends, splits, and merges, this yields a system of 160 equations in 160 unknowns. If we were given p1, T1, and **m1** (as well as Q, of course), this would still be a straightforward forward dynamics calculation for **p2**. The equations could be solved sequentially, starting with the first component, and marching downstream through the entire heat exchanger, with the outlet conditions calculated from each component serving as the inlet conditions used for the next component.

Figure 4
Six Leg Panel

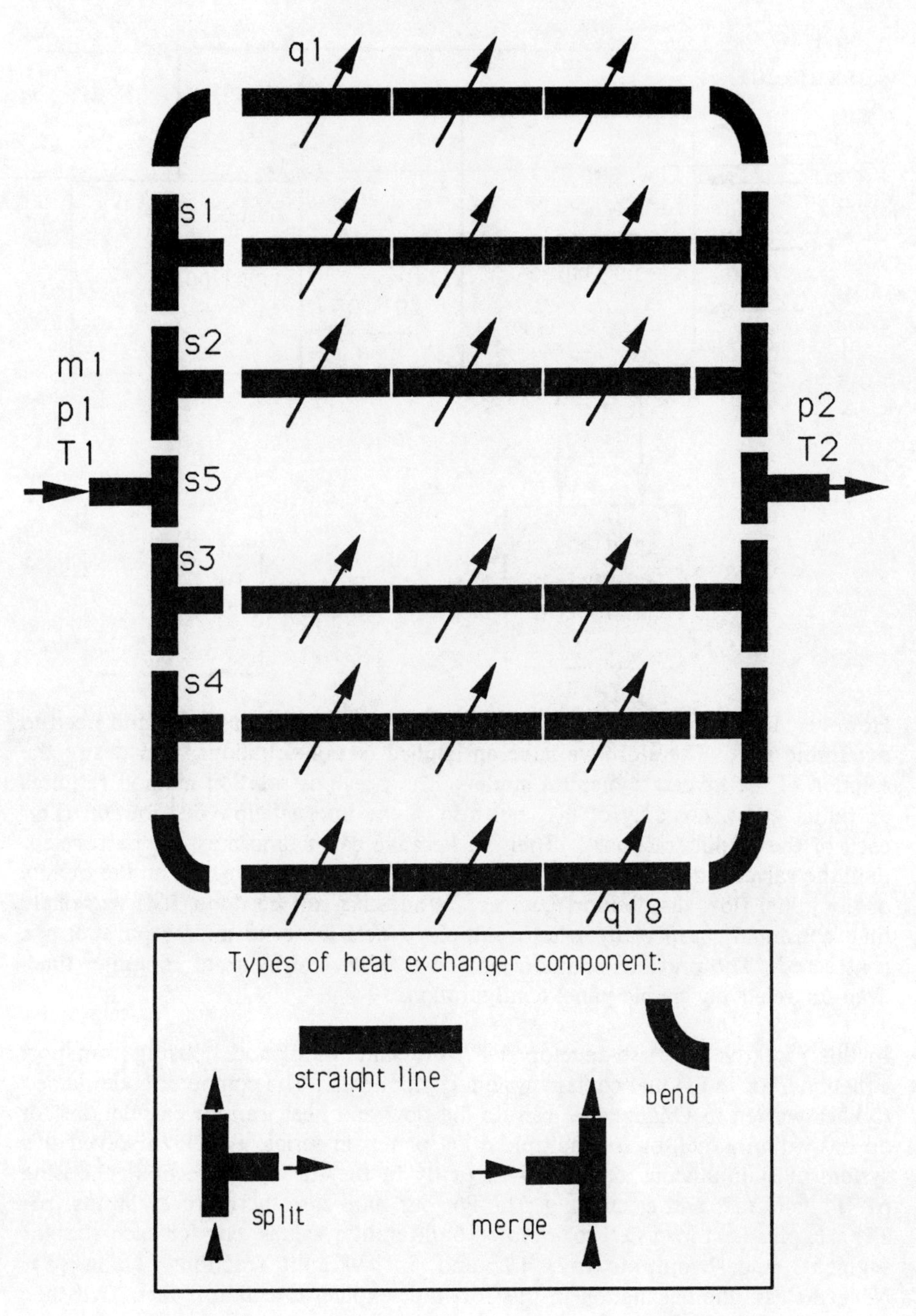

## Figure 5
## Iterative Marching Solution

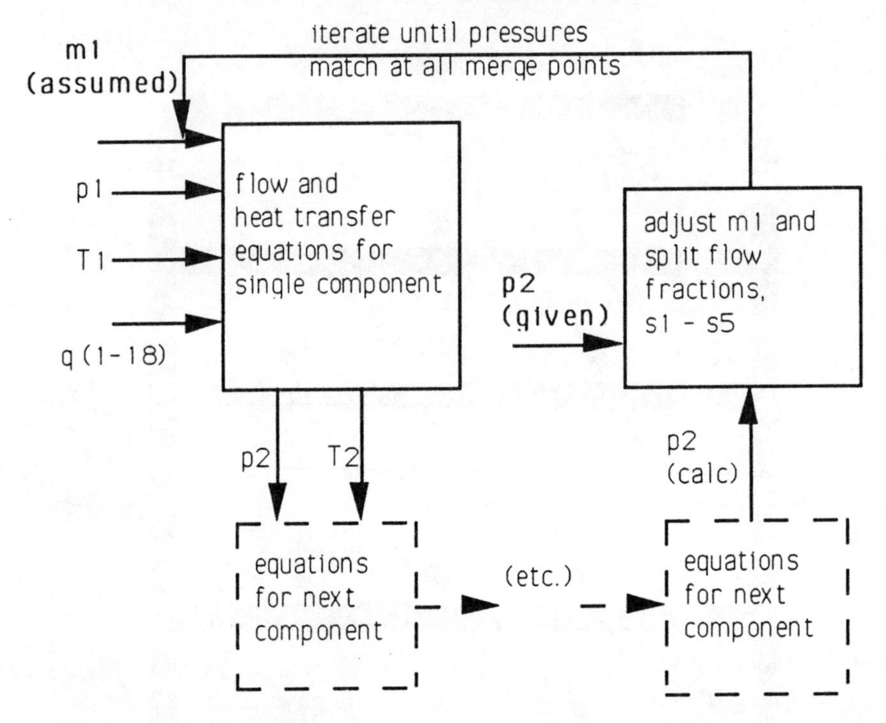


However, in this application, we are given p1, T1, and **p2** (and Q), and need to determine **m1**. Therefore we have an implicit set of equations, and desire the solution of the inverse dynamics model. The previous solution method requires an initial guess, not only of m1, but also of the internal flow distribution (i.e., each of the 5 split fractions). Then an iterative calculation must be performed, until the calculated p2 equals the desired p2. Convergence depends on the quality of the initial flow distribution guesses. While this can be done, it is extremely time consuming, especially when multiple panels in series and/or parallel are considered. The previous solution method took several hours of computer time, even for relatively simple panel configurations.

In this research we first developed a flow simulation model, using transport equations, for individual coolant system components. The computer code named COOL, written in C language, can do the flow and heat transfer calculations for an individual panel, or for multiple 6 leg panels in series. We then solved this system of simultaneous equations explicitly in the forward direction, choosing p1, T1, and m1, and calculating p2. For our base case there are 21 inputs (p1, T1, m1, plus heat load Q, broken into 18 different q values, one for each straight segment) and 7 outputs (p2, T2, and 5 flow split fractions, s1 to s5). Nevertheless, the calculation in this forward, explicit direction is fast. We then ran COOL many times in this forward direction, using randomly selected values of all the input parameters, each spanning their expected operating range. This gave us a training set of data for use in our neural networks. However, for the neural network training and tests, we rearranged the data so that p2 was considered to be an input and m1 was an output, still leaving 21 inputs and 7 outputs, but now arranged to allow for direct calculation of the desired variables (Figure 6).

Figure 6
Direct Flow Calculation

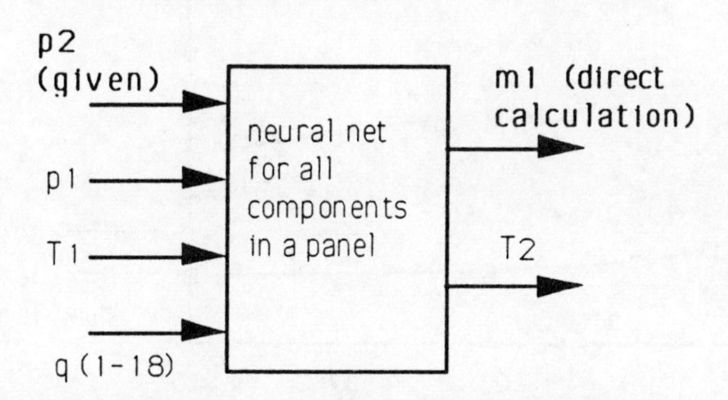


Training of an entire panel using a 6000 case data set can now be accomplished in less than 2 hours. This is less than the time that it takes to calculate a single case using the iterative method. Once the net is trained, subsequent calculation on any given set of input values can be accomplished in a fraction of a second.

We are frequently asked, "Since the forward dynamics model is available, why not just use it to generate a lookup table instead of using a neural network?" It must be remembered, however, that in this case a lookup "table" would not be a simple two-dimensional table, but 21-dimensional, one for each input variable. For even a very sparse table (e.g., 100 deg R spacing for temperature variables), to cover the range of variable values for this application would require a table of at least $10^{21}$ elements for each of the 7 output variables. It would require literally years of computer time to generate such a table, making it entirely impractical for this case. Therefore, the use of a neural network, which can generate satisfactory results with 6000 cases or less, is clearly justified.

## 4 Results for Single Panel

We first attempted to use traditional backpropagation neural nets, using either one or two hidden layers, for this project. We were unable to obtain satisfactory accuracy, regardless of the number of hidden nodes, learning rates, or number of epoches used. We then switched to using our competitive network architecture. This approach quickly yielded satisfactory results. The comparison results are summarized in Table 1. Figure 7a shows the training results for backpropagation with one and two hidden layers (LEARN1 and LEARN2, respectively), and for the

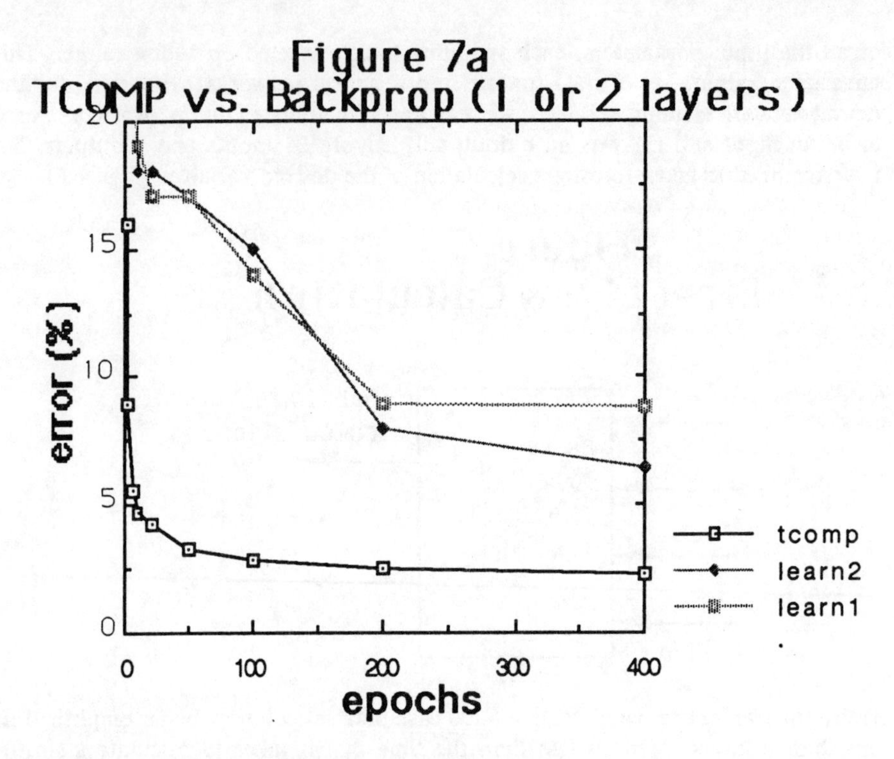
Figure 7a
TCOMP vs. Backprop (1 or 2 layers)
error (%)
epochs
0
5
10
15
20
0
100
200
300
400
tcomp
learn2
learn1

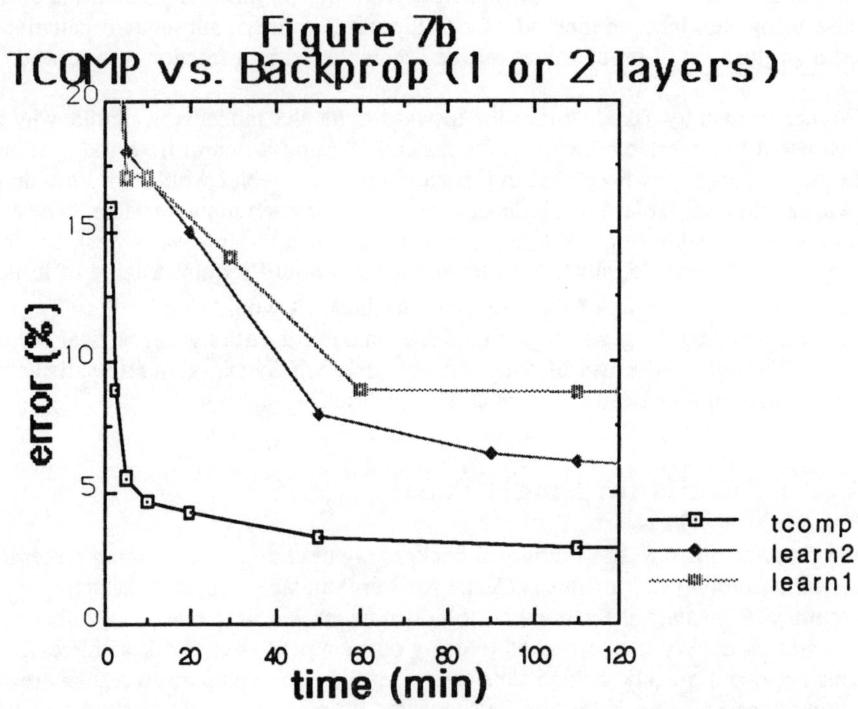
Figure 7b
TCOMP vs. Backprop (1 or 2 layers)
error (%)
time (min)
0
5
10
15
20
0
20
40
60
80
100
120
tcomp
learn2
learn1

competitive net (TCOMP), all plotted as average percent error vs. number of training epoches. To insure a fair comparison, the optimal number of hidden nodes and optimal learning rates were used for each type of network.

However, the competitive net requires more computer time per epoch. To insure a fair comparison, Figure 7b shows the same training results plotted as average percent error vs. cpu training time. Although many researchers report comparisons based on epoches, we believe that the comparison based on actual training time is more apropos for most practical cases of interest. But for our case, even this more harsh comparison demonstrates that TCOMP learns much more quickly in the first few minutes of training time (note the steep learning curve during the first ten minutes), and that it also attains a significantly lower error at the completion of training (2.8% after two hours, vs. 8.8% or 6.1%). Furthermore, Table 1 also lists the testing error, showing that TCOMP is also able to generalize well.

## Table 1. Competitive Net vs. Backpropagation Nets

| network | TCOMP | LEARN1 | LEARN2 |
|---|---|---|---|
| expert nets | 2 | N/A | N/A |
| hidden nodes | 9 | 10 | 6 & 6 |
| learning rates | 0.6 & 0.2 | 0.01 & 0.001 | 0.01, 0.01, 0.005 |
| training error after 200 epoches (%) | 2.5 | 9.0 | 8.0 |
| testing error after 200 epoches (%) | 2.3 | 9.0 | 7.8 |
| training error after 120 min. (%) | 2.8 | 8.8 | 6.1 |
| testing error after 120 min. (%) | 4.0 | 8.5 | 6.0 |

## Table 2. Data Range

| | min | max | |
|---|---|---|---|
| heat load Q | 50 | 600 | (BTU/ft2) |
| pressure p | 500 | 2000 | (psi) |
| temperature T | 100 | 1500 | (deg R) |
| delta p | 1 | 200 | (psi) |
| Reynolds # | 3000 | | |
| Mach # | | 0.3 | |

## Table 3. Sample Cases

| | case 1 uniform | case 2 uniform | case 3 uniform | case 4 shock | case 5 shock |
|---|---|---|---|---|---|
| q1 leg 1 | 191 | 139 | 91 | 99 | 92 |
| q2 | 181 | 150 | 95 | 90 | 92 |
| q3 | 193 | 143 | 94 | 97 | 94 |
| q4 leg 2 | 191 | 150 | 91 | 95 | 97 |
| q5 | 196 | 143 | 97 | 91 | 93 |
| q6 | 184 | 145 | 92 | 93 | 94 |
| q7 leg 3 | 188 | 146 | 95 | 95 | 93 |
| q8 | 192 | 149 | 91 | 96 | 93 |
| q9 | 179 | 138 | 97 | 92 | 94 |
| q10 leg 4 | 188 | 145 | 94 | 95 | **shock 278** |
| q11 | 188 | 141 | 93 | 99 | 99 |
| q12 | 182 | 148 | 97 | 96 | 96 |
| q13 leg 5 | 181 | 143 | 95 | 95 | 96 |
| q14 | 182 | 146 | 96 | 91 | 99 |
| q15 | 183 | 144 | 91 | **shock 295** | 98 |
| q16 leg 6 | 181 | 149 | 93 | 89 | 97 |
| q17 | 194 | 143 | 96 | 92 | 97 |
| q18 | 196 | 149 | 95 | 90 | 100 |
| Qavg (BTU/ft) | 187 | 145 | 94 | 95 | 95 |
| p1 (psi) | 1644 | 1005 | 1017 | 1764 | 1248 |
| T1 (deg R) | 265 | 225 | 634 | 590 | 612 |
| p2 (psi) | 1502 | 941 | 870 | 1600 | 1091 |
| T2 (deg R) | 778 | 1041 | 1073 | 912 | 1031 |
| m1 (lbm/s) | 0.0156 | 0.0762 | 0.0092 | 0.0139 | 0.0108 |
| **% m1 error** | **2.24%** | **1.29%** | **2.09%** | **3.66%** | **1.38%** |

Table 2 lists the data ranges that were used for the input parameters. These values were selected based on consultations with industry. Table 3 lists the results obtained for 5 typical sample cases taken from the test data set. The first three sample cases have relatively uniform heat loads, and the last 2 have a shock heat load on one of the straight leg components. These samples were chosen to illustrate typical performance; some cases were better, some worse. Percent error listed is for prediction of mass flow rate. This project was done on a Sun Sparc 2 Workstation, and all computing times listed are cpu times for this machine. Although training and prediction times are reasonable time frames for design use, even faster computers could be used, opening the possibility of real time prediction and control.

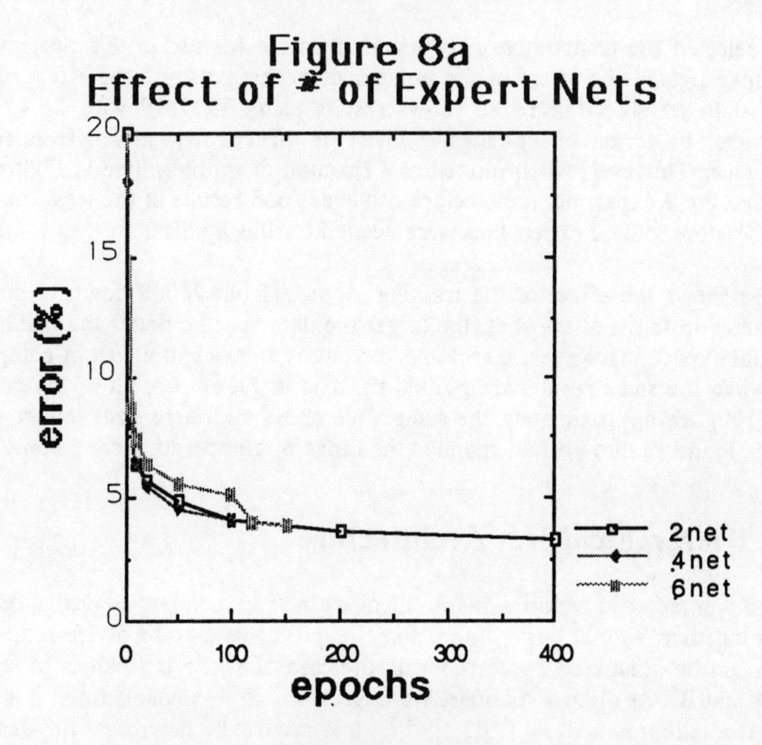
Figure 8a
Effect of # of Expert Nets
error (%)
epochs
2net
4net
6net
0
5
10
15
20
0
100
200
300
400

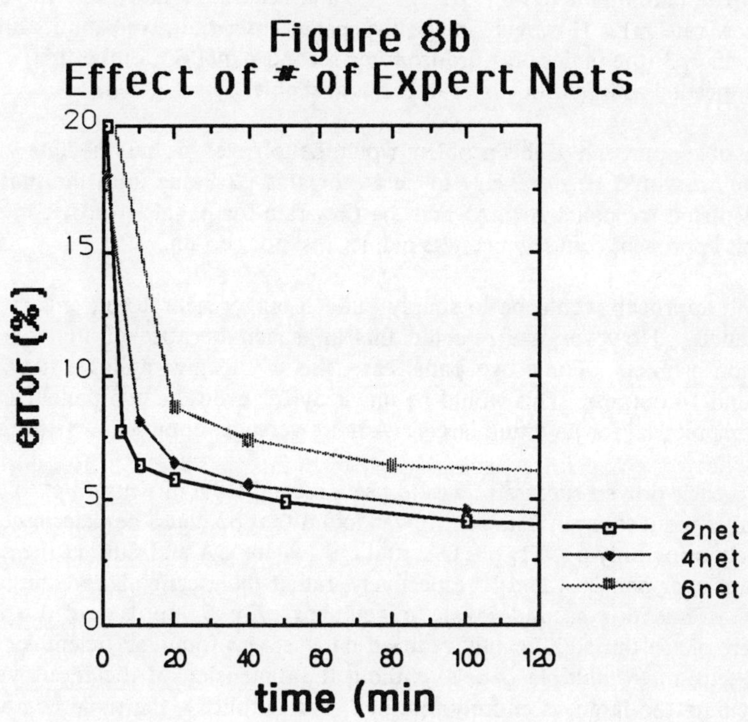
Figure 8b
Effect of # of Expert Nets
error (%)
time (min
2net
4net
6net
0
5
10
15
20
0
20
40
60
80
100
120

Having selected the competitive network architecture for use in this project, we further investigated the effect of the number of expert nets and the effect of the training data set size. Figure 8a shows results using TCOMP with 2, 4, or 6 expert nets, in terms of epoches. There is practically no difference in performance. However, when plotted as a function of training time in Figure 8b, we see that the 2 expert net architecture achieves good results in the least training time. Therefore, only 2 expert nets were needed for this application.

Figure 9a shows the effect of the training set size, from 20000 down to merely 100. Plotted in terms of epoches, the larger the data set, the better the results, as one would expect. However, increasing the data set comes at a cost in computer time. When the same results are plotted vs. time in Figure 9b, all of the curves (except 100) are approximately the same. We chose midrange sizes, either 6000 or 10000, to insure that we had spanned the range of all operating conditions.

## 5 Hierarchical Net Architecture

Next, we wanted to develop a hierarchical method to connect neural network modules together, so that larger, more complex problems may be addressed. The problem can be illustrated by considering the case of two 6 leg panels in series, panels A and B. As discussed above, we have successfully demonstrated a neural net that can take inputs p1, T1, p2, and Q, and accurately determine the desired mass flow rate m1. However, in the two panel case, the given inputs are now p1, T1, and p3 (the outlet pressure from the second panel B, not the first). In this case p2, needed as input for panel A, is not available.

One possible approach to this problem would be to guess an intermediate value of p2. Then it would be necessary to iterate on this p2 value until the mass flow rate calculated for panel A (mA) and the flow rate for panel B (mB) converged. With this approach, convergence was neither fast nor certain.

A second approach would be to simply build a bigger neural net, encompassing both panels. However, we rejected this approach because of the increasing dimension problem. For a two panel case, this would give a single net with 39 inputs and 14 outputs. This would be unwieldy for even the two panel case, and clearly impractical for anything larger. A more versatile approach was sought.

The approach proven successful was to use a second level of neural net to predict p2, using a reduced input data set. It was found that p2 could be determined very accurately using only p1, T1, p3, QA, and QB (where QA and QB are the average heat loads over panels A and B respectively, rather than using all 36 component q values). Thus this second level "p net" has only 5 inputs and 1 output. Therefore, since this significantly reduced data set was found sufficient for the "p net," extension to multiple panels, without the dimension of the input variable set becoming too large, is entirely feasible. For simplicity, the same competitive net architecture was used for this "p net" as previously used for the "panel net", although it is not necessary to use the same architecture for both levels.

Figure 9a
Effect of Training Set Size

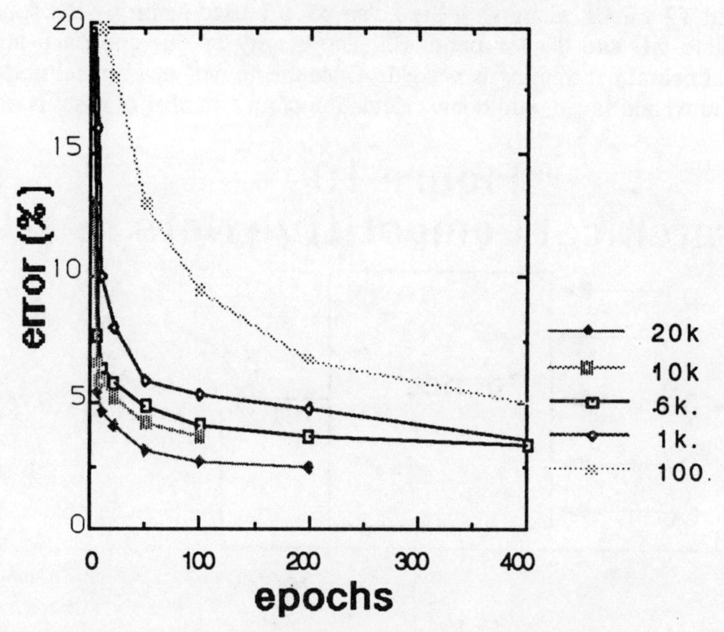


Figure 9b
Effect of Training Set Size

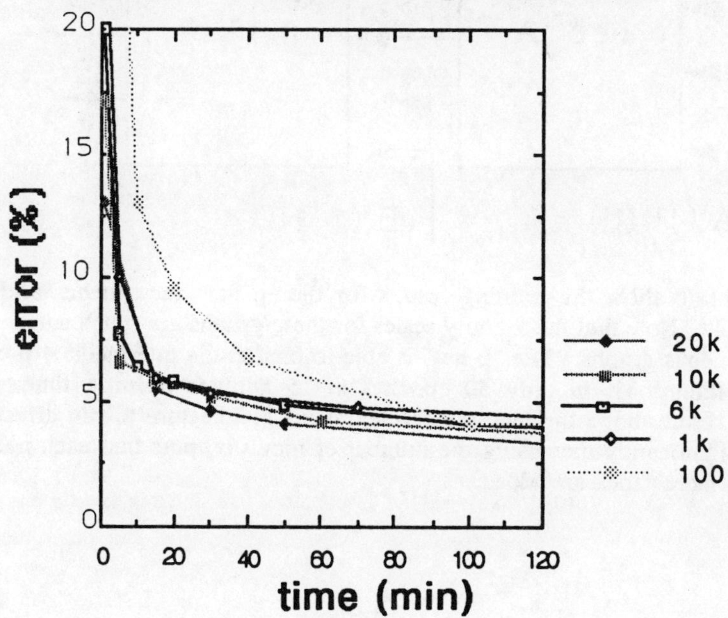

Thus we have created a hierarchical neural net structure (Figure 10). The "p net" first estimates the pressure between panels. Then this p2 value is used as an input variable by the "panel net" to calculate mA and T2 for panel A. Next the p2 and calculated T2 values, along with the given p3, are used again by the "panel net" to calculate mB and T3 for panel B. The beauty of this approach is that no additional neural net training is needed. Once the "p net" and "panel net" weights are computed and saved, additional calculation of any number of cases is rapid.

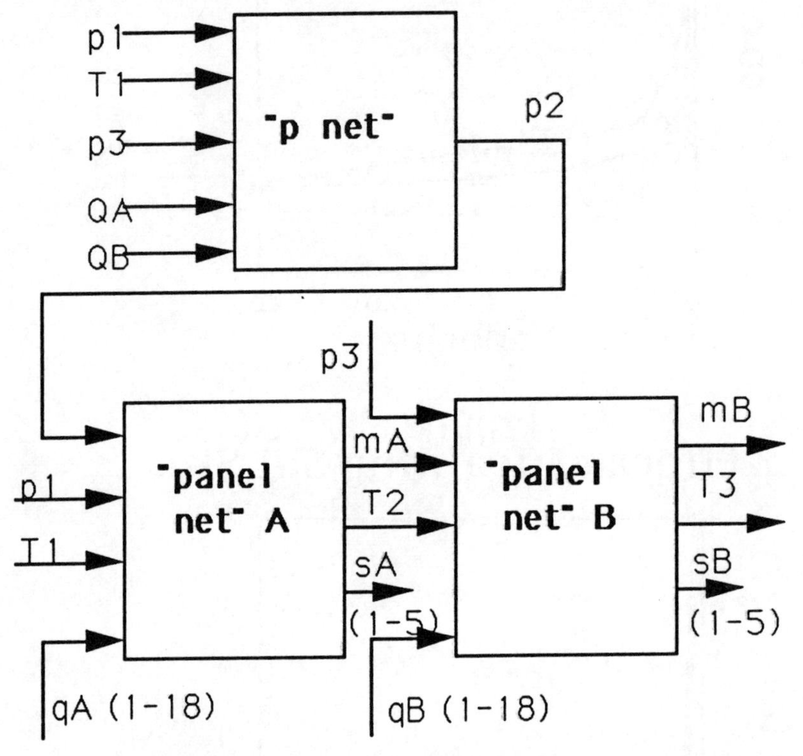


Figures 11a/b show the training results for the "p net" vs. epoches and time, respectively. Note that the x and y scales for these graphs are much smaller than in the previous graphs. The "p net" is able to predict the intermediate pressure, p2, to within 0.3% in only 50 epoches in 15 minutes training time. This excellent result allows the hierarchical neural net architecture to run effectively, without significantly increasing the number of inputs/outputs that each net must handle, as more panels are added.

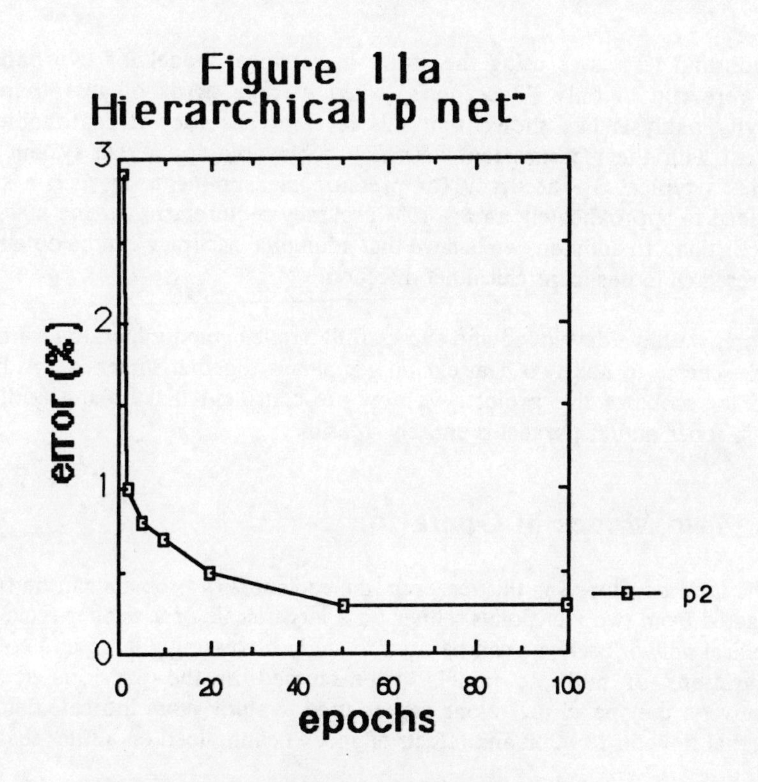
Figure 11a
Hierarchical "p net"
error (%)
epochs
p2
0
1
2
3
0
20
40
60
80
100

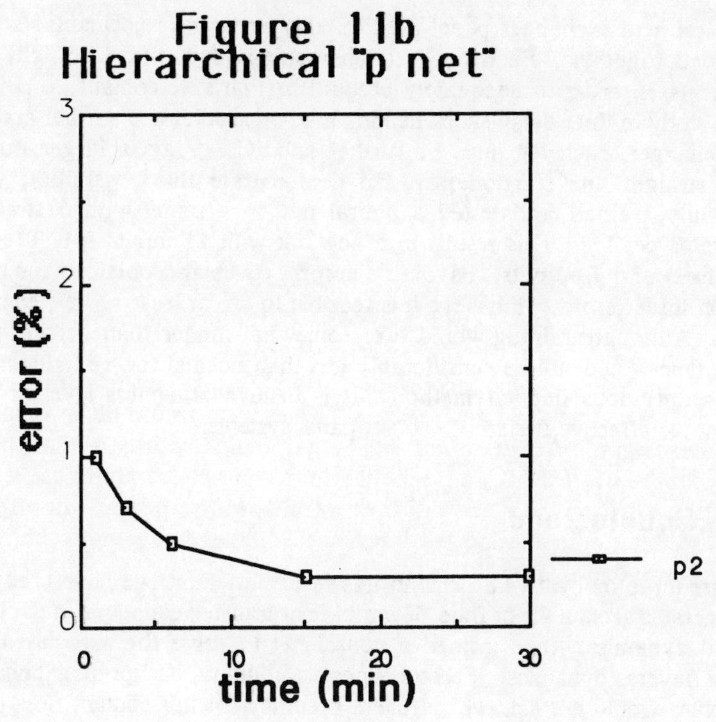
Figure 11b
Hierarchical "p net"
error (%)
time (min)
p2
0
1
2
3
0
10
20
30

One thousand test cases using the entire hierarchical model for two panels in series were run in only 33 seconds, with average error of less than 7%. Sensitivity analysis has shown that this level of accuracy is reasonable and consistent with the measurement accuracy achievable for a real system. For example, a typical 3% accuracy for pressure measurements in such a system would lead to approximately an 8 - 10% accuracy requirement for the mass flow rate prediction. In addition, we believe that additional accuracy can be obtained in future research in this hierarchical net direction.

Therefore, we have developed and successfully implemented a hierarchical neural network scheme to link two heat exchanger panels together in series. Although beyond the scope of this project, we now are confident that linking additional panels in series and/or parallel is entirely feasible.

# 6 Two Modes of Operation

With the tools developed in this research, the coolant flow problem can be further investigated from two viewpoints, either on a larger scale or a smaller scale. The hierarchical network scheme can be used to study increasingly large and complex configurations of multiple panels, as discussed in the previous section. Alternatively, the "panel net" alone can be used to study more intricate details of the internal flow distribution and effects of shock heating load on a finer scale.

The typical heat exchanger panel used for this research project consists of 6 legs manifolded together. Each leg is further subdivided into 3 straight leg line components, in order to adequately account for variable coolant properties. In order to study in finer detail shock heating load on smaller and smaller areas of the heat exchanger, each leg must be further subdivided into a larger number of smaller straight line components. To demonstrate this possibility, we have successfully trained and tested a neural net on a panel with 6 straight line components per leg. This results in a network with 21 inputs (p1, T1, p2, and 18 q values) and 7 outputs. Favorable accuracy was also obtained for this case. Based on these results, we believe it is feasible to use at least 12 line components per leg. Although training would take somewhat longer than before, the total training time would still be considerably less than needed for a single calculation using the previous implicit method. It is believed that this level of detail is sufficient for effective design of coolant panel systems.

# 7 Conclusions

There are three innovative aspects to this research. First, we have used training data generated from a set of fluid flow and heat transfer equations in explicit form (forward dynamics), then trained a neural net to solve the associated implicit problem (inverse dynamics). Second, since a traditional backpropagation network architecture would not achieve sufficient accuracy for this coolant flow problem, so we used an improved version of a competitive net architecture, conceived and developed in this project, to solve the problem. Finally, we have developed and successfully implemented a hierarchical neural network scheme to link multiple heat exchanger panels together. The possibility of using hierarchical neural networks is often mentioned, but there are few actual successful implementations reported for this type of engineering application (different versions of hierarchical nets are used in pattern recognition applications).

The immediate application to hypersonic vehicles, such as NASP is the rapid evaluation of coolant distribution performance. Once trained to simulate performance on a sectional basis, quick evaluation of multiple flight conditions or even multiple flight profiles will be possible. The coolant flow impact on vehicle size and performance can then be evaluated to an extent not permitted by current time consuming iterative calculations.

Rapid, direct solution of complex fluid flow problems has significant value in design simulations. Neural networks offer a potentially powerful tool for the expedient computation of complex non-linear flow dynamics. This capability would allow a designer to evaluate performance over full mission or multi-mission operating conditions, and to optimize system designs more effectively. Therefore, designs can be accomplished both faster and better.

## References

[1] Jacobs, R.A., M.I. Jordan, S.J. Nowlan, & G.E. Hinton, "Adaptive Mixtures of Local Experts", Neural Computation, 3, 79-87, 1991.

[2] Lange, T.E., "Dynamically-Adaptive Winner-Take-All Networks", J. Moody, S. Hanson, & R. Lippmann, Neural Information Systems,4, pp. 341-348, San Mateo, CA, Morgan Kaufmann, 1992.

[3] Jacobs, R.A., M.I. Jordan, "Learning Piecewise Control Strategies In a Modular Connectionist Architecture", in preparation, 1992.

[4] Pao, Y.H., "Adaptive Pattern Recognition and Neural Networks", Addison Wesley, pp.57-82, 1989.

[5] Bridle, J., "Probabilistic Interpretation of Feedforward Classification Network Outputs with Relationships to Statistical Pattern Recognition", F. Fogelman-Soulie & J. Herault (eds.), Neuro-computing: Algorithms Architectures, and Applications, New York, Springer-Verlag, 1989.

[6] Jordan, M.I., R.A. Jacobs, "Hierarchies of Adaptive Experts,", J. Moody, S. Hanson, & R. Lippmann, Neural Information Systems,4, pp.985-992, San Mateo, CA, Morgan Kaufmann, 1992.

[7] White, D.A., A. Bowers, K. Iliff, G. Noffz, M. Gonda, and J. Menousek, "Flight, Propulsion, and Thermal Control of Advanced Aircraft and Hypersonic Vehicles," D.A. White and D.A. Sofge, (eds.), Handbook of Intelligent Control, New York, Van Nostrand Reinhold, 1992.

## Acknowledgement

We gratefully acknowledge the National Science Foundation for research support.

# Index